V. W. Cairns, P. V. Hodson, J. O. Nria
Contaminant Effects on Fisheries 4

J. O. Nriagu and C. I. Davidson
Toxic Metals in the Atmosphere Volume 17 1986

Allan H. Legge and Sagar V. Krupa
Air Pollutants and Their Effects on the Terrestrial Ecosystem Volume 18 1986

J. O. Nriagu and J. B. Sprague
Cadmium in the Aquatic Environment Volume 19 1987

J. O. Nriagu and E. Nieboer
Chromium in the Natural and Human Environments Volume 20 1988

Marlene S. Evans
Toxic Contaminants and Ecosystem Health: A Great Lakes Focus Volume 21 1988

J. O. Nriagu and J. S. S. Lakshminarayana
Aquatic Toxicology and Water Quality Management Volume 22 1989

AQUATIC TOXICOLOGY AND WATER QUALITY MANAGEMENT

Volume

22

in the Wiley Series in

Advances in Environmental Science and Technology

JEROME O. NRIAGU, Series Editor

AQUATIC TOXICOLOGY AND WATER QUALITY MANAGEMENT

Edited by

Jerome O. Nriagu
National Water Research Institute
Burlington, Ontario, Canada

J. S. S. Lakshminarayana
University of Moncton
Moncton, New Brunswick, Canada

WILEY

A WILEY-INTERSCIENCE PUBLICATION

JOHN WILEY & SONS

New York • Chichester • Brisbane • Toronto • Singapore

Library of Congress Cataloging-in-Publication Data:
Aquatic toxicology and water quality management / edited by Jerome O.
Nriagu, J. S. S. Lakshminarayana.
p. cm.—(Wiley series in Advances in environmental science
and technology; v. 22)
"This volume is based on refereed contributions from the
Thirteenth Annual Aquatic Toxicology Workshop held in Moncton, New
Brunswick, Canada"—P.
"A Wiley-Interscience publication."
ISBN 0-471-61551-X
1. Water quality management—Congresses. 2. Organic water
pollutants—Toxicology—Congresses. 3. Water chemistry—Congresses.
4. Fishes—Effect of water pollution on—Congresses. I. Nriagu,
Jerome O. II. Lakshminarayana, J. S. S. III. Workshop on Aquatic
Toxicology (13th : 1987 : Moncton, N.B.) IV. Series: Advances in
environmental science and technology; v. 22.
TD180.A38 vol. 22
[TD365]
628 s—dc19 88-7282
[363.7′394] CIP

Printed in the United States of America

10 9 8 7 6 5 4 3 2 1

CONTRIBUTORS

ARSENEAU, A., Department of Chemistry and Biochemistry, University of Moncton, Moncton, New Brunswick, Canada

BERNSTEIN, JOAN W., Department of Applied Microbiology and Food Science, University of Saskatchewan, Saskatoon, Saskatchewan, Canada

BERMINGHAM, N., Environment Canada, Conservation and Protection Science, Quebec Region, Protection Directorate, Montreal, Québec, Canada

BISSON S., Sciences de l'Environnement, Université du Québec à Montréal, Montreal, Quebec, Canada

BLAISE, C., Environment Canada, Conservation and Protection Service, Quebec Region Directorate, Montreal, Quebec, Canada

BOURQUE, CHARLES L., Department of Chemistry and Biochemistry, University of Moncton, Moncton, New Brunswick, Canada

BURRIDGE, L. E., Department of Fisheries and Oceans, Marine Chemistry Division, Biological Station, St. Andrews, New Brunswick, Canada

CHAN, K. M., Marine Sciences Research Laboratory, Memorial University of Newfoundland, St. John's, Newfoundland, Canada

DAVIDSON, W. S., Department of Biochemistry, Memorial University of Newfoundland, St. John's, Newfoundland, Canada

EIDT, D. C., Canadian Forestry Service, Maritimes, Fredericton, New Brunswick, Canada

ERNST W. R., Environmental Protection Service, Environment Canada, Dartmouth, Nova Scotia, Canada

FLETCHER, G. L., Marine Sciences Research Laboratory, Memorial University of Newfoundland, St. John's, Newfoundland, Canada

GADSBY, M. C., Hoechst Canada, Inc., Regina, Saskatchewan, Canada

GUNN, JOHN M., Fisheries Branch, Ontario Ministry of Natural Resources, Toronto, Ontario, Canada and Department of Zoology, University of Guelph, Guelph, Ontario, Canada

GUY, ROBERT D., Trace Analysis Research Center, Department of Chemistry, Dalhousie University, Halifax, Nova Scotia, Canada

HADJINICOLAOU, J., Department of Civil Engineering and Applied Mechanics, McGill University, Montreal, Canada

HAMMER, U. T., Department of Biology, University of Saskatchewan, Saskatoon, Saskatchewan, Canada

HAYA, K., Department of Fisheries and Oceans, Marine Chemistry Division, Biological Station, St. Andrews, New Brunswick, Canada

HOLLEBONE, J. E., Planning and Priorities Division, Pesticides Directorate, Agriculture Canada, Ottawa, Ontario, Canada

HUANG, P. M., Department of Soil Science, University of Saskatchewan, Saskatoon, Saskatchewan, Canada

KINGSBURY, P. D., Forest Pest Management Institute, Canadian Forestry Service, Sault Ste. Marie, Ontario, Canada

LAKSHMINARAYANA, J. S. S., Biology Department, Faculty of Science and Engineering, University of Moncton, Moncton, New Brunswick, Canada

LALLIER, RÉAL, Faculty of Veterinary Medicine, University of Montreal, St. Hyacinthe, Quebec, Canada

LEE, KENNETH, Department of Fisheries and Oceans, Bedford Institute of Oceanography, Dartmouth, Nova Scotia, Canada

LÉGER, RACHEL, Aquarium de Montréal, Ile Ste-Hélène, Montréal, Québec, Canada and Faculty of Veterinary Medicine, University of Montreal, St. Hyacinthe, Quebec, Canada

LEVY, ERIC M., Department of Fisheries and Oceans, Bedford Institute of Oceanography, Dartmouth, Nova Scotia, Canada

LIAW, W. K., Saskatchewan Fisheries Laboratory, Department of Parks and Renewable Resources, Saskatoon, Saskatchewan, Canada

LOCKHART, W. L., Freshwater Institute Fisheries and Ocean Canada, Winnipeg, Manitoba, Canada

MEHRA, M. C., Department of Chemistry and Biochemistry, University of Moncton, Moncton, New Brunswick, Canada

REGIER, HENRY A., Institute for Environmental Studies, University of Toronto, Toronto, Ontario, Canada

SAMOILOFF, MARTIN R., Bioquest International Inc., Winnipeg, Manitoba, Canada

SPRAGGS, L. D., Department of Civil Engineering and Applied Mechanics, McGill University, Montreal, Canada

SWANSON, STELLA M., Saskatchewan Research Council, Saskatoon, Saskatchewan, Canada

VAN COILLIE, R., Environment Canada, Conservation and Protection Service Quebec Region, Protection Directorate, Montreal, Quebec, Canada

VEZEAU, R., Environment Canada, Conservation and Protection Service Quebec Region, Protection Directorate, Montreal, Quebec, Canada

Waldichuk, Michael, Department of Fisheries and Oceans, Fisheries Research Branch, West Vancouver Laboratory, West Vancouver, British Columbia, Canada

Wang, J. S., Department of Soil Science, University of Saskatchewan, Saskatoon, Saskatchewan, Canada

INTRODUCTION TO THE SERIES

The deterioration of environmental quality, which began when mankind first congregated into villages, has existed as a serious problem since the industrial revolution. In the second half of the twentieth century, under the ever-increasing impacts of exponentially growing population and of industrializing society, environmental contamination of the air, water, soil, and food has become a threat to the continued existence of many plant and animal communities of various ecosystems and may ultimately threaten the very survival of the human race. Understandably, many scientific, industrial, and governmental communities have recently committed large resources of money and human power to the problems of environmental pollution and pollution abatement by effective control measures.

Advances in Evironmental Sciences and Technology deals with creative reviews and critical assessments of all studies pertaining to the quality of the environment and to the technology of its conservation. The volumes published in the series are expected to service several objectives: (1) stimulate interdisciplinary cooperation and understanding among the environmental scientists; (2) provide the scientists with a periodic overview of environmental developments that are of general concern or that are of relevance to their own work or interests; (3) provide the graduate student with a critical assessment of past accomplishment, which may help stimulate him or her toward the career opportunities in this vital area; and (4) provide the research manager and the legislative or administrative official with an assured awareness of newly developing research work on the critical pollutants and with the background information important to their responsibility.

As the skills and techniques of many scientific disciplines are brought to bear on the fundamental and applied aspects of the environmental issues, there is a heightened need to draw together the numerous threads and to present a coherent picture of the various research endeavors. This need and the recent tremendous growth in the field of environmental studies have clearly made some editorial adjustments necessary. Apart from the changes in style and format, each future volume in the series will focus on one particular theme or timely topic, starting with Volume 12. The author(s) of each pertinent section will be expected to critically review the literature and the most important recent developments in the particular field; to critically evaluate new concepts, methods, and data; and to focus attention on important unresolved or controversial questions and on probable future trends. Monographs embodying the results of unusually extensive and well-rounded investigations will also be published in the series. The net result of the new editorial policy should be more integrative and comprehensive volumes on key environmental issues and pollutants. Indeed, the development of realistic standards of environmental quality for many pollutants often entails such a holistic treatment.

JEROME O. NRIAGU, Series Editor

PREFACE

This volume contains state-of-the-art surveys of the toxicology of important aquatic pollutants and provides an overview of key research in a rapidly expanding scientific field. The topics covered range from the principles of aquatic toxicology to environmental effects monitoring and the development of water quality guidelines for the protection of aquatic resources. The importance of the linkages between the aquatic and terrestrial ecosystems in environmental effects assessment has also been highlighted. It is expected that the volume will be of interest to professionals and graduate students in environmental toxicology, marine biology, limnology, ecology and systematics, chemical oceanography, and water management.

This volume is based on refereed contributions from the Thirteenth Annual Aquatic Toxicology Workshop held in Moncton, New Brunswick, Canada. We are grateful to the many people who contributed to the success of the workshop and to this publication. We particularly thank the following reviewers of the manuscripts: P. E. Belliveau, D. Besner, M. J. A. Butler, R. Cote, P. Couture, K. G. Doe, D. C. Eidt, W. R. Ernst, M. C. Gadsby, J. E. Hellebone, P. D. Kingsbury, P. J. LeBlanc, F. Leduc, W. L. Lockhart, P. Maltais, M. C. Mehra, A. S. Menon, C. Morry, T. Pollock, H. S. Samant, D. V. Subbarao, R. Van Coillie, and V. Zitko. Appreciation is also extended to the authors, to the members of the National and Workshop Steering Committees for their interest and encouragement, and to Marie Therese Babineau for secretarial assistance.

JEROME O. NRIAGU
J. S. S. LAKSHMINARAYANA

Burlington, Ontario, Canada
Moncton, New Brunswick, Canada
September 1988

CONTENTS

1. Will We Ever Get Ahead of the Problems? 1
Henry A. Regier

2. Aquatic Toxicology in Management of Marine Environmental Quality: Present Trends and Future Prospects 7
Michael Waldichuk

3. Survival of Lake Charr (*Salvelinus namaycush*) Embryos Under Pulse Exposure to Acidic Runoff Water 23
John M. Gunn

4. Hematological Parameters and Parasite Load in Wild Fish with Elevated Radionuclide Levels 47
Joan W. Bernstein and Stella M. Swanson

5. Colorimetric Determination of Cyanide in Aquatic Systems 65
M. C. Mehra and A. Arseneau

6. Nafion Dialysis Procedure for Speciation of Metal Cations 73
Charles L. Bourque and Robert D. Guy

7. Metallothionein Messenger RNA: Potential Molecular Indicator of Metal Exposure 89
K. M. Chan, W. S. Davidson, and G. L. Fletcher

8. Avoidance Tests with a Plating Industrial Effluent 111
J. Hadjinicolaou and L. D. Spraggs

9. Development of Multiresistance Patterns in the Bacterial Flora of Trout following an Antibiotic Therapy 125
Rachel Léger and Réal Lallier

10. Toxicity Testing of Sediments: Problems, Trends, and Solutions 143
Martin R. Samoiloff

11. Role of Dissolved Oxygen in the Desorption of Mercury from Freshwater Sediment 153
J. S. Wang, P. M. Huang, U. T. Hammer, and W. K. Liaw

12. Integrated Ecotoxicological Evaluation of Effluents from Dumpsites 161
R. Van Coillie, N. Bermingham, C. Blaise, R. Vezeau, and J. S. S. Lakshminarayana

13. The Use of a Fugacity Model to Assess the Risk of Pesticides to the Aquatic Environment on Prince Edward Island 193
L. E. Burridge and K. Haya

14. Assessment of the Inorganic Bioaccumulation Potential of Aqueous Samples with Two Algal Bioassays 205
S. Bisson, C. Blaise, and N. Bermingham

15. Biodegradation of Petroleum in the Marine Environment and Its Enhancement 217
Kenneth Lee and Eric M. Levy

16. Pesticides in Forestry and Agriculture: Effects on Aquatic Habitats 245
D. C. Eidt, J. E. Hollebone, W. L. Lockhart, P. D. Kingsbury, M. C. Gadsby and W. R. Ernst

Index 285

AQUATIC TOXICOLOGY AND WATER QUALITY MANAGEMENT

1

WILL WE EVER GET AHEAD OF THE PROBLEMS?

Henry A. Regier

Institute for Environmental Studies, University of Toronto, Toronto, Canada

1. **New domains of ignorance**
2. **Some roles of science**
3. **Concluding comments**
   References

1. NEW DOMAINS OF IGNORANCE

Fundamental methods of analyzing complex systems (which cannot be decomposed into simple systems) are currently being studied in many fields of science .... The evidence to date, however, indicates a continuing difficulty in understanding complex systems. There will, no doubt, be significant advances in the environmental sciences. The relevant point, however, ... is that it is unreasonable to expect that this advance can keep pace with more rapid expansion of man-made systems. Man is faced with an environmental predicament which can be stated as follows: Man's ability to modify the environment will increase faster than his ability to foresee the effects of his activities. ... If it is impossible to eliminate catastrophic outcomes by anticipating them, then it is necessary to adopt a strategy which will eliminate such outcomes without the requirements of anticipation. ... A principle of preserved diversity upon which this strategy is based can be stated: The chances for irreversible consequences and environmental catastrophy will be reduced by maintaining environmental diversity in two dimensions, spatial and organizational, and by elimination or constraint of those actions which have a persistent effect.

This rather long quotation from Bella and Overton (1972) provides a perspective from which to address the question posed in my title.

An oft-quoted rule of thumb states that scientific knowledge doubles every decade. Can such log-phase growth continue indefinitely; i.e., is the kind of ignorance that can be resolved through scientific methods infinitely large? If not, when will such shortage of ignorance become apparent so as to reduce and eventually stop this growth in scientific understanding?

Suppose that the store of resolvable ignorance is not a constant through time. In our biosphere, for example, it may have been growing since time immemorial as a consequence of natural evolution. One of the ways of characterizing evolution is in terms of increased diversity or increased "information." The more advanced the evolutionary process, the more there is to know about its consequences, and the greater the store of resolvable ignorance.

Is this also the case with creative evolution due to human processes operating within our noosphere (Serafin, 1988)? Partly as a result of the growth of scientific understanding of biophysical processes, humans invent new phenomena that have not previously occurred within the biosphere or have only occurred at much lower intensity. When such an invention is imposed on the biosphere a new domain of ignorance is created by us. In a complex system like our biosphere it is impossible to know, before the fact of such an imposition, what all the consequences will be. It is important to note that ignorance is created, and not that some old domain of preexistent ignorance is uncovered. Prior to the imposition on the biosphere of the invention in the noosphere, the relevant domain of ignorance simply did not exist (Ravetz, 1986).

It follows from the above that the store of ignorance (as perceived by a scientist) grows with the natural evolution of the biosphere and also with the cultural evolution of the noosphere—at least as that culture is expressed in the scientific–industrial world.

In the paradigm of Darwinian evolution, progress resulted from chance phenomena (e.g., gene mutations), most of which were disadvantageous to the organisms involved. By analogy, does something like this occur with respect to innovations in the noosphere? For example, are the various adverse consequences of persistent contaminants an unfortunate but necessary consequence of noospheric evolution? Some protagonists of conventional "progress" would so argue. *In any case, the Darwinian analogy is consistent with the idea that new domains of ignorance can and do emerge.*

Many of our human inventions act so as to harm, cripple, or destroy unique products of biospheric evolution. Biospheric phenomena are driven into highly unnatural systemic fields. A simple reversal of the processes of natural evolution is not what usually happens. It is not a matter of systemic rejuvenation, but rather of a generation of new types of system pathology. But such pathologies, say at a level of ecosystemic organization, often show some symptoms that resemble those of natural rejuvenation at first glance (Rapport et al., 1985). A point to be made here is that emergence of biospheric/ecosystemic pathology due to noospheric inventions does result in the creation of new ignorance, i.e., new pathologies.

According to Bella and Overton (see quotation above) it is unreasonable to expect that advances in environmental sciences can keep pace with more rapid expansion of man-made systems. In Ravetz's terms this would imply that we are generating ignorance more rapidly than understanding. Much of this ignorance is associated with undesirable phenomena. Hence the need for normative principles to be imposed on human inventiveness, such as "elimination...of those actions which have a persistent effect." To ecotoxicologists this may imply a policy of zero discharge of persistent contaminants. Within the noosphere it implies the need for clarification and broad acceptance of the principles of biospheric ethics.

2. SOME ROLES OF SCIENCE

From the perspective of the quotation from Bella and Overton, consider some roles of science as they relate to the fields of aquatic toxicity and ecotoxicology.

1. Full understanding, in the usual analytical sense, appears still to be the implicit goal of basic and/or academic science. We recognize that this goal is not attainable in real-life situations. It can be approached only in highly abstracted laboratory settings where life processes have been severely curtailed, to the point of a quasi-life state. But understanding of such abstracted phenomena can prove very useful, as in the establishment of testing protocols in which reproducible phenomena in a laboratory setting become surrogates of important, partially understood, real-life phenomena outside the laboratory. The surrogates are, at best, quite imperfect and we must guard against a tendency that would invest the laboratory protocols with more meaning and importance than the real-life phenomena themselves.

2. After the fact of some environmental harm, science may be used to identify a cause and to indict the agent of that cause. In jurisdictions in which such indictment frequently occurs in courts of law (as in the United States), the relevant science can become very "stylized" or contorted (Regier, 1985). The issue may be decided on the degrees of authority that may be exuded by scientists supporting two sides of the issue. As every freshman knows, reliance on authority is one of the most basic "thou shalt nots" of science. The long fight about the effects of tobacco smoke on health is a well-known case of what the struggle against contaminants is all about. A powerful interest group in our society can stall resolution of such an issue for a quarter century. Some chemical industries have become quite expert at such stalling, and have well-paid ecotoxicologists on staff to help extend the process. A kind of adversarial pseudoscience comes to be practised.

3. Science has come to be accepted as a kind of "process language" in interjurisdictional negotiation. Apparently it has something to offer in this respect now that the usual conventions of diplomacy and mililtary might are so thoroughly discredited. With respect to interjurisdictional differences, the greater the differences between the contending jurisdictions, the more elementary and primitive the kind of science that is invoked to serve as a process language in the negotiations. Initially, the phenomena that come to be studied jointly are far removed from the real phenomena underlying the interjurisdictional conflict (see Loucks and Regier, 1988). Eventually as relations improve, perhaps over decades, the issues under joint study approach more closely the real matters. With respect to persistent contaminants, toxic fall-out, acid rain, etc., it may be noted that Canada and the United State are progressing only very slowly toward joint study of even the most obvious aspects of the problems. In a situation where the more powerful jurisdiction is also the primary malefactor, such stalling can continue for a long time. Reinhold Niebuhr, an American theologian, is often quoted to have stated that there is no relevant ethic that applies to international relations. The disinterested scientific study of phenomena, however distantly related to the real issues involved, will eventually demonstrate that it is in each jurisdiction's self-interest to correct the abuses. That, apparently, is the hope invested in science.

4. Informed syntheses of scientific knowledge, however partial, may provide a basis for normative principles such as that of preserved diversity of Bella and Overton, see above. If such principles serve only as constraints or prohibitions after the fact of some deleterious activities, then growth of ignorance may outpace growth in understanding. Under such circumstances humans cannot become effective stewards of the biosphere because they exercise insufficient responsibility or self-control within their noosphere. Forward-looking scientists may help to create normative principles or goals that would foster desirable natural biospheric and cultural noospheric evolution. These would still lead to emergence of new ignorance but perhaps this ignorance would more frequently be associated with pleasant surprises rather than with undesirable catastrophes. Wouldn't it be fun to undertake scientific studies to explain the emergence of a desirable biospheric phenomenon? It may be that we need to foster unconventional science for such purposes—conventional science may be severely biased so as to foster the emergence of undesirable phenomena in our biosphere, and in our noosphere (Regier and Grima, 1985: see especially pp. 853–863).

3. CONCLUDING COMMENTS

The following excerpt is from a nongovernmental review of the 1978 Great Lakes Water Quality Agreement (NRC/RSC, 1985, p. 107):

The committee finds that the past century presents a record of resource degradation expanding in area, extending in the duration of the impairment, and intruding more deeply into ecosystemic processes. The causes of the impairments are now more complex with respect to causal linkages and have become less evident to the public as well as to scientists. Risks seem to affect much larger populations. The sequence of environmental degradation, corrective measures, and then the creation of new environmental threats has not been brought under control by the 1978 Agreement. Three types of future events are likely to occur in the Great Lakes:

- The certain ones that we can predict and understand;
- The uncertain ones that we cannot predict but can understand; and
- The surprising ones that we can neither predict nor understand.

The first allows the possibility for traditional control. The second requires flexibility in order to adapt and design for the uncertain. The third requires an atmosphere for learning. Looking to the future, water diversions and the possible consequences of climate change have the potential to produce additional region-wide alterations of the system, and another round of reactive measures will be needed.

Our current policies and programs related to science will not get us ahead of the second and third types of problem. Some critics of science blame it all on conventional science anyway. In order to get ahead of our problems we may have to step back and look for an alternate route than that of the dominant conventions of science. The latter may still be useful, in a subservient role.

REFERENCES

Bella, D.A., and Overton, W.S. (1972). Environmental planning and ecological possibilities. *J. Sanitary Eng. Div. ASCE.* **98** (SA3): 579–592.

Loucks, O.L., and Regier, H.A. (1988). International agreements and strategies for controlling toxic contaminants. In N.W. Schmidtke (Ed.), *Toxic Contamination in Large Lakes*, Vol. IV, *Prevention of Toxic Contamination in Large Lakes*, Lewis Publishers, Chelsea, Michigan, pp. 235–247.

NRC/RSC (1985). The Great Lakes Water Quality Agreement, an evolving instrument for ecosystem management. Binational review by the U.S. National Research Council and the Royal Society of Canada. National Academy Press, Washington, DC.

Rapport, D.J., Regier, H.A., and Hutchinson, T.C. (1985). Ecosystem behavior under stress. *Am. Nat.* **125**: 617–640.

Ravetz, J.R. (1986). Usable knowledge, usable ignorance: incomplete science with policy implications. In W.C. Clark and R.E. Munn (Eds.), *Sustainable Development of the Biosphere*, Cambridge Univ. Press, Cambridge, U.K., pp. 415–434.

Regier, H.A. (1985). Commentary on "Scientific perspective, freshwater systems" by J. Cairns, Jr., pp. 49–52. In G.E. Beanlands et al. (Eds). Proceedings of a Workshop on Cumulative Environmental Effects: A Binational Perspective. Canadian Environmental Assessment Research Council and U.S. National Research Council, Ottawa. Min. Supply and Services Canada, Cat. No. En. 106-2/1985.

Regier, H.A., and Grima, A.P. (1985). Fishery resource allocation: An exploratory essay. *Can. J. Aquat. Sci.* **42:** 845–859.

Serafin, R. (1988). Noosphere, Gaia and science of the biosphere. *Environmental Ethics* **10(2):** 121–137.

2

AQUATIC TOXICOLOGY IN MANAGEMENT OF MARINE ENVIRONMENTAL QUALITY: Present Trends and Future Prospects

Michael Waldichuk

Department of Fisheries and Oceans, Biological Sciences Branch, West Vancouver Laboratory, West Vancouver, British Columbia, Canada

1. Introduction
2. Acute bioassays
3. Sublethal effects testing
3.1. Physiological effects
3.2. Biochemical effects
3.3. Morphological and pathological effects
3.4. Genetic effects
3.5. Behavioral effects
3.6. Ecological effects
3.7. Bioassay
4. International initiatives
4.1. International Council for the Exploration of the Sea
4.2. Intergovernmental Oceanographic Commission
4.3. Convention on the Prevention of Marine Pollution by Dumping of Wastes and Other Matter (London Dumping Convention)
5. Discussion
6. Summary and Conclusions
References

Prepared for the Panel on Aquatic Toxicology and the Management of Marine Environmental Quality—Its Role, Opportunities and Constraints, at the 13th Annual Aquatic Toxicity Workshop, Hotel Beausejour, Moncton, New Brunswick, Canada, 11–14 November 1986.

1. INTRODUCTION

During the 1960s and 1970s, there was a flurry of activity into research on sublethal effects of contaminants, or, as it is more commonly expressed in Europe, *biological effects techniques in pollution-monitoring programs*. The old acute bioassay was becoming passé, inadequate for providing the kind of information needed to properly manage marine environmental quality, particularly that of the marine ecosystem. So various investigators were reporting in the scientific literature on their particular techniques for measuring the effect of pollution and often promoting each technique as the way to go in management of marine environmental quality.

Unfortunately, there has been little agreement among investigators, let alone among regulatory agencies, about which techniques are best for what and how they might be used for controlling marine pollution. Some workers in the field (e.g., Monk, 1983) have questioned certain applications of ecotoxicology to a number of problems. Workshops have been organized (see, e.g., McIntyre and Pearce, 1980) to review, evaluate, and essentially catalog various techniques for measuring effects of pollution. International conventions on marine pollution control, such as the London Dumping Convention, look for guidance to national laboratories on how best to measure the effects of contaminants in the marine environment. There really is none forthcoming. The system seems to have stalled. The next step required in standardizing methodology in biological effects monitoring in the marine environment, or, as we prefer to call it, *aquatic toxicology,* has not been taken (McIntyre, 1984). We are told by certain quarters (Bayne, 1985), however, that this halt in proceedings is only temporary. The next step, a practical workshop and field trials to apply as many of the available techniques as possible, was in fact taken in Norway during August 1986. More will be said on this later.

Even after evaluation of results from the foregoing workshop and field trials, it will be some time before agreement will be reached nationally and internationally on acceptable techniques for measuring effects of contaminants on the marine biota and ecosystem for the purpose of marine environmental impact assessments and pollution control. In the meantime, the much maligned acute bioassay will continue to be used for the immediate problems of waste disposal into the sea and for pollution control.

In this paper, I propose to examine once again acute bioassays and how they could be improved for management of marine environmental quality; to review briefly some of the techniques available for measuring effects of contaminants at the sublethal level and to single out those that appear to have most promise; and, finally, to explore international initiatives on the subject of aquatic toxicology (biological effects measurement) and the prospects for arriving at a consensus on universally acceptable techniques.

2. ACUTE BIOASSAYS

Engineers have often been criticized for their approach to pollution control, because they have sought design parameters without attempting to understand the bases of those parameters. They have, however, insisted on numbers, because only with numbers can they achieve a suitable design in a treatment facility or an outfall that will meet the criteria required for maintaining environmental quality. This has forced biologists, ecologists, chemists, and even oceanographers to be quantitative, even though this may be highly qualified, in arriving at the requirements for maintaining a viable aquatic living resource or ecosystem. Thus, we provide the engineer with an LC_{50} for a particular contaminant or waste stream. We may then apply an *application factor* for safety, which may range from 0.1 to 0.001, usually given in decades, often based on value judgement. Some will argue that the application factor is highly arbitrary and should be replaced by something more quantitative based on toxicological data. Others will actually attempt to quantify an application factor by pointing to some threshold of a measurable effect that can be used in association with an LC_{50}.

With all its shortcomings, the LC_{50} has been the mainstay for developing water quality criteria (NAS, 1972). The big problem with acute bioassays has been the lack of standardization in procedures from one laboratory to another. Not only does the test organism vary from one laboratory to another, but the characteristics of the water used for testing a given substance may differ widely among laboratories. Consequently, bioassay results may be greatly different for the same substance. Standardization is needed. It has been demonstrated in interlaboratory comparisons of bioassays, where conditions for the bioassays were carefully specified, that bioassays results need not be widely different (Davis and Hoos, 1975). An effort is being made now, at least internationally (e.g., OECD, 1981), to standardize techniques and test organisms for testing chemicals and other substances.

Standard acute bioassays have been conducted generally over a period of 96 hours to provide a 96-h LC_{50}. Longer exposures can provide more information on the toxicity of a substance through the shape of the toxicity curve, as the British have shown (Norton and Lloyd, 1981). This is particularly applicable to substances for which the toxic effects are not fully realized during short exposure periods. Full life-cycle bioassays have been conducted on organisms having a short life cycle to determine the effect of a contaminant on the different life stages, particularly on reproduction. Usually invertebrates are used for this purpose, although some fishes also have comparatively brief life cycles.

One of the big advances made in acute bioassays over the last three decades is the development of the flow-through system, as opposed to the static bioassay. This allows for a steady supply of fresh toxicant at a constant concentration without the problems of decreasing concentration due to adsorption, absorption, decomposition, and volatilization that are characteristic of the static bioassay.

The use of bioassays can be extended from the concept of two-dimensional dose–response curves to three-dimensional dose–response surfaces by multifactorial design of bioassay experiments (Alderdice, 1972). This can take into account the variability of environmental conditions, such as temperature, salinity, and dissolved oxygen. Such an approach is seldom taken for management purposes, however, and usually "average" environmental conditions are chosen for bioassays.

3. SUBLETHAL EFFECTS TESTING

Because toxicologists have always recognized the deficiencies in acute bioassays, they have endeavored individually to develop techniques for measuring effects of contaminants on aquatic organisms that would more fully describe the toxic effects, particularly at the sublethal level. As a result, there has been a plethora of techniques described in the scientific literature. When the Joint Group of Experts on the Scientific Aspects of Marine Pollution (GESAMP, 1980) reviewed the various biological variables related to marine pollution that could be monitored, it had a total of 36 to evaluate based on existing techniques. The real need, of course, has been to evaluate and discuss the techniques in various fora in order that some basis could be developed for selection of suitable techniques for particular purposes.

The first major discussion of sublethal techniques was carried out under the auspices of the Royal Society, London, in May 1978. The problems of marine pollution and the needs for sublethal studies were reviewed (Waldichuk, 1979; Perkins, 1979), and individual techniques utilized at that time were described (e.g., Steele, 1979). The proceedings of this discussion were published in both the *Philosophical Transactions* of the Royal Society of London and in book form (Cole, 1979).

On another front, the International Council for the Exploration of the Sea (ICES) was wrestling with the problem of measuring effects of contamination in the marine environment. A working group of that body examined available techniques and eventually published its views on the feasibility of using these techniques in monitoring biological effects of contaminants in the sea (ICES, 1978). Arising out of a recommendation of this working group, a Workshop on Biological Effects of Marine Pollution and the Problems of Monitoring was organized by ICES at Duke University in Beaufort, North Carolina, February/March 1979. The deliberations of the workshop were divided into seven panels to deal with the various techniques designed to measure seven different biological aspects of pollution: (1) physiology, (2) biochemistry, (3) morphology and pathology, (4) genetics, (5) behavior, (6) ecology, and (7) bioassay. Each panel evaluated the different techniques and listed the advantages and disadvantages of each. The resulting publication from this workshop (McIntyre and Pearce, 1980) is the most complete description of different toxicological techniques available to date. The potential application of these techniques for marine pollution monitoring has been discussed by GESAMP (1980); measuring the effects of chemicals on aquatic animals, as indicators of ecological damage, has been reviewed by Waldichuk (1985). As noted earlier, the next step, application of a wide sample of available techniques through practical workshops and field trials, has just been taken (Bayne, 1985; IOC, 1986).

GESAMP (1980) ranked the various biological variables given in McIntyre and Pearce (1980) that can be measured for an estimate of pollution effects. The ranking system was based on the following criteria:

1. Highly recommended for immediate use in monitoring programs in all regions.
2. Recommended only for selective use, because it is more costly or requires further field testing before it can be used routinely.
3. Potentially useful but not recommended at present, because the approach and techniques require further development.

The rankings given to the biological variables under each of the categories represented by panels into which the Beaufort Workshop was subdivided are given as follows.

3.1. Physiological Effects

Rank 1: Feeding rate; body condition index; scope for growth (+ growth efficiency); oxygen/nitrogen ratio.
Rank 2: Nil.
Rank 3: Respiration.

It is of interest that the recent international exercise (IOC, 1986) identified "scope of growth" in mussels as an index that may be used successfully to assess the biological effect of pollution at the organismic or "whole animal" level. Scope for growth is an analysis of energy intake and energy losses by individual animals, set in an equation that considers the resultant energy balance (Widdows, 1985). In the same exercise (IOC, 1986), measurements of tissue (gill) respiration rate showed mixed success: Crab tissues from the field situation were responsive to the contamination gradient, but neither mussels in the field or a mesocosm, nor crabs in the mesocosm showed significant responses. On the other hand, biochemical composition of the digestive glands of both crabs and mussels showed significant changes in response to contaminant levels and provided a physiological expression of changes.

3.2. Biochemical Effects

Rank 1: Lysosomal stability; taurine/glycine ratio.

Rank 2: Mixed function oxidase; metallothionein; energy charge; blood chemistry; primary production.

Rank 3: Steroids.

In the IOC Workshop on Biological Effects Measurements (IOC, 1986), mixed function oxidase in flounder liver, as assayed by the activity of the enzyme known as EROD, gave the most sensitive index to the expected pollution gradient in the field, with a 15-fold increase in activity from the reference site to the most contaminated site. This measurement could be recommended as an indicator of marine contamination by polychlorinated biphenyls and polycyclic aromatic hydrocarbons. Metallothionein production was also measured as a biochemical response in marine organisms, but at the time the preliminary report (IOC, 1986) was prepared, the data on metallothionein could not yet be assessed. Variability in the reproductive state of exposed animals confounded interpretation of results of cytochemical analysis of the functional state of subcellular organelles, such as the lysosomes, which proved to be responsive to the contamination gradients.

3.3. Morphological and Pathological Effects

Rank 1: Liver as percentage body weight; ulcers; fin erosion; assymetry.

Rank 2: Liver structure; gametogenic cycle; neoplasia/tumors; early developmental stage.

Rank 3: Gill deformity.

In the IOC Workshop on Biological Effects Measurements (IOC, 1986), the more descriptive approaches of histopathology did not prove sensitive to the contaminant levels encountered.

3.4. Genetic Effects

Rank 1: Nil.

Rank 2: Nil.

Rank 3: Chromosomal abnormalities; mutagenicity assay.

No measurements of the foregoing type were conducted in the IOC Workshop on Biological Effects Measurements (IOC, 1986), but both quantitative and more classical descriptive approaches to cellular pathology were evaluated. A sensitive measure of cellular condition was the accumulation of lipid residues within some cell types (lipidosis). Quantitative steriological tests were useful in discriminating all damage in mussels from different sites. Some of the approaches that can be taken to cellular responses to pollutants have been described by Moore (1985).

3.5. Behavioral Effects

Rank 1: Nil.

Rank 2: Torque test.

Rank 3: Nil.

The ICES Beaufort Workshop (McIntyre and Pearce, 1980) did not recommend any specific behavioral tests, partly because behavioral effects are often difficult to measure in the field. GESAMP (1980) considered the "torque test" as a behavioral response, because it has been demonstrated (Lindahl and Schwanbom, 1971) as a useful measurement of pollution effects in fish. Being moderately sensitive, it has been applied to a limited extent in a field monitoring program. It has the disadvantage, however, of being applicable to only a few fish species, and requiring fairly expensive equipment.

Effects of abnormal behavior may be more readily observed and used as a response in marine organisms than immediate behavioral response. Thus avoidance in fish may be reflected in altered migration and distribution.

3.6. Ecological Effects

Rank 1: Community biomass; abundance; diversity; alterations in distribution; species density; growth rate; reproduction (gonad as % body wt); population structure.

Rank 2: Nil.

Rank 3: Nil.

An analysis of benthic communities was undertaken at the IOC Workshop on Biological Effects Measurements (IOC, 1986). The aims were, first, to discriminate among sampling stations, irrespective of the causes of any differences recorded, and second, to link differences to pollution, bearing in mind the many confounding variables that may be present in any natural system such as a fjord. The Workshop succeeded in the first aim by employing advanced statistical analyses. Success or failure in the second aim will be unclear until more chemical data are available. The results of applying multidimensional scaling to numbers or biomass distribution of fauna in the field clearly discriminated the reference (control) station from others along the contaminant gradient, and also identified differences among contaminated sites. It was not possible, however, to entirely reject the hypothesis that these differences were due, at least in part, to differences in depth between sites. Benthic community structure is known to be sensitive to depth.

Two important insights into fjord benthos were derived from application of multivariate statistical tests to the benthic faunal data obtained in the above program. First, it was possible to distinguish between stations at various levels of taxonomic discrimination (species, order, family, etc.), suggesting that in some situations a time-consuming, full, taxonomic analysis to the species level may not be necessary to identify within-station differences. Second, it was discovered that copepods are more sensitive in their species distributions than other components of the marine fauna to differences in environmental contamination.

3.7. Bioassay

Rank 1: Bivalve/echinoderm larvae; microalgae bioassay; hydroid bioassay.
Rank 2: Nil.
Rank 3: Nil.

GESAMP (1980) noted that bioassay techniques are used primarily for assessment of water quality and have the advantage of being highly quantitative, sensitive, and precise, with a high signal/noise ratio and response rate, combined with a low cost. In monitoring, however, GESAMP suggested that bioassays should be limited to identifying the presence of "hot spots," because the response measured in isolation has little ecological significance. Nevertheless, for predicting biological impact from a given waste discharge, bioassays in combination with oceanographic information on currents and flushing rates still provide an effective route in management of marine environmental quality.

4. INTERNATIONAL INITIATIVES

International activities in aquatic toxicology or, as it is usually referred to in Europe, *biological effects measurements,* provide a useful barometer on the needs and available means for assessing marine environmental quality.

4.1. International Council for the Exploration of the Sea

The International Council for the Exploration of the Sea (ICES) has led the way in these endeavors (ICES, 1978; McIntyre and Pearce, 1980), because of its traditional role in advising member states on management of fish and shellfish stocks, as well as in maintaining marine environmental quality in fisheries waters of the North Sea and the northeast Atlantic. This role extends into monitoring to determine trends in concentrations of various constituents of anthropogenic origin in fisheries products, seawater, and marine sediments. Thus, the ICES Advisory Committee on Marine Pollution (ACMP) reviews all aspects of pollution arising in waters and affecting living marine resources of the area under its jurisdiction.

At the 1984 ICES Statutory Meeting, a Study Group on Biological Effects Techniques was established to review and evaluate the extent to which existing biological effects techniques serve (1) to identify and quantify the presence and effects of potentially harmful anthropogenic inputs or activities, including at the population or community level; and (2) to identify the causes of the effects detected. One of the first things this group noted was that biological effects techniques in monitoring programs face all the problems associated with chemical monitoring programs, but additionally, there are problems due to the inherent variability of the biological material used for assays. Moreover, biological methods are highly time- and labor-consuming, require highly trained personnel, and are more difficult to standardize and automate than analytical chemical methods.

4.2. Intergovernmental Oceanographic Commission

In another international arena, the Intergovernmental Oceanographic Commission (IOC) of UNESCO, action is underway on biological effects measurements. The role of IOC in marine pollution investigations has been thoroughly reviewed recently by Kullenberg (1986). Under the umbrella of its Scientific Committee (formerly Working Committee) on GIPME (Global Investigation of Pollution in the Marine Environment), a Group of Experts on the Effects of Pollutants (GEEP) was formed in 1982. Among its five terms of reference was one "to formulate international cooperative research proposals for the study of the effects of pollutants on marine organisms and at different levels in the marine ecosystem." Along this line, the practical Workshop on Biological Effects Measurements was held at the University of Oslo, Norway, 11–29 August 1986, when 31 scientists from 12 countries met to evaluate various techniques for measuring the effects of pollutants in the sea (IOC, 1986). This was done by applying available procedures to material collected along a contamination gradient in Frierfjord and from experimental exposures to different levels of a contaminant "cocktail" (a mixture of a water-accommodated fraction of diesel oil and soluble copper) within the mesocosm facilities at Solbergstrand on Oslofjord. Careful consideration was given to the statistical demands of the sampling strategy and to rigorous statistical analysis of the results. All samples underwent blind analysis, the biologists being unaware of the source of their material until after all the analyses were completed. Only preliminary results of this Workshop are available at this time (IOC, 1986), as discussed in the previous section, but it is anticipated that some useful recommendations on biological effects techniques will arise from the various evaluations.

There is a good prospect for useful collaboration in GEEP activities from another body under IOC's Scientific Committee on GIPME, the Group of Experts on Methods, Standards and Intercalibration (GEMSI). This group has made great advances on sampling and analytical techniques for different materials in the marine environment since it was formed in 1977.

4.3. Convention on the Prevention of Marine Pollution by Dumping of Wastes and Other Matter (London Dumping Convention)

The need to be able to relate laboratory tests to field conditions has been identified in various regional ocean dumping conventions (e.g., Convention for the Prevention of Marine Pollution by Dumping from Ships and Aircraft of 1972, commonly known as the Oslo Convention), and in the global Convention on the Prevention of Marine Pollution by Dumping of Wastes and Other Matter (commonly referred to as the London Dumping Convention (LDC)).

In the Scientific Group of the LDC, delegations from different member states have been requested to submit information on techniques used to test materials to be dumped at sea and on the relation between laboratory tests and field assessments. The United Kingdom delegation outlined its routine toxicological testing program, in operation since 1968, which has been used on wastes proposed for dumping at sea. The program has been based largely on acute bioassays, and the British investigators point out that much can be learned about the characteristics of a waste from longer exposures of test organisms, e.g, 20 days, than in the conventional 96-h acute bioassay. Moreover, the shape of the toxicity curve obtained from a standardized bioassay can indicate whether changes in the characteristics of a waste have occurred in the periods between bioassays conducted for the purpose of issuing a dumping permit.

The United States delegation presented preliminary information from the 6-year joint U.S. Corps of Engineers/Environmental Protection Agency Field Verification Program initiated in 1982, in relation to dumping of dredged materials. This program includes a comparison of land, estuary, and sea disposal of highly contaminated dredged material (Dillon, 1986). Objectives of the program are to (1) document in the laboratory predictive methods for assessing the effects of disposal of contaminated dredged material; (2) verify these methods in the field following disposal; (3) examine the association between tissue contaminant residue and observed biological effect; and (4) evaluate the findings of the first three objectives for aquatic, intertidal, and confined upland disposal alternatives.

The dredging project in Black Rock Harbor near Bridgeport, Connecticut, is being used as a case study. The potential for aquatic organisms to accumulate contaminants has been examined in two species, the mussel *Mytilus edulis* and the polychaete *Nereis virens*. Bioenergetics, adenylate energy charge, population dynamics, and histopathology have been studied using a number of benthic organisms. The preliminary results indicate "scope for growth" (Nelson et al., 1985) offers good potential for measuring the effects of contaminated dredged material on benthic organisms (R.M. Engler, personal communication).

5. DISCUSSION

Choosing the right type of measurement(s) for management of marine environmental quality is constrained by a host of requirements that are difficult to meet with one or even a combination of techniques. Investigators usually work with individual organisms or a group of organisms, and it has often been said that what happens to an individual organism on exposure to a toxicant may be insignificant for a population of those organisms. Depending on the choice of test organism, any measurement of a biological effect, acute or sublethal, may have little relation to the impact on the marine ecosystem. Modern ecological theory may be able to relate the measured effect of a particular waste on an individual or group of organisms to a population of those organisms, but this tells us nothing about the effect on communities. There is still a need to determine the effect of a waste discharge on marine communities by carrying out appropriate field investigations, or by examining the impact in large enclosures, such as CEPEX bags, or in laboratory mesocosms or microcosms. The exercise carried out by GEEP in Norway during August 1986 (IOC, 1986) should cast a great deal of light on various approaches that can be taken to examine anthropogenic impacts on marine ecosystems.

An effect that cannot be ascertained through any short-term acute or sublethal bioassay is the cumulative effect of a variety of wastes or of even one specific waste. For example, the cumulative effect of a pulp and paper mill effluent, containing wood solids that may settle out, does not become apparent for months or even years (e.g., Waldichuk, 1988). A variety of different discharges into coastal water will have cumulative effects that must be taken into account in any waste disposal plan (Waldichuk, 1986). In evaluating the assimilative capacity of a given body of water for a proposed substance, based on physical and chemical observations, existing inputs must be recognized and accounted for in any mass-balance evaluation (GESAMP 1986).

Management of marine environmental quality must involve monitoring to determine trends in seawater quality, and in concentrations of certain constituents in marine organisms and sediments. It is of some interest to examine the views of the ICES Advisory Committee on Marine Pollution concerning the use of biological effects techniques in pollution monitoring programs (ICES, 1986). It was noted that GESAMP (1980) recognized three phases in biological monitoring: Phase I, *identification*—detecting a change in time and/or space; Phase II, *quantification*—establishing the degree or extent of the change; and Phase III, *causation*—determining the cause of the observed change.

It was agreed that Phases I and II are closely interconnected and require a similar approach involving a baseline and monitoring studies. Phase III usually requires an experimental research approach. ACMP considered that a fourth phase is required to assess the consequences of observed effects. Most biological techniques used in the identification and quantification phases provide only comparative scales of measurement and cannot be immediately or simply related to consequences of direct concern to regulatory agencies. Techniques utilizing measurements at the individual level are often difficult to interpret in terms of population or community. The assessment phase would provide the essential link between the observation of biological change and management action.

Cause–effect relations are often difficult to establish. ACMP (ICES, 1986) suggested that even if clear evidence of causal relations is lacking, but an assessment of the observed change indicates that either the change itself or some consequence of the change is deleterious and significant, this should be sufficient indication for regulatory agencies to stop further inputs into the area. To assess the consequences of measured effects, ACMP proposed the following criteria as relevant.

1. The reversibility, and its time scale, of the effect.
2. The immediate ecological significance of the measured effect, particularly to valued species or communities.
3. The relevance of the effect to other levels of biological organization, i.e., the possibility of extrapolation to population and/or community levels.
4. The relevance to a range of taxa.
5. Trends in recorded effects.

It was emphasized that no single technique could fulfil the requirements of all phases and that usually a suite of techniques must be chosen. These techniques should be selected on the basis of their ability to facilitate assessment of the consequences of observed effects and to determine the causes of observed effects. The selected suite of techniques must be appropriate to the actual phase of the monitoring strategy, the aims and needs of management, and the indigenous biota.

Clearly, further advances in monitoring biological effects of contaminants in the marine environment can be achieved only through a better understanding of the interactions among physical, chemical, and biological components of the marine ecosystem. Both basic and applied interdisciplinary ecosystem research should continue and even increase. Some investigators (e.g., Bayne, 1985) suggest that biological studies must be integrated with current chemical monitoring programs, inasmuch as the biological responses that are measured require information on contaminant levels in the environment and in the biota if they are to achieve full relevance for environmental quality assessment. Efforts in development of techniques for quantifying the effects of contaminants on marine communities, both in the water column and on the bottom, must be intensified.

The foregoing needs are mainly in the research arena. Research is an essential step in the direction of effective management of marine environmental quality. In the meantime, protection of the marine environment against the onslaught of pollutants and development must go on. The best techniques already available and proven must be used in a consistent way to provide the necessary tools for management. Not the least of these are acute bioassays. As new proven techniques become available, they must be standardized and subjected to comparisons among laboratories. An important first step in any standardization is the control of the natural variability of biological assay material. Monitoring of biological effects of contaminants in such programs as the "mussel watch" (Martin, 1985), carried out in association with chemical programs, should continue and be enhanced as more information and experience are acquired.

In terms of impact of contaminants on marine living resources, particularly commercially important fish stocks, effort should be focused on the effects of contamination on reproductive success and recruitment. So far, any such effects that may exist have been obscured by the effects of natural variability in the marine environment that may affect reproduction and recruitment, and by the effects of harvesting. There is a growing interest in the recruitment problem among stock assessment biologists, oceanographers, chemists, and fishery ecologists, so that the complex web of interaction among the various variables may be closer to becoming unraveled.

6. SUMMARY AND CONCLUSIONS

1. Acute bioassays, notwithstanding their many shortcomings, continue to be a mainstay among the tools used by regulatory agencies for management of marine environmental quality. Various refinements of bioassays have improved the information derived from this technique.

2. There is now a multitude of laboratory techniques to measure the impact of contaminants at the sublethal level (McIntyre and Pearce, 1980). These cover effects on marine organisms in physiology, biochemistry, morphology and pathology, genetics, behavior, and ecology. Few have been fully tested in the field in a monitoring program.

3. Until this year (1986), there has been no effort through practical workshops and field trials to apply as wide a sample of available techniques as possible to convincingly evaluate the effectiveness of various biological measurements. In August 1986, such a practical workshop was conducted in Norway, under the auspices of the Intergovernmental Oceanographic Commission's Group of Experts on the Effects of Pollution (GEEP) (Bayne, 1985; IOC, 1986).

4. Preliminary results of the foregoing workshop (IOC, 1986) suggest that scope for growth in mussels, among physiological responses, has promise for standardized, routine biological effects measurements. Biochemical compositions of the digestive glands of both crabs and mussels showed significant changes in response to contaminant levels (mixture of water-accommodated fraction of diesel oil and soluble copper). The accumulation of lipid residues within some cell types was shown to be a sensitive measure of cellular

condition. In analysis of benthic communities, it was demonstrated to be possible to discriminate the control station from others along the contaminant gradient, and also to identify differences among contaminated sites, by applying multidimensional scaling to numbers or biomass distributions of fauna in the field.

5. Information on functioning of certain cellular systems, which can now be measured quite reliably by a number of techniques, can be used to reasonably predict effects on the individual organism.

6. Effects on the individual can be related with some confidence to the impact on a population through modern ecological theory (Bayne, 1985).

7. Assessment of effects of a disturbance on communities cannot as yet be made based on either theory or on the body of available empirical knowledge. Large enclosed bags, mesocosms, and microcosms are serving in part to predict effects on communities.

8. As techniques for measuring effects of contaminants on marine organisms become widely accepted, they will need to be standardized through interlaboratory comparisons.

9. There are four phases now recognized in biological effects monitoring for management purposes: (I) *identification*—detecting a change in space and/or time; (II) *quantification*—establishing the degree or extent of the change; (III) *causation*—determining the cause of the observed change; (IV) *consequences*—assessment of the consequences of observed effects. It has been suggested that usually a combination of techniques for measurement of biological effects is needed to meet the requirements of each of the foregoing phases.

10. In management of marine environmental quality, cumulative effects of both individual waste materials discharged over the long term (e.g., deposition of particulate material) and the combination of substances from different sources must be borne in mind when making mass-balance evaluations.

REFERENCES

Alderdice, D.F. (1972). Responses of marine poikilotherms to environmental factors acting in concert. In O. Kinne (Ed.), *Ecology,* Vol. 1, *Environmental Factors*. Wiley-Interscience, New York, pp. 1659–1722.

Bayne, B.L. (1985). Biological effects monitoring. *Mar. Pollut. Bull.* **16**: 86.

Cole, H.A. (Ed.) (1979). *The Assessment of Sublethal Effects of Pollutants in the Sea*. The Royal Society, London.

Davis, J.C., and R.A.W. Hoos (1975). Use of sodium pentachlorophenate and dehydroabietic acid as reference toxicants for salmonid bioassays. *J. Fish. Res. Board Can.* **32**: 411–416.

Dillon, T.M. (1986). The field verification program—a progress report on the aquatic disposal alternative. Ecosystem Research and Simulation Division, Environmental Laboratory, Waterways Experiment Station, U.S. Army Corps of Engineers, Vicksburg, MS.

GESAMP (1980). Monitoring Biological Variables Related to Marine Pollution. IMCO/FAO/UNESCO/WMO/WHO/IAEA/UN/UNEP Joint Group of Experts on the Scientific Aspects of Marine Pollution. GESAMP Reports and Studies No. 12, UNESCO, Paris.

GESAMP (1986). Environmental capacity: An approach to marine pollution prevention. IMO/FAO/UNESCO/WMO/WHO/IAEA/UN/UNEP Joint Group of Experts on the Scientific Aspects of Marine Pollution. GESAMP Reports and Studies No. 30, FAO, Rome.

ICES (1978). On the feasibility of effects monitoring. ICES Coop. Res. Rep. Cons. perm. int. Explor. Mer, No. 75, International Council for the Exploration of the Sea, Charlottenlund, Denmark.

ICES (1986). Report of the ICES Advisory Committee on Marine Pollution (ACMP), 1985. ICES Coop. Res. Rep. Cons. perm. int. Explor. Mer, No. 135, International Council for the Exploration of the Sea, Copenhagen, Denmark.

IOC (1986). Preliminary Report on the IOC Workshop on Biological Effects Measurements (Oslo, Norway, 11–29 August 1986). Intergovernmental Oceanographic Commission IOC/GGE(EP)-III/3 Prov. Extract, Paris, 12 September 1986.

Kullenberg, G. (Ed.) (1986). The IOC programme on marine pollution. *Mar. Pollut. Bull.* **17**: 341–352.

Lindahl, P.E., and E. Schwanbom (1971). Rotary-flow technique as a means of detecting sublethal poisoning in fish populations. *Oikos* **22**: 354–357.

Martin, M. (1985). State Mussel Watch: Toxics surveillance in California. *Mar. Pollut. Bull.* **16**: 140–146.

McIntyre, A.D. (1984). What happened to biological effects monitoring? *Mar. Pollut. Bull.* **15**: 391–392.

McIntyre, A.D., and J.B. Pearce (Eds.) (1980). Biological effects of marine pollution and the problems of monitoring. *Rapp. P.-v. Rêun. Cons. perm. int. Explor. Mer* **179**: 1–346.

Monk, D.C. (1983). The uses and abuses of ecotoxicology. *Mar. Pollut. Bull.* **14**: 284–288.

Moore, M.N. (1985). Cellular responses to pollutants. *Mar. Pollut. Bull.* **16**: 134–139.

NAS (1972). *Water Quality Criteria 1972*. National Academy of Sciences/National Academy of Engineering, Washington, DC.

Nelson, W.G., D. Black, and D. Phelps (1985). Utility of scope for growth index to discuss the physiological impact of Black Rock Harbor sediment in the blue mussel, *Mytilus edulis*: A laboratory evaluation. Tech. Rept. D-85-6, prepared by the U.S. Environmental Protection Agency, Narragansett, R.I., for the U.S. Army Corps of Engineers, Waterways Experiment Station, Vicksburg, MS.

Norton, M.G., and R. Lloyd (1981). The role of laboratory testing in the United Kingdom's controls on the dumping of wastes at sea. *Chemosphere* **10**: 641–657.

OECD (1981). OECD Guidelines for Testing of Chemicals. Organization for Economic Co-operation and Development, Paris.

Perkins, E.J. (1979). The need for sublethal studies. *Philos. Trans. R. Soc. London Ser. B* **286**: 425–442.

Steele, J.H. (1979). The use of experimental ecosystems. *Philos. Trans. R. Soc. London Ser. B* **286**: 583–595.

Waldichuk, M. (1979). Review of the problems. *Philos. Trans. R. Soc. London Ser. B* **286**: 399–424.

Waldichuk, M. (1985). Methods for measuring the effects of chemicals on aquatic animals as indicators of ecological damage. In V.B. Vouk, G.C. Butler, D.G. Hoel, and D.B. Peakall (Eds.), *Methods for Estimating Risk of Chemical Injury: Human and Non-Human Biota and Ecosystems*. Wiley, Chichester, U.K.

Waldichuk, M. (1986). Management of the estuarine ecosystem against cumulative effects of pollution and development. In G.E. Beanlands, W.J. Erckmann, G.H. Orians, J. O'Riordan, D. Policansky, M.H. Sadar, and B. Sadler (Eds.), *Cumulative Environmental Effects: A Binational Perspective*. Canadian Environmental Assessment Research Council (CEARC), Ottawa, Canada; United States National Research Council (NRC), Washington, DC, pp. 93–105.

Waldichuk, M. (1988). Prediction of environmental effects based on invalid assumptions. *Mar. Pollut. Bull.* **19**: 45–46.

Widdows, J. (1985). Physiological responses to pollution. *Mar. Pollut. Bull.* **16**: 129–134.

3

SURVIVAL OF LAKE CHARR (*SALVELINUS NAMAYCUSH*) EMBRYOS UNDER PULSE EXPOSURE TO ACIDIC RUNOFF WATER

John M. Gunn

Fisheries Branch, Ontario Ministry of Natural Resources, Toronto, Ontario, Canada and Department of Zoology, University of Guelph, Guelph, Ontario, Canada

1. **Introduction**
2. **Fish species**
3. **Study sites**
4. **Materials and methods**
   - 4.1. Field toxicity tests
     - 4.1.1. Fertilization—Encapsulated embryo
     - 4.1.2. Overwinter test of site-specific mortality
   - 4.2. Laboratory toxicity tests
     - 4.2.1. H^+ and Al lethality tests
     - 4.2.2. Laboratory experiments using natural water at 1°C
   - 4.3. Data handling and statistical analysis
5. **Results and discussion**
   - 5.1. Field studies
   - 5.2. Laboratory studies
     - 5.2.1. Tests with synthetic solutions
     - 5.2.2. Experiments with natural lake water
6. **General discussion**
7. **Conclusion**

References

1. INTRODUCTION

Acidification of surface waters appears to cause the reduction or extinction of natural fish populations mainly through a failure to produce recruits rather than a mass mortality of juveniles or adults (Haines, 1981; Muniz and Leivestad, 1980; Rosseland, 1986). The adverse affects of low pH and associated chemistry on reproduction and recruitment in freshwater fish have been extensively studied. Some of the processes addressed include adult maturation (Beamish et al., 1975; Weiner et al., 1986), gametogenesis (Craig and Baksi, 1977; Hutchinson and Sprague, 1986; Ruby et al., 1978), embryo ion regulation (Peterson and Martin-Robichaud, 1986), development (Geen et al., 1985; Nelson, 1982), growth (Cleveland et al., 1986), early feeding (Lacroix et al., 1985) and survival (Baker and Schofield, 1982; Ingersoll, 1986; Rombough, 1983).

Peterson et al. (1982) summarized the results of many of the early studies on reproductive failure of freshwater fish in acidic water. They concluded that mortality occurred mainly at three sensitive developmental stages: (1) during early cleavage, (2) during or shortly after hatching, and (3) at the initiation of exogenous feeding. They also identified a variety of modifying factors that required further study. The role of such modifying factors in the lethal response to low pH has been the focus of much of the recent work. Included has been the study of the effects of metals: singly, in the case of Al (Baker and Schofield, 1982); Cleveland et al., 1986; Wood and McDonald, 1987), or as mixtures of several metals (Hutchinson and Sprague, 1986); concentrations of cations in the water (Brown, 1982; McDonald, 1983); season (Stuart and Morris, 1984); body condition prior to exposure (Barton et al., 1985; Kwain et al., 1984); activity level of the fish (Graham and Wood, 1981); water temperature (Korwin-Kossakowski and Jezierska, 1985); and exposure frequency and duration (Siddens et al., 1986).

Laboratory-derived findings have been used to make predictions about changes in natural populations of fish with increasing acidity of the water (Breck et al., 1986; Sadler, 1983). To date most of these mechanistic-type models use lethal threshold relations (eg. mortality versus H^+) to predict the presence or absence of a fish species. Mixed results have been achieved. For example, Sadler (1983) used laboratory toxicity data on survival of brown trout (*Salmo trutta*) under varying levels of hydrogen ions (H^+) and Ca^{2+} in a model that was tested against survey data of brown trout populations in culturally acidified lakes in southern Norway. His predictions generally underestimated observed population losses in these lakes. He concluded that effects on other unmeasured life stages and the additive or synergist effects of other toxicants, such as Al (Baker and Schofield, 1982), were probably important in natural systems. Use of data for average chemical conditions in the survey lakes, rather than extreme conditions observed during episodic events, may also have lead to underestimated effects.

Episodic acidification is a widespread phenomenon in areas with low alkalinity lakes that receive a high deposition of acidic precipitation (Henriksen et al., 1984; Jeffries et al., 1979; Kelso et al., 1986). These short-term chemical changes are usually associated with snowmelt or periods of heavy rain. In addition to the acid, large quantities of Al may be leached from the watershed soils and carried in solution to receiving lakes and streams (Cronan and Schofield, 1979). The combination of high levels of H^+ and inorganic Al in dilute runoff water can have severe effects on fish and other aquatic biota (Driscoll et al., 1980; Henriksen et al., 1984; Leivestad and Muniz, 1976).

In this study, *in situ* and laboratory toxicity tests were conducted to examine the effects of episodic acidification on survival of lake charr (*Salvelinus namaycush*) embryos. The objectives of the field study were to determine survival relative to measured chemical conditions (1) during fertilization and water hardening, (2) during encapsulation of embryos within egg membranes, and (3) as free embryos. Laboratory tests were used to assess survival of lake charr embryos under 5-d pulse exposure to various levels of H^+ and Al in low-Ca^{2+} water. Water collected from acid lakes was also used in laboratory tests.

2. FISH SPECIES

Lake charr (*Salvelinus namaycush*) is an important commercial and sport-fish species that has been lost from about 60 acidified lakes in Ontario (Beggs et al., 1985; Beggs and Gunn, 1986) and another 4–5 lakes in the Adirondack Mountains area of New York (Haines and Baker, 1986). Lake charr appear to be particularly vulnerable to damage from acidic runoff water because they spawn at shallow nearshore sites in the fall, and embryonic life stages are present and remain within the spawning substrate throughout the snowmelt period (Gunn, 1986). In a survey of 82 Ontario lakes, McMurtry (1986) found that more than 75% of lake charr spawning sites are in water less than 2 m deep and within 10 m from shore.

Lake charr are relatively well studied with regard to acidification effects. Netting surveys indicate that populations decline or become extinct when whole-lake average pH drops below 5.2–5.6 (Baker and Harvey, 1984; Beamish and Harvey, 1972; Beggs et al., 1985; Beggs and Gunn, 1986). Population assessments in lakes acidified through atmospheric inputs (Beggs et al., 1985), or artificially, through acid additions (Mills, 1984; Schindler et al., 1985), demonstrate that adults continue to spawn and produce viable gametes in affected lakes, but that recruitment to the juvenile stage ceases. The mechanism/s for this recruitment failure are still unknown, and wide discrepancies between laboratory and field findings exist (Hutchinson et al., 1987). High mortality during episodic acidification events is one possible explanation for the apparent discrepancy.

In this paper, terminology of lake charr life stages follows that of Balon (1980). "Embryo" refers to the period when lake charr are dependent only on yolk material for nutrition. "Encapsulated embryos" are embryos within the

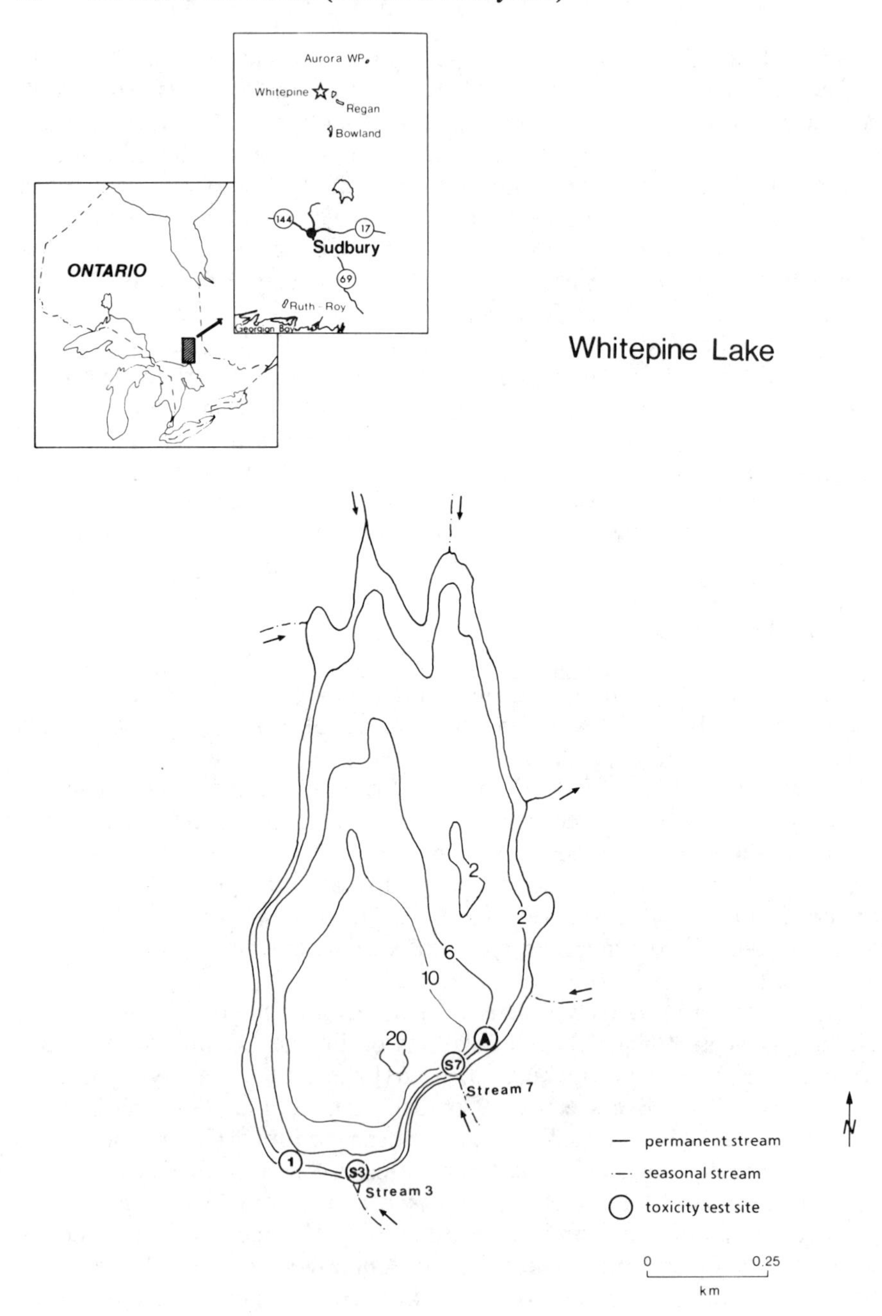

Figure 3.1. Location of the study lakes used for field toxicity tests (Bowland, Whitepine, Regan) and lakes used as sources of water for laboratory experiments (Aurora WP, Ruth-Roy). The toxicity test sites on the principal study lake, Whitepine Lake, are indicated. The depth contours are in meters.

egg membranes. "Free embryos" are embryos after hatching. "Alevins" are free-swimming fish in the transition stage between endogenous and exogenous feeding.

3. STUDY SITES

Field toxicity tests were conducted in three lakes known to have contained natural populations of lake charr: Bowland Lake, Whitepine Lake, and Regan Lake. The lakes were located within an area of boreal forest and granitic bedrock, 70–90 km north of Sudbury, Ontario (Fig. 3.1, Table 3.1).

Bowland Lake (pH 5.1) was the most acidic lake. Its native lake charr were extinct and only acid-tolerant yellow perch (*Perca flavescens*) were observed in the lake during the early 1980s (Kelso and Gunn, 1984). Experiments with hatchery-reared fish demonstrated that the ambient chemical conditions of Bowland were lethal to a variety of age classes (5 months–2 years) of juvenile

Table 3.1 Chemical Characteristics of Study Lakes and Laboratory Test Water[a]

	Field Toxicity Test Lakes			Water Collection Lakes for Laboratory Tests		
	Regan 47°14′N 80°47′W	Whitepine 47°17′N 80°50′W	Bowland 47°05′N 80°50′W	Aurora WP[b] 47°23′N 80°38′W	Ruth-Roy 46°06′N 80°14′W	Laboratory Soft Water
pH	6.5 (0.3)	5.6 (0.2)	5.1 (0.05)	5.0 (0.1)	4.6 (0.05)	7.0 (0.02)
Cond. $\mu S \cdot cm^{-1}$	40 (3)	35 (3)	35 (1)	32 (0.2)	34 (2)	21 (2)
DOC $mg \cdot L^{-1}$	2.1	2.3 (0.8)	1.9 (0.4)	2.3 (0.5)	1.5 (0.3)	0.1 (0.05)
Ca^{2+} $mg \cdot L^{-1}$	3.5 (0.6)	2.8 (0.3)	2.9 (0.2)	2.2 (0.2)	1.4 (0.1)	1.7 (0.2)
Mg^{2+} $mg \cdot L^{-1}$	1.0 (0.04)	0.8 (0.05)	0.7 (0.1)	0.7	0.4 (0.1)	0.4 (0.04)
Na^{+} $mg \cdot L^{-1}$	1.0	0.8 (0.1)	0.7 (0.02)	0.8 (0.1)	0.5 (0.1)	1.5 (0.1)
K^{+} $mg \cdot L^{-1}$	0.5	0.6 (0.1)	0.3 (0.03)	0.4	0.2 (0.05)	0.1 (0.004)
Cl^{-} $mg \cdot L^{-1}$	0.2	0.4 (0.1)	0.4 (0.0)	0.4	0.4 (0.01)	2.0 (0.2)
$SO_4^{=}$ $mg \cdot L^{-1}$	11.3 (0.2)	10.4 (1.1)	10.9 (0.4)	10.4	9.35 (0.4)	1.4 (0.2)
Cu $\mu g \cdot L^{-1}$	<1	<2	<2	<1	3 (1)	<1
Ni $\mu g \cdot L^{-1}$	2	2 (0.5)	4 (0.6)	<2	16 (2)	<1
Zn $\mu g \cdot L^{-1}$	2	4 (2)	6 (2)	10	25 (2)	2
Pb $\mu g \cdot L^{-1}$	<3	<3	<3	<3	<3	<3
Fe $\mu g \cdot L^{-1}$	11 (4)	42 (28)	63 (15)	65	118 (24)	29 (19)
Mn $\mu g \cdot L^{-1}$	5 (2)	30 (7)	84 (4)	116	96 (3)	<1
Al $\mu g \cdot L^{-1}$	21 (8)	49 (20)	143 (12)	158 (43)	610 (27)	<3
n	1–6	2–13	2–3	1–5	2–3	4–9

[a] Mean (SD) given. Lake water collected as epilimnetic composite samples in 1982 (Whitepine, Bowland), 1982-83 (Regan) and 1985 (Aurora WP and Ruth-Roy). Laboratory soft water was a dechlorinated mixture of deionized and untreated Toronto tap water.

[b] Also known as Whitepine Lake.

lake charr (Booth et al., 1986; Gunn et al., 1987). Bowland Lake was limed approximately 6 months after the completion of this field toxicity experiment, as part of a long-term study of lake neutralization (Booth *et al.*, 1986). The neutralization study used procedures for chemical monitoring and toxicity testing that were similar to mine, and therefore provide some useful data for comparison.

Whitepine Lake (pH 5.6) was the principal study site (Fig. 3.1). It, too, was dominated by yellow perch, but had remnant populations of adult lake charr and other acid-sensitive species, such as white sucker (*Catostomus commersoni*) and burbot (*Lota lota*). Netting assessments indicated that there was no recruitment of juvenile lake charr in Whitepine Lake during the 1970s (Gunn, unpubl. data). Recruits began to appear in 1982, presumably due to improved water quality resulting from reduced emissions from the Sudbury smelters (Keller and Pitblado, 1986). Thus, the chemical conditions in Whitepine during the period of this study (1982–1983) may have been less severe than in the past, but extensive areas of the shoreline still displayed large pH depressions during snowmelt (Gunn and Keller, 1984a).

Regan Lake was a circumneutral lake (Table 3.1). It contained a mixed fish community including a healthy population of lake charr (Beggs et al., 1985). It served as a source of gametes, and as a reference lake for toxicity tests.

In 1985 water was collected from two highly acidified lakes, Aurora WP and Ruth-Roy lakes (Fig. 3.1, Table 3.1), for use in laboratory tests of survival of embryos exposed to "natural" water from acid lakes. Aurora WP Lake was one of the native lakes of the aurora trout, a unique color phase of brook charr (*S. fontinalis*). Aurora trout disappeared from Aurora WP Lake and two other nearby lakes when the lakes acidifed in the 1950s and 1960s (Keller, 1978). Netting surveys in 1983 indicated that Aurora WP Lake was fishless (V. Liimatainen, Ontario Ministry of Natural Resources, unpubl. data). Ruth-Roy Lake was also a fishless, acidic (pH 4.6) lake with high levels of Al (ca. 600 μg/L total, 300–400 μg/L inorganic), Zn (25 μg/L), Mn (96 μg/L), and other metals in the surface water. In previous field toxicity tests, the waters of Ruth-Roy have proven to be lethal to a variety of fish species (Hulsman et al., 1983; Gunn and Keller, 1984b).

4. MATERIALS AND METHODS

4.1. Field Toxicity Tests

In situ experiments were conducted during the winter of 1982–83. Adult lake charr were captured in Regan Lake, the circumneutral lake, and transported alive to the field station on Whitepine Lake, to provide gametes for field toxicity tests.

4.1.1. Fertilization—Encapsulated Embryo

The first experiment was a test of the effects of acidic water on survival during the periods of (1) fertilization and water hardening, and (2) encapsulation of embryos within the egg capsule. Eggs were manually spawned from an anesthetized female (54.3 cm total length, 1.65 kg wet wt) and divided into 3 lots of 500 eggs each. Each lot was dry fertilized with 3 drops of sperm (52.0 cm., 1.65 kg male) and rinsed with 250 mL of water from one of the study lakes (Bowland, Whitepine, Regan). After 3 min the rinse water was decanted off and the fertilized eggs were transferred to a 10-L Nalgene pail of the same lake water. The eggs were left undisturbed for 3 h while they absorbed water, creating the perivitelline fluids (hardening). After hardening, eggs were loaded into sandwich-style incubators (Hulsman et al., 1983) with fifty 1.2-cm-diam chambers arranged in 5 rows of 10. The individual rows of chambers in an incubator were randomly assigned eggs from each fertilization treatment. This mixing of eggs within incubators was used to eliminate incubator-specific effects. Six incubators (i.e., a total of 10 rows of eggs from each treatment) were placed at a nearshore site on each lake. On Whitepine and Regan lakes, the incubation sites were known spawning sites of the native fish. On Bowland Lake the original spawning sites were unknown, therefore an incubation site similar in appearance to those on the other lakes was used. Mortality counts were conducted on all incubators when this experiment was terminated during the period 3–10 Feb. 83.

4.1.2. Overwinter Test of Site-Specific Mortality

Lake charr embryos were held at four sites along the shoreline of Whitepine Lake throughout the winter of 1982-83 to determine the timing of periods of high mortality. Sites were chosen that were known to differ chemically during snowmelt. The sites included spawning site A, shoreline site 1, and sites at the inlets of two streams, 3 and 7 (Fig. 3.1). Stream 3 had a relatively large drainage area and ran frequently throughout the year in response to both rainfall and snowmelt. Stream 7 was a tiny stream that ran only in the spring during snowmelt. The egg incubation sites were located within 1 m of the inlets of these intermittent streams. The other sites were also within 1 m of shore at a similar depth (0.6–1.0 m).

Fertilized eggs from a pair of Regan Lake adults (54.6 cm TL female, 47.6 cm TL male) were hardened in Whitepine Lake water on 24 Oct, placed in plexiglass incubators ($n = 3$ per site with 50 eggs per incubator), and suspended at each shoreline site. Mortality counts were conducted periodically throughout the winter (2 Feb, 23 Feb, 15 Mar, 29 Mar) and then on a daily basis beginning on 6 Apr. Just prior to snowmelt, the live free embryos from each incubator were carefully transferred into separate 1-L screened pens. The pens floated 30–50 cm below the ice at the exposure sites. Mortality counts and removal of dead individuals were conducted daily throughout the period of

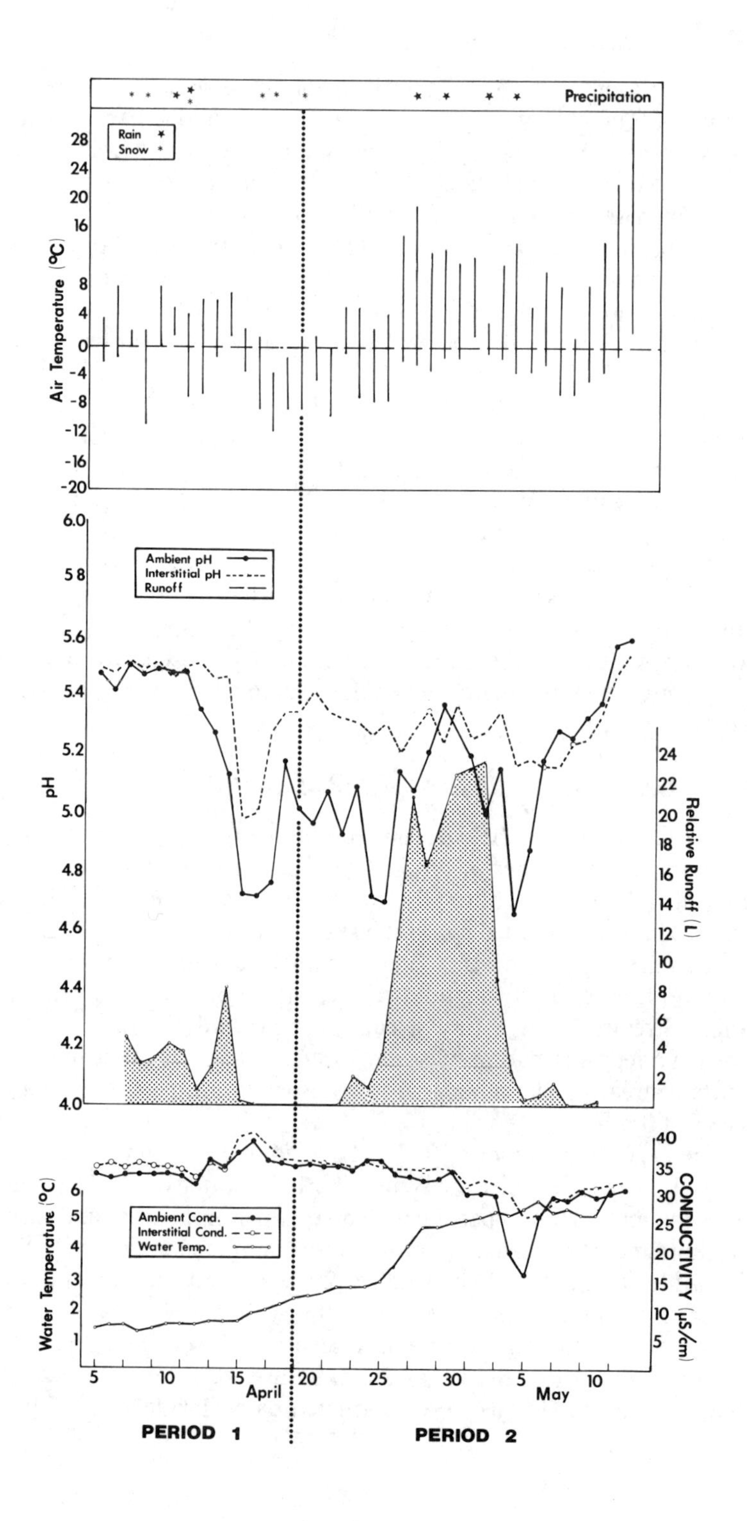

Precipitation
Rain
Snow
Air Temperature (°C)
28
24
20
16
8
4
0
-4
-8
-12
-16
-20
Ambient pH
Interstitial pH
Runoff
pH
6.0
5.8
5.6
5.4
5.2
5.0
4.8
4.6
4.4
4.2
4.0
Relative Runoff (L)
24
22
20
18
16
14
12
10
8
6
4
2
Ambient Cond.
Interstitial Cond.
Water Temp.
Water Temperature (°C)
6
5
4
3
2
1
CONDUCTIVITY (µS/cm)
40
35
30
25
20
15
10
5
5
10
15
April
20
25
30
5
May
10
PERIOD 1
PERIOD 2

snowmelt and ice breakup (30 d). Daily water samples were drawn from beside the pens. Dissolved O_2 and temperature were measured daily *in situ* using a YSI model 54 meter. Water samples were analyzed daily for pH and specific conductance at the field station on Whitepine Lake (Radiometer model PM84 pH meter, YSI model 32 conductivity meter). Duplicate water samples were collected 2–3 times weekly for analysis of metals and major ions. Inorganic Al was analyzed by the dialysis separation procedure of LaZerte (1984).

In addition to monitoring surface water chemistry at the toxicity test sites, interstitial chemical conditions at Spawning Site A on Whitepine Lake were also monitored during the period 5 Apr 83 to 11 May 83. Sampling methods of interstitial water are described in Gunn and Keller (1984a). The observed characteristics (duration of pH depressions, water temperatures, etc.) of runoff pulses in the interstitial environment (Fig. 3.2) were used to design laboratory experiments that closely simulated natural conditions.

4.2. Laboratory Toxicity Tests

4.2.1. H^+ and Al Lethality Tests

Laboratory experiments were designed to test survival of those life stages observed to be affected by acidic meltwater *in situ*. These were (1) encapsulated embryos just prior to hatch, and (2) free embryos. Different ages of free embryos were simultaneously tested to determine if sensitivity to H^+ and Al changed as the embryos absorbed yolk material and began branchial respiration. The test conditions were restricted to the chemical conditions observed in the field rather than a wide matrix of H^+ and Al combinations. An experiment consisted of a 5-d exposure to the test solution followed by a return to control conditions until completion of yolk absorption. This exposure regime closely matched conditions observed during snowmelt at a natural spawning site on Whitepine Lake (Fig. 3.2).

Lake charr eggs were obtained from Lake Manitou as part of a hatchery collection that consisted of pooled gametes from several males and females. Fertilized eggs or free embryos were transfered to the laboratory and held for a minimum of 19 d in soft water (Ca^{2+} 2 mg/L, hardness 5 mg/L as $CaCO_3$) before use in an experiment. Chemical characteristics of the laboratory soft water are presented in Table 3.1.

Figure 3.2. Effects of acidic runoff water on chemical conditions at lake charr spawning site (Site A) on Whitepine Lake. During periods of cold water temperature (1–2°C; Period 1) acidic runoff water (pH 4.1) penetrates to the substrate level causing changes in interstitial pH. After warming to 4°C (Period 2) little change in interstitial pH occurs even during periods of heavy runoff. The melting of the icepack produces the sharp decline in specific conductance of the water column in late April. Air temperature (range) and occurrence of precipitation events are indicated. Relative runoff is measured as daily volume of water obtained from a shoreline collector. Sampling methods follow those of Gunn and Keller (1984a).

Exposure of embryos to H^+ and Al occurred in 40-L static tanks in a 5°C water bath. Tanks were continually aerated and solutions changed daily. Test solutions were prepared by adding $Al_2(SO_4)$ and H_2SO_4 to soft water. Vigorous aeration and repeated titration with 0.02 N H_2SO_4 were used to produce stable pH conditions. New test solutions were prepared daily and cooled for approximately 24 h to equilibrate (±0.5°C) to temperatures in the exposure tanks. There were three replicate exposure tanks per treatment including controls. Equal numbers of randomly selected test animals (depending on availability, the number varied from 20 to 50) were placed in small floating pens (9.5 cm diam; Styrofoam with polyethylene mesh) in each tank. Fish were kept in the dark except during water sampling, water changes, and mortality checks. At the end of the 5-d exposure all fish were transferred to their identical floating pens to temperature-regulated (5.0°C) control water until yolk absorption was complete (21–70 d).

Temperature (±0.05°C) was measured daily in each tank using a NBS Hg thermometer. Daily water samples were collected in duplicate in polystyrene bottles from each tank for analysis of pH (± 0.01), specific conductance at 25 °C (±0.3 μS/cm), and total hardness (± 0.3 mg L^{-1} as $CaCO_3$). All pH readings were made within 12 h of sample collection using a Radiometer PM82 meter equipped with a low ionic strength electrode (Radiometer GK2401C). Water samples were submitted to Ontario Ministry of Environment laboratories for analysis of Na, Ca, and total Al by atomic absorption spectroscopy (Ontario Ministry of the Environment, 1981), and analysis of inorganic Al by the pyrocatechol violet colorimetric method (Rogeberg and Henriksen, 1985). On several occasions, samples were collected immediately before and after the daily water changes to assess loss of aqueous inorganic Al through precipitation or adherence to surfaces. Losses were less than 5%. The low loading rate (0.1 g L^{-1}) of nonfeeding fish, and the low levels of dissolved organic carbon (0.2 mg/L) in the test water were presumably responsible for the good agreement between nominal and measured inorganic aluminum concentrations in the static tanks.

4.2.2. *Laboratory Experiments Using Natural Water at 1°C*

Addition toxicity tests were conducted using natural lake water and a test temperature (1°C) that better matched the conditions observed *in situ* (Fig. 3.2).

Epilimnetic water was collected from the Ruth-Roy and Aurora WP lakes and transported in 200-L polyethylene barrels to the laboratory. The barrels of water were kept in the dark and refrigerated (8°C) until use. Toxicity tests began within 48 h of collection. Some additional experiments with synthetic solutions were also run for comparison. In these experiments test solutions (pH 4.0; pH 4.5 + 500 μg/L Al), were prepared by adding H_2SO_4 and $Al_2(SO_4)$ to soft water as described above.

Lake charr embryos were obtained from the Chatsworth fish hatchery, and held in cold (1°C) soft water for 60 d prior to exposure to acidic solutions. Static toxicity tests were performed in polyethylene-lined glass bottles containing 2.5 L of continuously aerated water. There were 6 replicate bottles per treatment and 10 fish per replicate. Temperatures were maintained at 1 + 0.1°C during the toxicity tests by suspending bottles in large (0.7 × 1.2 × 0.7 m), insulated, thermoregulator-controlled water baths. During the 2- to 7-d toxicity tests embryos were inspected several times a day to assess time of mortality. Loss of body color, cessation of movement, and lack of response to tactile stimuli were the criteria for assessing death. Temperature, pH, and specific conductance were measured daily for each replicate. Samples (2–3) collected from within the bottles were analyzed for major ions and metals as described earlier.

4.3. Data Handling and Statistical Analysis

An arcsine transformation was applied to percentage mortality data prior to statistical comparisons by ANOVA. Median survival times (MST) were determined by the methods of Litchfield (1949). In experiments with small numbers of replicates ($n \leq 3$), means ± 1 SD are presented in figures and tables without attempting further statistical tests.

5. RESULTS AND DISCUSSION

5.1. Field Studies

The fertilization medium had no effect ($F = 0.88$, $p > 0.05$) on long-term survival of encapsulated embryos in any of the test lakes (Table 3.2). There were significant differences ($p < 0.001$) between lakes, but the highest mortality during the 106-d toxicity test (13–20%) was in Regan Lake, the circumneutral reference lake, rather than in Bowland Lake, the site with the most acidic water. The observed mortality differences between lakes were presumably due to handling differences during transport, or to physical factors such as wind and

Table 3.2 Effects of Lake Water pH at the Time of Fertilization and Water Hardening (3 h) on Long-Term Survival of Lake Charr Eggs under Various Incubation Conditions[a]

			% Survival in each Fertilization Medium		
Fertilization Date	Rearing Site	Inspection Date	Bowland (pH 5.1)	Whitepine (pH 5.6)	Regan (pH 6.5)
20 Oct 82	Bowland	(10 Feb 83)	92.0 (6.3)	95.0 (10.8)	92.0 (7.9)
	Whitepine	(3 Feb 83)	99.0 (3.0)	92.0 (9.0)	95.0 (5.3)
	Regan	(10 Feb 83)	87.0 (12.5)	81.0 (17.9)	80.0 (21.1)

[a]Mean (SD) percentage survival and n given. Each sample contained 10 eggs.

wave action. The incubation site on Regan Lake was much further offshore than the sites on the other lakes and appeared to be more vulnerable to disturbance from wind-driven currents during the fall, when the lakes were still ice free. Overall, the survival in all lakes and treatment groups was high (80–99%) relative to values reported from similar studies (Booth et al., 1986; Gunn and Keller, 1984b; Johnson et al., 1987).

In the site-specific mortality experiment, the inlet stream sites exhibited the highest mortality, but the timing of observed mortality differed between the two inlet sites (Fig. 3.3). At the site in front of the larger stream, stream 3, mortality (79.5 ± 15.4%) occurred between fertilization and the first inspection of the site on 2 Feb. 83. No chemical data exist for the period of high mortality. However, sampling conducted at other times of the year, showed that stream 3 was consistently very acidic (pH 4.2–4.6), and that during periods of melt or heavy rain it kept the pH at <5.0 in the area where the egg incubators had been located (Gunn, unpubl. data). The other inlet site showed less than 10%

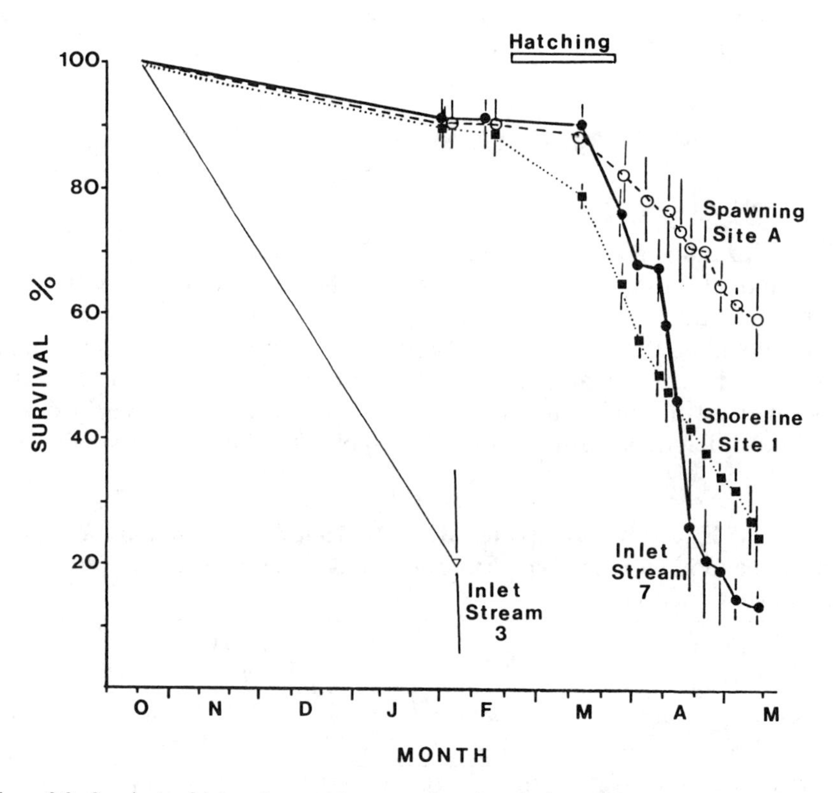

Figure 3.3. Survival of lake charr embryos at four incubation sites along the shoreline of Whitepine Lake, 1982–1983. Mean (± 1 SD) survival for groups of three incubators at each site are shown. In some cases standard deviation bars are omitted to increase clarity.

mortality during the fall and early winter, similar to that observed at sites away from inlets (Fig. 3.3). The initial 10% mortality appeared to be due to incomplete fertilization prior to placing the eggs in the lake. Most of the dead eggs contained no developing embryos.

Mortality began at the high-survival sites during or shortly after hatching (Fig. 3.3). Final survival was lowest at the second inlet site, 7 (13.3 ± 3.5%). The rapid mortality occurred at this site when embryos were free from the egg capsule (Fig. 3.4), and began during snowmelt as pH dropped to 4.5 and inorganic Al rose to 48 μg/L (temperature 0.5–0.8°C, DO 10 mg/L, Ca^{2+} 2.9–3.0 mg/L, Na^{+} 0.65–0.70 mg/L). Site 1 also had low survival (24.0 ± 5.3%), but

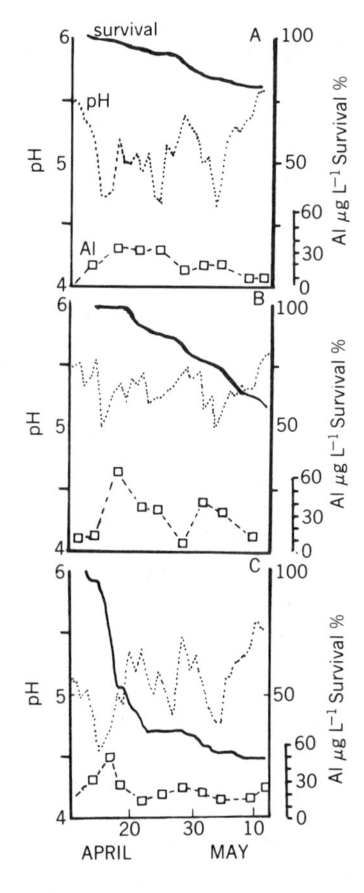

Figure 3.4. Survival of lake charr free embryos in 1 L screened pens, pH (daily measurements) and inorganic Al (2–3 measurements per week) at (A) spawning site A, (B) shoreline site 1, and (C) stream inlet 7 on Whitepine Lake, during the snowmelt period of 1983. Survival is expressed as a percentage of the initial number (n = 20–43) of embryos that were alive at the time of transfer from the plexiglass incubators to the screen pens used in this test.

mortality occurred steadily throughout the free embryo period, a period when pH was not observed to drop below 5.0 (Fig. 3.4). Relatively high survival (58.7 ± 6.1%) occurred at spawning site A in spite of frequent fluctuations in pH (min. 4.7) and other chemical parameters (inorganic Al 6–30 μg/L, Ca^{2+} 1.10–2.9 mg/L, Na^+ 0.30–0.65 mg/L) during snowmelt (Fig. 3.4; Gunn, 1987).

5.2. Laboratory Studies

5.2.1. Tests with Synthetic Solutions

Lake charr embryos were relatively tolerant of low pH. No significant mortality of free embryos occurred during 5-d exposures to pH 4.25–5.0 (Table 3.3). High mortality occurred only when pH was lowered to 4.0, well below the minimum observed pH in the interstitial water (pH 4.6) in Whitepine Lake, or in other study lakes (Gunn, 1987). The addition of 100 μg/L Al to low-pH (4.5, 5.0) water did not cause mortality of free embryos. Some increased mortality (13.3–16.7%) and delays in hatching (Fig. 3.5) occurred at realistic H^+ and Al concentrations in tests that were conducted using encapsulated embryos (Table 3.3).

Table 3.3 Survival of Lake Charr Embryos during 5-d Exposure to Low pH and Elevated Al, Followed by 10-d Recovery Period in Control Water (pH 7.0, 0 Al)[a]

Treatment		Overall Survival (%)			
		During Hatching[b]		Free Embryos	
pH	Al (μg · L^{-1})	5-d Exposure	10-d Recovery	5-d Exposure	10-d Recovery
(1) 4.0	0	na[e]	na	1.0 (1.9)[c]	na
(2) 4.25	0	61.1 (7.0)	32.2 (5.1)	99.4 (1.1)	98.7 (1.2)
(3) 4.5	0	93.7 (5.0)	86.7	100 (0)	100 (0)
(4) 4.5	100	87.3 (11.5)	na	100 (0)	na
(5) 5.0	0	na	na	100 (0)	100 (0)
(6) 5.0	100	97.8 (1.9)	83.3 (3.3)	100 (0)	100 (0)
Controls[d]					
7.0	0	98–100	96.7–100	99.4–100	99.4–100

[a] Toxicity tests conducted in laboratory soft water (specific conductance 21 μs cm^{-1}, Ca^{2+} 1.7 mg L^{-1}, Na^+ 1.5 mg L^{-1}, Cl^- 2.0 mg L^{-1}) at 4.8–6.5°C. There were three replicate control and treatment tanks for each pH and Al level, with 20–50 embryos (450–600 degree-days old) per tank. Mean (SD) survival indicated.

[b] Experiments begun with fully encapsulated embryos drawn from batch of eggs in which some hatching had already begun.

[c] Experiment terminated on Day 4.

[d] Controls (pH 7.0, 0 Al) run for each experiment 1–6. Ranges in survival of controls indicated.

[e] na, no data available.

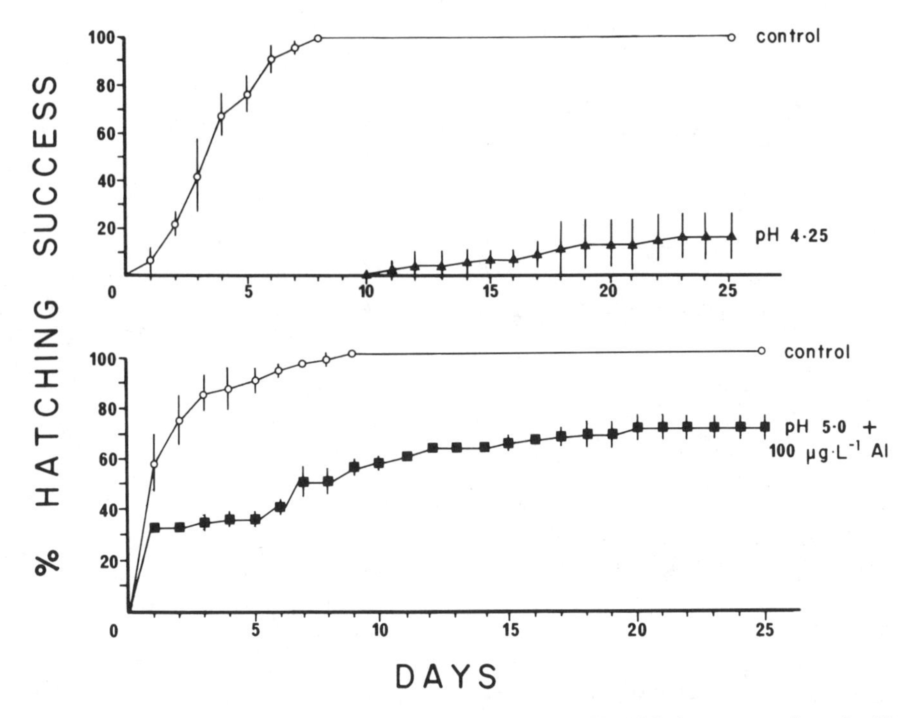

Figure 3.5. Delayed hatching of lake charr exposed to pH 4.25 (upper panel) and pH 5.0 + 100 ug/L Al (lower panel). The treatment groups were exposed to the acidic water during Days 1–5, then transferred to control water (pH 7.0) for the following 20 d. Control groups remained at pH 7.0 + 0 Al throughout. Mean (± 1 SD) percentage hatch indicated.

In one low-pH (4.25) test, hatching was delayed until 5 d after transfer to control water. Mortality continued throughout the control water period (Fig. 3.6). Embryos died while still partially enclosed in the chorion, a phenomenon commonly observed in low-pH water (Peterson et al., 1982). Peterson and Martin-Robichaud (1983) showed that the vigorous trunk movements needed to rupture the chorion are reduced at pH 4.0–4.5, while Haya and Waiwood (1981) suggested that low pH decreased the activity of chlorionase, the enzyme that partially dissolves the chorion prior to hatch. In this experiment, hatching ceased immediately upon exposure to the treatment water, suggesting that decreased embryo movement rather than reduced rate of chorion dissolution was responsible for the inability of the embryo to free itself from the egg membranes.

The embryos entered a second period of high sensitivity near the end of yolk absorption (Fig. 3.6). This increase in sensitivity may reflect the shift in the principal site of respiration from cutaneous to branchial membranes in the developing lake charr embryo (Balon, 1980). The development of the gills not only increases the surface area of the embryo over which ion loss can occur, but

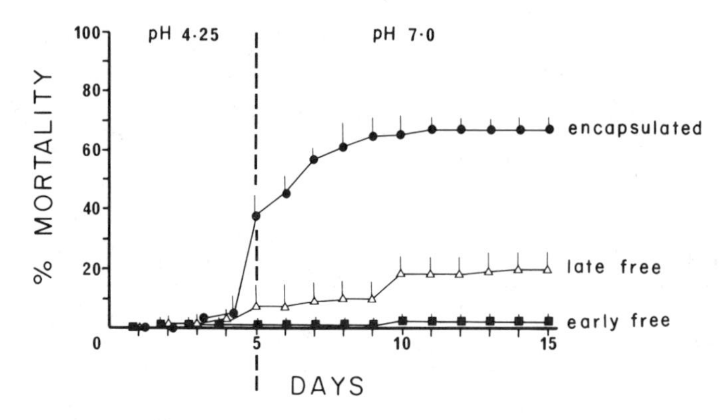

Figure 3.6. Mortality of different life stages of lake charr exposed to pH 4.25 for 5 d before transfer to control water. Mean (± 1 SD) percentage mortality indicated.

produces branchial membranes that are more permeable than epithelial tissues (McDonald, 1983; Talbot et al., 1982).

Lake charr embryos were less sensitive to acidic aluminum-rich water in laboratory test than in field tests. The high mortality of embryos at stream inlet 7 in Whitepine Lake, at pH 4.5 and 50 μg/L inorganic Al, did not occur in 5-d laboratory exposures of embryos to pH 4.5 and 0–200 μg/L Al. Differences in concentrations of other cations were not responsible for the differences in mortality. Calcium, which is known to ameliorate the toxic effects of H^+ and Al (Brown, 1982; 1983; McWilliams and Potts 1978), was at lower concentrations in the synthetic than in the natural water (Table 3.1). Sodium may also reduce the toxicity of H^+ (Brown and Lynam, 1981), but was in relatively low concentrations in the laboratory soft water (1.8 mg/L).

Hutchinson et al. (1987) also found that lake charr were relatively tolerant of low pH and elevated Al in soft water. Significant mortality occurred only when pH was <4.8. H^+ appeared to be the dominant toxicant. Aluminum produced few independent effects and increased mortality only at H^+ concentrations (pH 4.2–4.6) that were already causing some mortality. Free embryos were the most sensitive life stage to low pH and elevated Al. Sensitivity, in terms of immediate mortality during short-term exposure, declined in order of free embryos > alevins > cleavage eggs > encapsulated embryos. Some evidence of longer term effects was also obtained. Mortality of fish exposed to H^+ and Al during encapsulation continued after transfer to circumneutral control water.

5.2.2. Experiments with Natural Lake Water

Free embryos did not respond with rapid mortality in tests conducted at a realistic temperature (1°C) using natural lake water (Fig. 3.7), with its mixtures of associated metals and low levels of cations (Ca^{2+}, Na^+). No mortality occurred during 192-h exposures to Aurora WP Lake water (pH 5.1, inorganic

Al 56 ± 10 μg/L). Ruth-Roy water with its very low pH (4.7) and high Al (inorganic Al 332 ± 22 μg/L) was lethal to embryos after exposure periods of 105–120 h. Median survival time (MST) of embryos in Ruth-Roy water was 155 h. The prepared solution with a pH of 4.6 and measured inorganic Al of 376 ± 12 μg/L, but without the other metals of a "natural" acidic lake, produced a mortality pattern that was quite similar to that of Ruth-Roy Lake water. Mortality began between 52 and 72 h, with an estimated MST of 113 h. As in previous tests, mortality was very rapid (MST = 18 h) in pH 4.0 water.

6. GENERAL DISCUSSION

The low survival of lake charr embryos at the stream inlet sites cannot be used to predict population effects in acidifying lakes. Spawning sites of lake charr in lakes on the Precambrian shield are rarely located near streams (McMurtry, 1986). In Whitepine Lake, the highest survival of embryos occurred at one of the natural spawning sites, which, like most described spawning sites, was located along a windswept section of shoreline away from stream inlets.

Heavy fall rains can cause pH depressions in lakes and streams (Jeffries et al., 1979), but it is unlikely that fertilization or early cleavage stages of development will be affected in natural lake charr populations. Mortality was less than 20% prior to hatching in all lakes and treatments. This low mortality appeared to represent the initial level of fertilization achieved when mixing gametes, combined with some handling effects. Daye and Glebe (1984) found in Atlantic salmon (*Salmo salar*) that sperm mobility and fertilization success declined with pH, but only at levels below pH 5.0. In this experiment with lake charr under realistic exposure conditions (including lake water where population losses have occurred), brief exposure of eggs to acidic water during

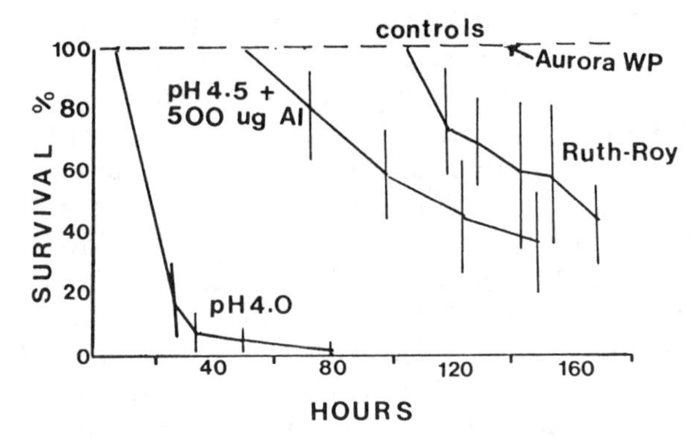

Figure 3.7. Survival of lake charr free embryos in laboratory tests using water collected from two acid lakes (Ruth-Roy Lake, Aurora WP) and additional prepared solutions. Mean survival (%) and standard deviations (vertical bars) during the course of the 5- to 6-d toxicity test are indicated.

fertilization and water hardening produced little or no mortality. The high pH of semen and ovarian fluids (pH > 8.0; Daye and Glebe, 1984; Inaba et al., 1958), may partially explain these results. Even after rinsing with site water, the pH in the 10-L containers of lake water rose rapidly during the hardening process. The pH of Bowland Lake water rose from 5.0 to 5.9. However, under natural conditions lake charr spawn in mass at shallow sites, and there may be similar local changes in water chemistry through the release of reproductive products that protect the earliest life stages from acidic water.

The mortality of free embryos, observed during the overwinter incubation study, is consistent with the reported high sensitivity of this life stage to acid water in laboratory tests (Hutchinson et al., 1987), and in other field studies. Booth et al. (1986) conducted egg incubation experiments in the nearshore waters of acidic Bowland Lake, and in a circumneutral control lake. Mortality was high in the acid lake (95.5%) and occurred mainly at free embryo stage. In the control, mortality was significantly lower (44.3%) and occurred early in development. After liming, both lakes had similar low mortality, confined generally to early stages.

These findings suggest that mortality of free embryos *in situ* may not be due to pulse exposure to acutely lethal runoff water. Mortality occurred at sites where pulse events were neither frequent nor extreme. The once-daily water sampling may have missed many short pulses; however, the laboratory tests with natural water and/or a synthetic solution of H^+ and Al (Table 3.3), demonstrated that free embryos can tolerate severe chemical changes for periods of 1–5 d. Mortality may therefore be the result of more gradual processes, at concentrations of toxicants that were sublethal in short-term tests. Some suggestions to account for observed mortality include exposure to a combination of chronic and episodic toxic conditions (Hutchinson et al., 1987), and toxic responses through exposure to frequent fluctuations in H^+ and Al (Siddens et al. 1986).

Some findings from recent field studies suggest that chronic effects of acid and/or metals on older life stages may produce the observed recruitment failures. In toxicity tests conducted in acidified lakes in the Adirondacks, Johnson et al. (1987) compared mortality among four stages of lake charr. They found that alevins during the initiation of exogenous feeding and 10- to 16-cm yearlings were more sensitive to low pH and elevated levels of Al than juveniles (4–8 cm) in the first year of life, or free embryos, during snowmelt. In Lake 223, a lake located in an area with low deposition of acidic precipitation and less acidic runoff water than in these study area, lake charr suffered recruitment failure when the pH was artificially lowered to 5.6 (total Al 7–36 μg/L) (Schindler et al., 1985). No significant mortality of early life stages occurred in toxicity tests (Mills, 1984) that were similar in design to those described here. Mortality appeared to occur between emergence from the substrate in April–May and the fall of the first year.

A variety of mechanisms for recruitment failure under chronically acidic conditions have been suggested. Weiner et al. (1986) showed that exposure of adult rainbow trout to sublethal pH conditions reduced the viability of progeny, even when raised in circumneutral control water. Reduced growth under sublethal stress (Gunn and Noakes, 1987; Kwain and Rose, 1985) may reduce survival potential of alevins and juveniles. Natural activity patterns of fish, required for feeding and predator avoidance, may also increase their sensitivity to sublethal chemical conditions (Barton et al., 1985, Graham and Wood, 1981). Few laboratory experiments have attempted to test responses of active fish to toxins. Finally, the interactive effects of low pH and a mixture of metals (principally Al, Cu, Zn) may produce high mortality in some acid lakes, particularly those lakes near smelters or other sources of metals (Hutchinson and Sprague, 1986).

7. CONCLUSION

This study demonstrates that mortality of early life stages of lake charr can occur through exposure to acidic runoff events, but the findings suggest that chronic whole-lake acidification precedes and causes the recruitment failure that leads to extinction of lake charr populations. With the observed variability in water chemistry and biological response, simple empirical models dependent only on whole-lake data (e.g. Reckhow, 1984; Beggs et al., 1985) may be more useful for predicting population loss than mechanistic models that attempt to relate seasonal water chemistry to laboratory-derived lethal relations.

ACKNOWLEDGMENTS

W. Keller and P. Gale of Ontario Ministry of Environment, in Sudbury, provided chemical analyses and assisted in the field studies. V. Liimatainen and R. Furchner of Ontario Ministry of Natural Resources assisted with the field toxicity tests. Laboratory space and equipment at the Ontario Ministry of the Environment Laboratory, at Rexdale, Ontario were provided by G. Westlake. W. Kerr and R. Angelo assisted in the laboratory studies. Funds were provided by Ontario Ministry of Natural Resources and National Research Council of Canada (Contract DSS 20ST 31048-4-200 to D. Noakes, University of Guelph). L. Beattie and K. Kovacs helped prepare the figures. L. Deacon, N. Hutchinson, P. Ihssen, W. Keller, D. Noakes, and J. Sprague provided helpful review comments.

REFERENCES

Baker, J.P., and Harvey, T.B. (1984). Critique of acid lakes and fish population status in the Adirondack region of New York State. U.S. Envir. Prot. Agency (NAPAP Project E3-25).

Baker, J.P., and Schofield, C.L. (1982). Aluminum toxicity to fish in acidic waters. *Water Air Soil Pollut.* **18**: 289–309.

Balon, E.K. (1980). Early ontogeny of the lake charr, *Salvelinus (Cristivomer) namaycush*. In E.K. Balon (Ed.), *Charrs, Salmonid Fishes of the Genus Salvelinus* Junk, The Hague, pp. 485–562.

Barton, B.A., Weiner, G.S., and Schreck, C.B. (1985). Effects of prior acid exposure on physiological responses of juvenile rainbow (*Salmo gairdneri*) to acute handling stress. *Can. J. Fish. Aquat. Sci.* **42**: 710–717

Beamish, R.J., and Harvey, H.H. (1972). Acidification of the LaCloche Mountain lakes, Ontario, and resulting fish mortalities. *J. Fish. Res. Board Can.* **29**: 1131–1143.

Beamish, R.J., Lockhart, W., Van Loon, J., and Harvey, H.H. (1975). Long-term acidification of a lake and resulting effects on fishes. *Ambio* **4**: 98–102.

Beggs, G.L., and Gunn, J.M. (1986). Responses of brook trout and lake trout to surface water acidification in Ontario. *Water Air Soil Pollut.* **30**: 711–718.

Beggs, G.L., Gunn, J.M., and Olver, C.H. (1985). The sensitivity of Ontario Lake trout (*Salvelinus namaycush*) and lake trout lakes to acidification. *Ont. Fish. Tech. Rept.* **17**: 1–24.

Booth, G.M., Hamilton, J.G., and Molot, L.A. (1986). Liming in Ontario: Short-term biological and chemical changes. *Water Air Soil Pollut.* **31**: 709–720.

Breck, J.E., Beauchamp, J.J., and Ingersoll, C.G. (1986). Phalca: A microcomputer model for estimating the survival of brook trout early life stages in different combinations of pH, aluminum and calcium. Paper presented at 116th Annual Meeting Am. Fish. Soc., Providence, RI, Sep. 17.

Brown, D.J.A. (1982). The influence of calcium on the survival of eggs and fry of brown trout at 4.5. *Bull. Envir. Contam. Toxicol.* **28**: 664–668.

Brown, D.J.A. (1983). Effect of calcium and aluminum concentrations on the survival of brown trout (*Salmo trutta*) at low pH. *Bull. Environ. Contam. Toxicol.* **30** 582–587.

Brown, D.J.A., and Lynam, S. (1981). The effect of sodium and calcium concentrations on the hatching of eggs and yolk-sac fry of brown trout, *Salmo trutta* L. at low pH. *J. Fish Biol.* **19**: 205–211.

Cleveland, L., Little, E.E., Hamilton, S.J., Buckler, D.R., and Hunn, J.B. (1986). Interactive toxicity of aluminum and acidity to early life stages of brook trout. *Trans. Am. Fish. Soc.* **115**: 610–620.

Craig, G.R., and Baksi, W.F. (1977). The effects of depressed pH on flagfish reproduction, growth and survival. *Water Res.* **11**: 621–626.

Cronan, C.S., and Schofield, C.L., (1979). Aluminum leaching response to acidic precipitation: Effects on high elevation watersheds in the northeast. *Science* **204**: 304–306.

Daye, P.G., and Glebe, B.D. (1984). Fertilization success and sperm motility of Atlantic salmon (*Salmo salar* L.) in acidified water. *Aquaculture* **43**: 307–312.

Driscoll, C.T. Jr., Baker, J.P., Bisogni, J.J., Jr., and Schofield, C.L. (1980). Effects of aluminum speciation on fish in dilute acidified waters. *Nature (London)* **284**: 161–164.

Geen, G.H., Neilsen, J.P., and Hobden, B. (1985). Effects of pH on the early development and growth and otolith microstructure of Chinook salmon, *Oncorhynchus tshawytscha*. *Can. J. Zool.* **63**: 22–27.

Graham, M.S., and Wood, C.M. (1981). Toxicity of environmental acid to the rainbow trout: Interactions of water hardness, acid type, and exercise. *Can J. Zool.* **59**: 1518–1526.

Gunn, J.M. (1986). Behaviour and ecology of salmonid fishes exposed to episodic pH depressions. *Envir. Biol. Fish.* **17**: 241–252.

Gunn, J.M. (1987). The role of episodic acidification in the extinction of lake charr (*Salvelinus namaycush*) populations. Ph.D. Thesis, Guelph University, Guelph, Ontario.

Gunn, J.M. and Keller, W. (1984a). Spawning site chemistry and lake trout (*Salvelinus namaycush*) sac fry survival during spring snowmelt. *Can. J. Fish. Aquat. Sci.* **41**: 319–329.

Gunn, J.M., and Keller, W. (1984b). In situ manipulation of water chemistry using crushed limestone and observed effects on fish. *Fisheries* **9**: 19–24.

Gunn, J.M., and Noakes, D.L.G. (1987). Latent effects of pulse exposure to aluminum and low pH on size, ionic composition and feeding efficiency of lake trout (*Salvelinus namaycush*) alevins. *Can. J. Fish. Aquat. Sci.,* **45**: 1418–1424.

Gunn, J.M., McMurtry, M.J., Bowlby, J.N., Casselman, J.M., and Liimatainen, V.A. (1987). Survival and growth of stocked lake trout in relation to body size, stocking season, lake acidity and biomass of competitors. *Trans. Am. Fish. Soc.,* **116**: 618–627.

Haines, T.A. (1981). Acidic precipitation and its consequences for aquatic ecosystems: a review. *Trans. Am. Fish. Soc.* **110**: 669–707.

Haines, T.A., and Baker, J.P. (1986). Evidence of fish population responses to acidification in the eastern United States. *Water Air Soil Pollut.* **31**: 605–629.

Haya, K., and Waiwood, B. (1981). Acid pH and chorionase activity of Atlantic salmon (*Salmo salar*) eggs. *Bull. Envir. Contam. Toxicol.* **27**: 7–12.

Henriksen, A., Skogheim, O.K., and Rosseland, B.O. (1984). Episodic changes in pH and aluminum-speciation kill fish in a Norwegian salmon river. *Vatten* **40**: 255–260.

Hulsman, P., Powles, P., and Gunn, J.M. (1983). Mortality of walleye eggs and rainbow trout yolk-sac larvae in low-pH waters of the LaCloche mountain area, Ontario. *Trans. Am. Fish. Soc.* **112**: 680–688.

Hutchinson, N.J., and Sprague, J.B. (1986). Toxicity of trace metal mixtures to American flagfish (*Jordanella floridae*) in soft, acidic water and implications for cultural acidification. *Can. J. Fish. Aquat. Sci.* **43**: 647–655.

Hutchinson, N.J., Holtze, K.E., Munro, J.R., and Pawson, T.W. (1988). Lethality of H^+ and Al to eggs and fry of lake trout and brook trout as modified by ionic strength, episodic exposure and post-exposure mortality. *Aquat. Toxicol.*, in press.

Inaba, D., Nomura, M., and Suyama, M. (1958). Studies on the improvement of artificial propogation in trout culture, II. On the pH values of eggs, milt, coelomic fluid and others. *Bull. Jpn. Soc. Sci. Fish.* **23**: 762–765.

Ingersoll, C.G. (1986). The effects of pH, aluminum, and calcium on survival and growth of brook trout (*Salvelinus fontinalis*) early life stages. Ph. D. thesis. University of Wyoming, Laramie.

Jeffries, D.S., Cox, C.M., and Dillon, P.J. (1979). Depression of pH in lakes and streams in central Ontario during snowmelt. *J. Fish. Res. Board Can.* **36**: 640–646.

Johnson, D.W., Simonin, H.A., Colquhoun, J.R., and Flack, F.M. (1987). *In situ* toxicity tests of fishes in acid waters. *Biogeochemistry*, **3**: 181–208.

Keller, W. (1978). Limnological observations on the aurora trout lakes. Ont. Min. Environment Tech. Rept., Sudbury, Ontario.

Keller, W., and Pitblado, J.R. Water quality changes in Sudbury area lakes: A comparison of synoptic surveys in 1974-76 and 1981-83. *Water Air Soil Pollut.* **29**: 285–296.

Kelso, J.R.M., and Gunn, J.M. (1984). Responses of fish communities to acid waters in Ontario. In G.E. Hendrey (Ed.), *Early Biotic Responses to Advancing Lake Acidification*. Butterworth, Stoneham, pp. 105–115.

Kelso, J.R.M., Minns, C.K., Lipsit, J.H., and Jeffries, D.S. (1986). Headwater lake chemistry during the spring freshet in north-central Ontario. *Water Air Soil Pollut.* **29** 245–259.

Korwin-Kossakowski, M., and Jezierska, B. (1985). The effect of temperature on the survival of carp fry, *Cyprinus carpio* L., in acidic water. *J. Fish Biol.* **26**: 43–47.

Kwain, W., and Rose, G.A. (1985). Growth of brook trout *Salvelinus fontinalis* subject to sudden reductions of pH during their early life history. *Trans. Am. Fish. Soc.* **114**: 564–570.

Kwain, W., McCauley, R.W., and MacLean, J.A. (1984). Susceptibility of starved, juvenile smallmouth bass, *Micropterus dolomieui* (Lacepede) to low pH. *J. Fish Biol.* **25**: 501–504.

Lacroix, G.L., Gordon, D.J., and Johnston, D.L. (1985). Effects of low environmental pH on the survival, growth and ionic composition of post-emergent Atlantic salmon (*Salmo salar*). *Can. J. Fish. Aquat. Sci.* **42**: 768–775.

LaZerte, B.D. (1984). Forms of aqueous aluminum in acidified catchments of central Ontario: A methodological analysis. *Can. J. Fish. Aquat. Sci.* **41**: 766–776.

Leivestad, H., and Muniz, I.P. (1976). Fish kill at low pH in a Norwegian river. *Nature (London)* **259**: 391–392.

Litchfield, J.T., Jr. (1949). A method of rapid graphic solution of time–per cent curve. *J. Pharm. Exp. Ther.* **97**: 399–408.

McDonald, D.G. (1983). The effects of H^+ upon the gills of freshwater fish. *Can. J. Zool.* **61**: 691–703.

McMurtry, M.J. (1986) Susceptibility of lake trout (*Salvelinus namaycush*) spawning sites in Ontario to acidic meltwater. Ontario Ministry of Natural Resources Unpubl. Rept., Ont. Fish. Acid. Rept. Series 86-01. Toronto, Ontario.

McWilliams, P.G., and Potts, W.T.W. (1978). The effects of pH and calcium concentrations on gill potentials in the brown trout, *Salmo trutta*. *J. Comp. Physiol.* **126**: 277–286.

Mills, K.H. (1984). Fish populations responses during the experimental acidification of a small lake. *In* G.R. Hendry (Ed.), *Early Biotic Responses to Advancing Lake Acidification*. Butterworth, Stoneham, pp. 117–131.

Muniz, I., and Leivestad, H., (1980). Acidification—Effects on freshwater fish. *In* Drablos and Tollan (Eds.), *Ecological Impact of Acid Precipitation*. Proceedings, Int. Conf. Ecol. Impact Acid Precip., Sandefjord, Norway, 1980, pp. 84–92.

Nelson, J.A. (1982). Physiological observations on developing rainbow trout (*Salmo gairdneri*, Richardson) exposed to low pH and varied calcium ion concentrations. *J. Fish Biol.* **20**: 359–372.

Ontario Ministry of the Environment (OMOE). (1981). Outline of analytical methods. Laboratory Service Branch Report, Toronto, Ontario.

Peterson, R.H., and Martin-Robichaud, D.J. (1983). Embryo movements of Atlantic salmon (*Salmo salar*) as influenced by pH, temperature, and state of development. *Can. J. Fish. Aquat. Sci.* **40**: 777–782.

Peterson, R.H., and Martin-Robichaud, D.J., (1986). Growth and major cation budgets of Atlantic salmon alevins at three ambient acidities. *Trans. Am. Fish. Soc.* **115**: 220–226.

Peterson, R.H., Daye, P.G., LaCroix, G.L., and Garside, E.T. (1982). Reproduction of fish experiencing acid and metal stress. In R.E. Johnson (Ed.), *Acid Rain/Fisheries*, Am. Fish. Soc., Bethesda, MD, pp. 177–196.

Recknow, K. (1984). Empirical models of fish reponse to acidification. *In* Aquatic Effects Task Group (E) and Terrestrial Effects Task Group (F) Peer Review, Research Summaries. EPA/NCSU Acid Deposition Program, North Carolina State University, Raleigh, pp. 255–263.

Rogeberg, E.J.S., and Henriksen, A. (1985). An automatic method for fractionation and determination of aluminum species in fresh-water. *Vatten* **41**: 48–53.

Rombough, P.J. (1983). Effects of low pH on eyed embryos and alevins of Pacific salmon. *Can. J. Fish. Aquat. Sci.* **40**: 1575–1582.

Rosseland, B.O. (1986). Ecological effects of acidification on tertiary consumers. Fish population responses. *Water Air Soil Pollut.* **30**: 451–460.

Ruby, S.M., Aczel J., and Craig, G.R. (1978). The effects of depressed pH on spermatogenesis in flagfish, *Jordanella floridae*. *Water Res.* **12**: 621–626.

Sadler, K. (1983). A model relating the results of low pH bioassay experiments to the fishery status of Norwegian lakes. *Freshwater Biol.* **13**: 453–463.

Schindler, D.W., Mills, K.H., Malley, D.F., Findley, D.L., Shearer, J.A., Davis, I.J., Turner, M.A., Linsey, G.A., and Cruikshank, D.R. (1985). Long-term ecosystem stress: the effects of years of experimental acidification on a small lake. *Science* **228**: 1395–1401.

Siddens, L.K., Seims, W.K., Curtis, L.R., and Chapman, G.A. (1986). Comparison of continuous and episodic exposure to acidic, aluminum-contaminated water of brook trout (*Salvelinus fontinalis*). *Can. J. Fish. Aquat. Sci.* **43**: 2036–2040.

Stuart, S., and Morris, R. (1984). The effects of season and exposure to reduced pH (abrupt and gradual) on some physiological parameters in brown trout (*Salmo trutta*). *Can. J. Zool.* **63**: 1078–1083.

Talbot, C., Eddy, F.B., and J. Johnston, J. (1982). Osmoregulation in salmon and sea trout alevins. *J. Exp. Biol.* **101**: 61–70.

Weiner, G.S., Schreck, C.B., and Li, H.W. (1986). Effects of low pH on reproduction of rainbow trout. *Trans. Am. Fish. Soc.* **115**: 75–82.

Wood, C.M., and McDonald, D.G. (1987). The physiology of acid/aluminum stress in trout. *Ann. Soc. R. Zool. Belgium* **117**: 399–410.

4

HEMATOLOGICAL PARAMETERS AND PARASITE LOAD IN WILD FISH WITH ELEVATED RADIONUCLIDE LEVELS

Joan W. Bernstein

Department of Applied Microbiology and Food Science, University of Saskatchewan, Saskatoon, Saskatchewan, Canada

Stella M. Swanson

Saskatchewan Research Council, Saskatoon, Saskatchewan, Canada

1. Introduction
- 1.1. Background
- 1.2. Literature review
- 1.3. Objectives

2. Methods
- 2.1. Study area
- 2.2. Field methods
- 2.3. Laboratory methods
- 2.4. Data analysis

3. Results
- 3.1. Tumor and parasite survey
- 3.2. Hematology

4. Discussion

5. Conclusion

References

1. INTRODUCTION

Studies of accumulation of uranium- (U) series radionuclides in aquatic systems have shown elevated levels in plants and animals near uranium mines or high natural background areas (Anderson et al., 1963; Ruggles and Rowley, 1978; Dean et al., 1982; Swanson, 1983). The sources of these radionuclides include effluent discharge from mill tailings treatment areas or groundwater movement and surface drainage from waste sites (Morin et al., 1982).

U-series radionuclides have relatively high biological mobility (relative to fission products) (Whicker and Schultz, 1982). They primarily emit α and β radiations, which pose a hazard if deposited internally in organisms (Casarett,

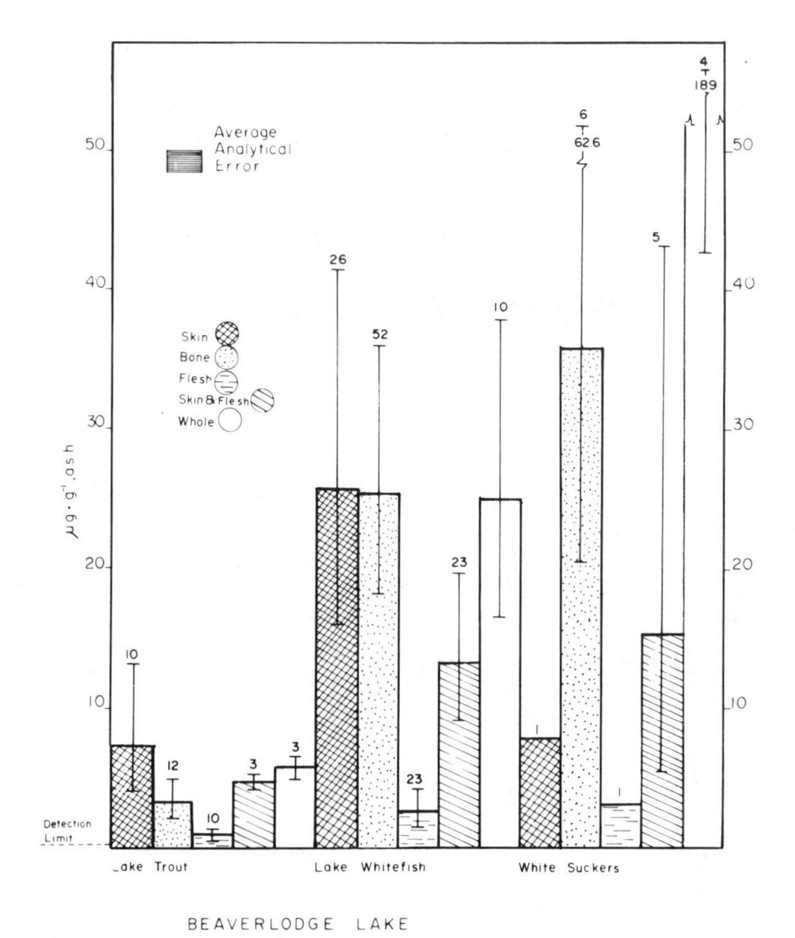

Figure 4.1. Mean U-238 content in white suckers, lake whitefish, and lake trout from the contaminated Beaverlodge Lake. The number of samples and the 95% confidence limits are indicated for each tissue of each fish species.

1980). Investigations of the effects of these radionuclides on aquatic animals in natural systems have predominantly focused on population dynamics such as growth and fecundity (Blaylock, 1969; Blaylock and Trabalka, 1978). This study was initiated to assess the chronic effects of low-level U-series radionuclides on wild fish. A tumor and parasite survey and measurement of standard blood parameters were used to evaluate potential U-series radionuclide effects on fish health.

1.1. Background

Saskatchewan has a relatively long history of uranium mining. Over the past decade, large deposits of high-grade uranium ore have been found and three new mines have been developed (Dean et al., 1982). From 1952 to 1982 a uranium mine and mill located in northern Saskatchewan near Uranium City was operated by Eldorado Nuclear Ltd. The mill tailings were discharged into a natural drainage system that ultimately drains into Beaverlodge Lake. Two nearby lakes, Milliken and Fredette, were used as controls in this study. They do not receive runoff from any operating or abandoned mines, although they are part of the same drainage basin (Beck, 1969; Ruggles and Rowley, 1978; Swanson, 1983).

Beaverlodge Lake water shows some effects from input of the tailings system effluent: sodium sulfate, radium-226 (Ra), and U levels are elevated. Sediment levels of lead-210 (Pb), Ra-226, and U-238 are elevated. Fredette Lake has the largest overall plankton population; Milliken Lake has the most diverse and abundant benthic fauna. Fredette and Beaverlodge Lakes have relatively poor benthic populations (Swanson, 1982, 1983).

Three fish species in Beaverlodge Lake were found to have significantly elevated U-series radionuclide levels ($p < 0.001$). U-238, Ra-226, and Pb-210 were significantly elevated in these fish. Figure 4.1 shows U-238 levels found in these species. A similar pattern was observed with Ra-226 and Pb-210: concentrations were highest in white suckers (*Catostomus commersoni*), intermediate in lake whitefish (*Coregonus clupeaformis*), and lowest in lake trout (*Salvelinus namaycush*). Lake trout levels were sometimes similar to background levels seen in control fish. Skin and bone contained the highest levels, while levels in the flesh were much lower (Swanson, 1983). This is similar to distributions found in other studies (Justyn and Lusk, 1976; Dean et al., 1982) since elements such as Pb and Ra are known to have much longer biological half-lives in the bones than in the whole body. Lake whitefish and white suckers had Ra-226 and Pb-210 levels 2–63 times those found in nearby control lakes while U concentrations were 3–115 times higher than control levels. Lake trout had lower radionuclide levels than the other two species, 1.4–7 times control values. These radionuclide levels in Beaverlodge fish represent low dose rates compared with those shown to have effects in laboratory experiments. They resemble levels found in fish in other field studies (Anderson et al., 1963; Ruggles and Rowley 1978; Eldorado Nuclear Ltd., 1978). Levels were uniformly low in control lake species and similar to background levels in fish from other studies (Holtzman, 1969; Lucas et al., 1970). No significant differences occurred in other chemical concentrations (Zn, Mn, Fe, Cu, Cr, Al, Mo, Ni, Pb, Ti, and V) measured in fish in the three study lakes (Swanson and Tones, personal communication, 1984).

1.2. Literature Review

Both Ra-226 and Pb-210 are naturally produced during U decay; Ra-226 primarily emits α radiation while Pb-210 primarily emits β radiation (Eisenbud, 1973). Both are deposited in the skeleton. Radium is a metabolic analog of calcium and when retained in bone serves as a source of α radiation. Radium has been shown to induce tumor formation, e.g., osteosarcoma (Raabe, 1984). Radium that becomes incorporated in the skeleton could also potentially emit α particles into adjacent hematopoietic tissue located subdorsal to the vertebral column in fish.

Chronic radiation studies in fish have dealt almost exclusively with γ radiation. The length of "chronic" exposure in these studies is also relatively short, e.g, several months. Results show that both acute and chronic radiation exposure cause hematopoietic damage (Cosgrove and Blaylock, 1972; Bonham et al., 1948).

The hematopoietic system has been shown to be extremely sensitive to radiation and has been used to detect radiation syndromes in several species of fish (Patel and Patel, 1979). Acute, high-dose γ or X radiation results in decreases in blood cells (both red and white), hematocrit, serum protein, and hemoglobin concentrations in fish (Ulrikson, 1973; Patel and Patel, 1979). Chronic γ-radiation exposure at much lower doses was found to cause less hematopoietic damage than acute higher dose exposure (Cosgrove and Blaylock, 1972).

Effects of chronic, low-level α and β radiation, the type primarily emitted by the U-series radionuclides, are largely unknown. Alpha radiation has a much higher "relative biological effectiveness" than γ radiation and produces more cellular damage per disintegration (Casarett, 1980). Alpha and β radiation delivered over the entire lifespan of a fish may result in large culmulative doses.

Effects of low-level β radiation from strontium-90 were studied in carp (*Cyprinus carpio*) in the USSR (Shleifer and Shekhanova, 1980). Cumulative doses above 4–5 rads produced significant hematological changes involving decreased numbers of white blood cells, changes in differential blood counts, and suppression of phagocytic activity and antibody formation. Morphological changes in erythrocytes were noted.

The hematopoietic system is a major target system of lead (chemical toxicity). Exposure to lead can lead to anemia due to both red blood cell lifespan reduction and inhibition of hemoglobin synthesis (Casarett, 1980). Lead has been shown to cause changes in fish blood parameters. Dawson

(1935) first reported that catfish developed anemia after lead exposure. This response was documented in other species (Demayo et al., 1980; Hodson et al., 1978).

1.3. Objectives

The discovery of elevated radionuclide levels in fish from Beaverlodge Lake, Saskatchewan, and the lack of much information on the effects of low-level radiation on fish prompted a field investigation of the general health of the Beaverlodge Lake and two control lake fish populations. The objectives were to (1) document tumors, lesions, and other abnormalities, and inventory parasite levels; and (2) evaluate standard blood parameters based on the known relation between lead, radiation, and the hematopoietic system.

2. METHODS

2.1. Study Area

The study area was located on the northeast shore of Lake Athabasca, Saskatchewan (59° 30′ N, 108° 35′ W). Three lakes were sampled, Beaverlodge, Milliken, and Fredette (Fig. 4.2). Beaverlodge is affected by drainage from the now decommissioned Eldorado Nuclear uranium mill tailings treatment area and also receives contaminated drainage from Ace Creek, which runs through old tailings deposits and the abandoned Lorado uranium mill. The two control lakes, Milliken and Fredette, do not receive runoff from any operating or abandoned uranium mines, although minor uranium deposits are located near both.

The three study lakes, Beaverlodge, Milliken, and Fredette, are typical shield lakes in that they are oligotrophic. Beaverlodge Lake is the largest, with a total area of 4771 hectares (ha) and a maximum depth of 70 m. Milliken and Fredette Lakes have areas of 975 and 566 ha and maximum depths of 55 and 30 m, respectively. Beaverlodge and Milliken Lake support populations of all three fish species investigated in this study, while Fredette Lake contains only lake trout and lake whitefish.

2.2. Field Methods

Fish were gill netted in June 1982 and August 1983 and 1984. Live fish were bled in the boat by severing the tail and collecting free flowing blood in 10-mL heparinized tubes. Fish were then placed in covered holding containers and kept cool to await necropsy which was carried out on shore.

Immediately on return to shore, microhematocrit tubes were filled with blood and centrifuged (International Model MB, Canlab) at 1500 rpm for 3 min. The packed cell volume (International Capillary Tube Reader Model CR,

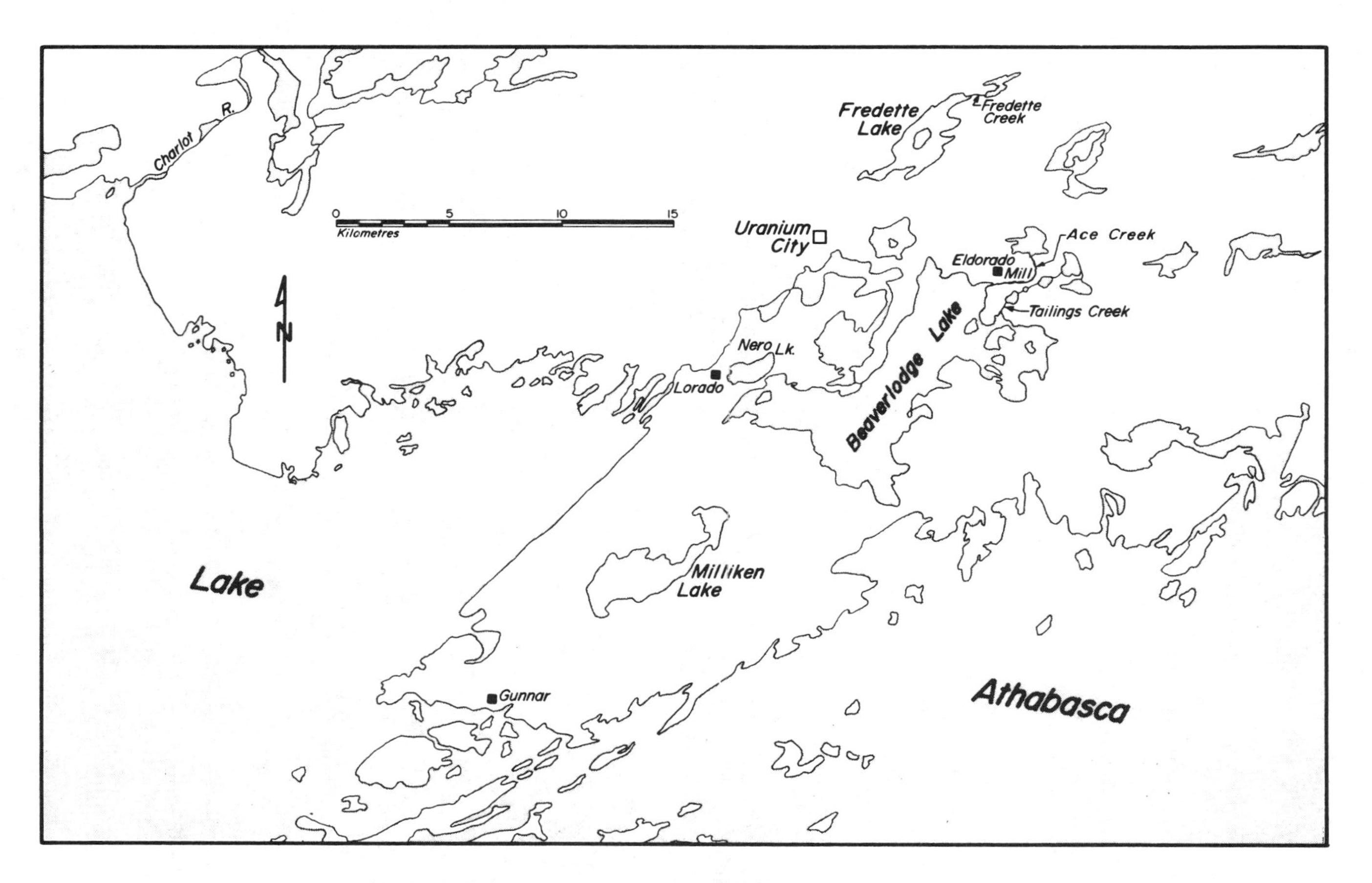

Figure 4.2. The location of Beaverlodge, the contaminated lake, and Milliken and Fredette, the control lakes.

Canlab) and the total protein content of the plasma (American Optical Goldberg Refractometer, Canlab) were determined. Individual fish blood samples were stained (Lehmann and Sturenberg, 1974) and set aside for later cell counts to be carried out indoors. Blood counts were read directly using a Neubauer chamber and phase-contrast microscope. Blood smears were made and fixed in methanol for differential counts to be completed at a later time. Bled fish were weighed, measured (fork length), and sexed, and the stage of sexual maturity was noted. Otoliths were taken from whitefish and lake trout and pectoral fin rays were taken from suckers for aging. A gross necropsy performed on each fish examined the general health condition and noted parasites, lesions, and any abnormalities. Samples of ventral skin, left gill, left eye tissue, and parasites were collected and fixed in 10% formalin for later histological examination. All fish were then frozen for further detailed laboratory examination.

Additional fish netted at the same time, were collected for radionuclide analysis to confirm that levels were still similar to those recorded in 1979–1981. These fish were gutted and filleted, and the skin, bone, and flesh were bagged separately and frozen.

2.3. Laboratory Methods

Blood smears prepared and fixed in the field were stained with Wright's Giemsa on an automated stainer (Ames Hema-Tek, Canlab). Differential cell counts were made of two groups of 100 cells each. The identity of the smear was unknown at the time of counting. The average of the two results was taken as the final reading. White blood cell types for white suckers were classified as eosinophils, lymphocytes, neutrophils, and finely reticulated cells. Whitefish and lake trout white blood cells were classified as monocytes, lymphocytes, and neutrophils. Fifty cells were examined in those few cases where very low overall numbers were present.

Uranium analysis was performed on ashed fish bone using the fluorometric method (HASL, 1972).

2.4. Data Analysis

Data were analyzed using the Statistical Analysis System (SAS) software package. Data were tested for heterogeneity and normality. Log transformation was then performed. Differences in parameters among lakes and species were tested with a factorial ANOVA design with interaction. The Student-Newman-Keuls test was used to test individual differences within species and lakes for each parameter.

3. RESULTS

3.1. Tumor and Parasite Survey

Approximately 60 fish of each species in each lake were examined and gross necropsies revealed no evidence of tumors, lesions, or abnormalities. Histological results revealed lesions primarily associated with parasite infestation.

There were differences in number and types of parasites observed. Although the following description of the parasite results is preliminary and a more in-depth analysis is in preparation, the general trends are clear. Lake whitefish exhibited higher levels and a greater variety of parasite types than did common suckers in both Beaverlodge and Milliken Lakes (Figs. 4.3 and 4.4). Parasite burdens in lake whitefish did not differ greatly between all three lakes. In contrast, parasite number and level in Beaverlodge common suckers were significantly lower than that seen in Milliken Lake. Overall, Milliken Lake appeared to support a more abundant parasite fauna than the other two lakes. Similar trends were revealed with histological parasite evaluation of the eye and gill, while no unusual lesions or differences were observed in the skin samples.

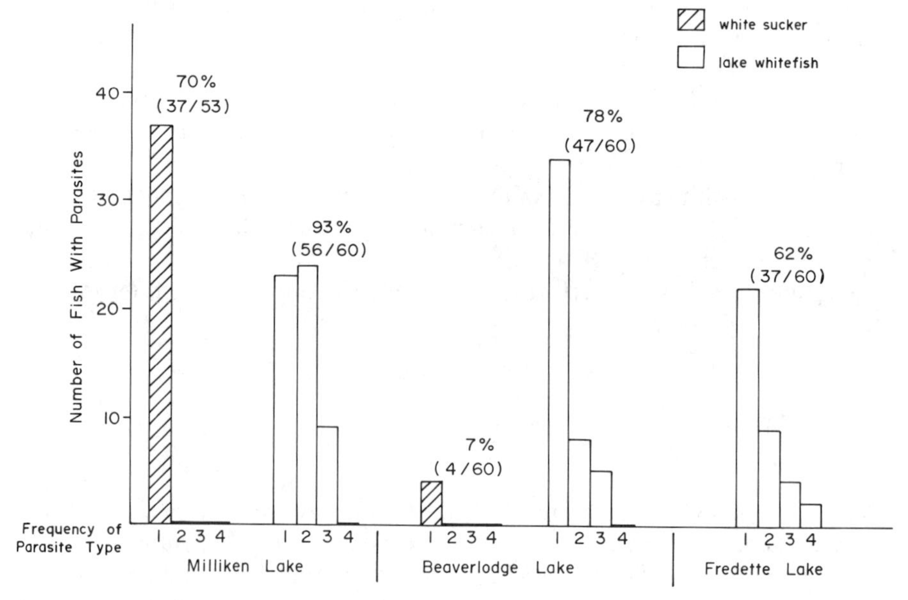

Figure 4.3. Parasite occurrence based on gross examination. Frequency of parasite type (1, 2, 3, or 4) refers to different types of parasites observed in the fish, e.g., cestode, nematode, adult, larva.

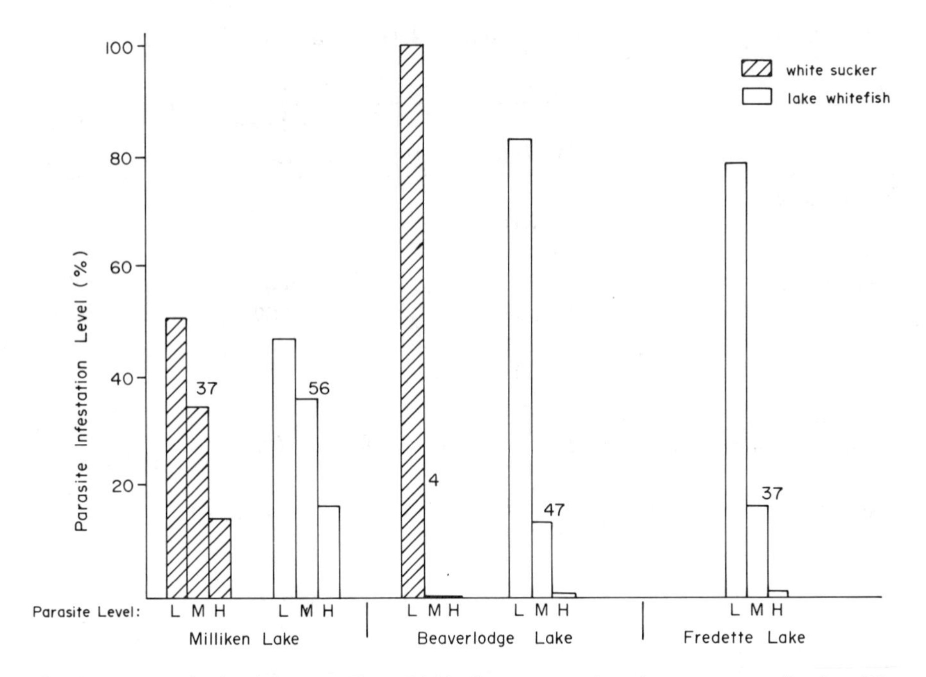

Figure 4.4. Parasite level (low, medium, high) of occurrence based on gross examination. The sample size is indicated for each species in each lake.

3.2. Hematology

Twenty fish of each species in each lake were bled for determination of packed cell volume, plasma protein, and red and white blood cell counts and differentials. The ANOVA of fish blood parameters versus lake and species showed significant differences in all four blood parameters both between lakes and between species ($p < 0.001$). The Student-Newman-Keuls test specified where the differences occurred.

Packed cell volume, total protein, and red blood cell counts were significantly lower in common suckers and lake whitefish from Beaverlodge Lake in comparison with those from Milliken and Fredette Lakes (Tables 4.1–4.3). White blood cell counts in Beaverlodge suckers and whitefish, however, were significantly higher than those in Milliken Lake, while lake whitefish white blood cell counts, although higher, were not significantly different from those in Fredette Lake (Table 4.4).

Lake trout blood parameters did not exhibit the same pattern seen in suckers and whitefish. Either no differences were found or the pattern was opposite to that found in the other two species. Total protein and red blood cell counts in Beaverlodge trout did not differ significantly from those in Milliken

Table 4.1 Mean Packed Cell Volume (L/L) and Standard Deviations[a]

	Species		
Lake	Lake Whitefish	White Sucker	Lake Trout
Beaverlodge Lake	[A]0.42 ±0.06	[C]0.30 ±0.06	[A]0.46 ±0.07
Milliken Lake	[B]0.50 ±0.07	[D]0.45 ±0.09	[B]0.41 ±0.08
Fredette Lake	[B]0.46 ±0.08		

[a] In rows and columns different letters indicate significant differences ($p < 0.001$).

Table 4.2 Mean Total Protein (g/L) and Standard Deviations[a]

	Species		
Lake	Lake Whitefish	White Sucker	Lake Trout
Beaverlodge Lake	[A]58.00 ±10.49	[D]31.63 ±06.66	[F]82.00 ±09.40
Milliken Lake	[B]70.06 ±17.35	[G]58.95 ±14.49	[F]79.54 ±10.77
Fredette Lake	[C]63.60 ±10.08		

[a] In rows and columns different letters indicate significant differences ($p < 0.001$).

Table 4.3 Mean Number of Erythrocytes ($\times 10^{12}$/L) and Standard Deviations[a]

	Species		
Lake	Lake Whitefish	White Sucker	Lake Trout
Beaverlodge Lake	[A]0.72 ±0.09	[D]0.61 ±0.12	[E]0.65 ±0.16
Milliken Lake	[B]0.88 ±0.15	[F]0.78 ±0.23	[E]0.61 ±0.10
Fredette Lake	[C]0.87 ±0.14		

[a] In rows and columns different letters indicate significant differences ($p < 0.001$).

Table 4.4 Mean Number of White Blood Cells ($\times 10^9$/L) and Standard Deviations[a]

	Species		
Lake	Lake Whitefish	White Sucker	Lake Trout
Beaverlodge Lake	^{A}12.80 ±4.69	^{A}11.41 ±4.41	^{C}1.84 ±1.20
Milliken Lake	^{C}7.29 ±1.69	^{B}5.67 ±3.19	^{D}3.15 ±2.51
Fredette Lake	^{A}9.06 ±4.31		

[a] In rows and columns different letters indicate significant differences ($p < 0.001$).

trout (Tables 4.2 and 4.3). Packed cell volume was higher in Beaverlodge trout than in Milliken trout and leucocytes were lower in Beaverlodge trout than in Milliken trout. This is opposite to what was found in suckers and whitefish.

There was a significant interaction between lake and species effects on blood parameters. Differences among lakes were seen to be significantly affected by the species being examined.

Higher leucocyte counts in Beaverlodge suckers were primarily due to greater numbers of neutrophils and to a lesser extent, finely reticulated cells (Table 4.5). Higher leucocyte numbers in Beaverlodge whitefish were due to greater numbers of lymphocytes and, to a lesser extent, neutrophils. Lower leucocyte counts in Beaverlodge lake trout were primarily due to fewer lymphocytes.

A check on uranium levels in fish bone revealed that between 1981 and 1983 sucker and whitefish levels in Beaverlodge and Fredette Lakes remained the same. Between 1983 and 1984 there was a significant decline in uranium levels in Milliken whitefish (Table 4.6). It is assumed that Ra-226, Pb-210, and Po-210 levels follow the same trends as uranium. Therefore, we concluded that radionuclide levels were still elevated in Beaverlodge fish in 1983.

4. DISCUSSION

Cell proliferation is often related to radiation damage (Casarett, 1980). Therefore, one might expect tumor development in wild fish chronically exposed to radiation. The lack of tumor involvement in this study indicates that long-term exposure of these fish species to the low levels of radionuclides is not manifested teratogenically, but in some other way.

Table 4.5 Differential White Blood Cell Counts (Means ($\times 10^9$/L) ± Standard Deviation)

Lake and Species	WBC	Eosinophils	Lymphocytes	Finely Reticulated Cells	Neutrophils
Beaverlodge					
White suckers	11.41 ± 4.41	0.11 ± 0.25	4.1 ± 2.8	1.0 ± 1.2	6.2 ± 4.9
Milliken					
White suckers	5.67 ± 3.19	0.09 ± 0.16	4.16 ± 2.8	0.60 ± 1.16	0.82 ± 1.9

Lake and Species	WBC	Monocytes	Lymphocytes	Neutrophils
Beaverlodge				
Lake whitefish	12.80 ± 4.69	0	10.82 ± 6.0	1.98 ± 2.0
Lake trout	1.84 ± 1.20	0.001 ± 0.007	1.83 ± 1.2	0.01 ± 0.02
Milliken				
Lake whitefish	7.29 ± 1.69	0.01 ± 0.04	6.47 ± 1.7	0.81 ± 0.46
Lake trout	3.15 ± 2.51	0	3.06 ± 2.47	0.10 ± 0.14
Fredette				
Lake whitefish	9.06 ± 4.31	0.08 ± 0.26	8.43 ± 4.75	0.54 ± 0.42

Table 4.6 Uranium Levels in Bone of Fish from the Three Study Lakes

Species	Location	Mean 1983 U Level (μg/g ash)	Mean 1984 U Level (μg/g ash)	Significant Difference
White sucker	Beaverlodge Lake	58.1 ± 6.8 ($n = 10$)	47.0 ± 1.8 ($n = 20$)	No
Lake whitefish	Fredette Lake	1.9 ± 0.5 ($n = 10$)	4.1 ± 1.5 ($n = 20$)	No
Lake whitefish	Milliken Lake	0.89 ± 0.17 ($n = 10$)	2.0 ± 0.35 ($n = 20$)	Yes $p < 0.05$

Lake whitefish and common suckers did not exhibit the same trends in parasite load. Beaverlodge Lake whitefish appeared to exhibit generally the same levels as populations in the control lakes.

Differences in types and infestation levels of parasites between lakes and fish species probably depend on feeding habits, available food, and presence of suitable intermediate hosts. Previous work has shown Milliken Lake to have a greater and more diverse benthic fauna than Beaverlodge and Fredette Lakes (Swanson, 1982), thus potentially indicating a greater abundance of intermediate hosts. This corresponds to the more abundant parasite fauna found in both species in Milliken Lake. In contrast, the reduced parasitism in Beaverlodge common suckers in the contaminated lake may be due to radionuclide levels reducing benthic faunal diversity and thus intermediate host populations. However this does not explain the similarity in parasite infestation levels of lake whitefish populations in Beaverlodge and control lakes. Potential effects on intermediate host populations should be investigated further.

Blood parameters in Beaverlodge Lake fish showed significant differences in several respects from fish in Milliken and Fredette Lakes. Since the method of capture was standard and time of capture the same the population component sampled should be comparable between lakes. Variances in blood parameters in this study were lower than those usually attributed to differences in fish strains (Barnhart, 1969; Wedemeyer and Nelson, 1975; Christensen et al., 1978). Therefore, the differences observed in this study may be related to environmental factors, e.g., radionuclide levels, and they may also be influenced by other factors such as size, age, and/or sex differences of the fish samples taken from each lake.

Total red and white blood cells, total protein, and hemoglobin have been shown to increase with length and weight in some fish species (primarily during the early years) (Joshi and Tandon, 1977) and total protein has been shown to increase with weight in rainbow trout (Hille, 1982). Gill net mesh size was the same for all sample collections; however, since Beaverlodge fish are small (Table 4.7), size may be a contributing factor.

Although age has been correlated with blood parameters, these differences are usually seen between very young, sexually immature, and old mature fish (McCarthy et al., 1975; Hardig and Hoglund, 1984). Fish from all three lakes in this study were mainly mature, older fish. Therefore, age (i.e., sexual maturity) is not a likely cause of the observed blood differences.

Sexual differences have been noted to contribute to differences in blood parameters (Hille, 1982). Specifically, red blood cells and packed cell volume have been shown to be higher in males than females and total protein higher in females than males. There was a higher female-to-male ratio in the Beaverlodge common sucker sample, while lake whitefish samples had very similar sex ratios (Table 4.7) and yet the blood parameters for both these species exhibited similar trends. The greater number of females may have influenced the sucker blood parameter results. However, total protein values would be expected to be

Table 4.7 Size, Sex Ratio, Maturity, and Age of Fish Sampled for Blood Parameters (Means ± Standard Error)

Species	Lake	Size Range		Sex Ratio	Sexual	
		Mean Fork Length (cm)	Mean Weight (g)	Female : Male	Maturity	Mean Age
White suckers	Beaverlodge	33.1 ± 0.5	482 ± 23	20 : 8	87% mature	13.0[a]
	Milliken	46.0 ± 0.7	1565 ± 68	12 : 15	89% mature	11.9[a]
Lake whitefish	Beaverlodge	36.5 ± 0.7	697 ± 37	8 : 10	82% mature	13.2[b]
	Milliken	44.9 ± 1.3	1376 ± 118	8 : 10	84% mature	15.2[b]
	Fredette	30.6 ± 0.7	454 ± 43	16 : 18	82% mature	7.7[b]
Lake trout	Beaverlodge	52.2 ± 1.4	2140 ± 188	7 : 15	82% mature	7.1[b]
	Milliken	51.4 ± 0.5	1599 ± 63	8 : 15	82% mature	12.8[b]

[a] Pectoral fin ray.
[b] Otolith.

higher in the Beaverlodge sucker sample if the sex ratio was a significant factor and this was not the case.

Amount and quality of food affects a variety of blood parameters (Barnhart, 1969). Starvation causes anemia (Lane et al., 1982) and a decline in packed cell volume and total protein (Cairns and Christian, 1978). The lower benthic fauna population and smaller size and lower fecundity of Beaverlodge Lake fish populations in comparison to Milliken and Fredette Lakes suggest a restricted food supply. Although gross necropsies did not reveal any evidence of starvation, a limited food supply in Beaverlodge Lake could be contributing to the blood parameter differences in Beaverlodge fish.

Bacterial or viral infections and parasite load can affect blood parameters in fish. Gross necropsy in the field and more detailed studies and histological examinations in the laboratory revealed that all fish appeared to be normal and in good condition with adequate fat reserves. The blood parameter results exhibited by Beaverlodge fish would be expected from a population with a high parasite load rather than one with the low levels found in the common suckers. This suggests that factors other than disease or parasitism may have caused these blood changes.

Beaverlodge fish, e.g., common suckers and lake whitefish, exhibited higher white blood cell numbers in comparison to fish from control lakes. Beaverlodge common suckers exhibited a higher level of neutrophils and lake whitefish exhibited a higher level of lymphocytes and neutrophils than fish in the control lakes. Chronic infections can cause antigenic stimulation resulting in rising lymphocyte counts and although no overt disease symptoms were found, there is still the possibility that one was present but undetected. Early stages of leukemia may be associated with or cause lymphocytosis and a temporary increase in the neutrophil count has been seen at low doses in some cases of radiation exposure (Casarett, 1980). Chronic exposure to copper (Dick and Dixon, 1985) and chlorinated hydrocarbons (McLeay, 1973) has also been shown to result in neutrophilia. Radiation effects on white blood cells are generally manifested as a decline in the number of peripheral lymphocytes (Roberts, 1978); however, it may be possible that at the low levels in the Beaverlodge fish, radiation causes stimulation of white cells rather than a decline. Additionally, an increase in white blood cells can occur during periods of stress; possibly the Beaverlodge fish were either under a greater stress or exhibiting a greater response to the stress of being gill netted. All fish were sampled in the same manner and all were presumably stressed in the same manner.

General water quality (major ions, nutrients, pH) is very similar among the three lakes. The only elements that are significantly higher in Beaverlodge water and fish are the radionuclides; specifically uranium, Ra-226, Pb-210, and polonium-210 (Swanson, 1983, 1984). Medium- to high-level radiation is known to cause severe anemia and damage to blood-forming organs in fish resulting in a decline in blood parameters (Roberts, 1978). Chronic low-level radiation may also cause declines in red blood cells and other blood parameters. Thus radionuclides are a possible cause of the lower packed cell volume, total protein, and red blood cell numbers in Beaverlodge fish. Lead-210 is also known to act directly on the hemopoietic system causing anemia.

If radionuclides are a significant factor affecting blood parameters, differences in lake trout parameters between lakes would be expected to be less pronounced or different since radionuclide levels are lower in this species. There were fewer differences in lake trout results, and the differences that did occur were the reverse of those found in whitefish and suckers. Thus, based on lake trout results, radionuclides are a possible factor affecting blood parameters.

5. CONCLUSION

The data presented show consistent differences in blood parameters in the two species with the highest levels of radionuclides, but the physiological significance of these data is unclear. In field studies of this type many other factors can influence the results. The blood parameters, measured although nonspecific indicators, potentially demonstrate differences between the experimental and control groups that may be due to the presence of radionuclides in these fish. The fact that tumors and other changes characteristic of radiation damage were not observed may indicate several possibilities. Sample sizes may need to be much larger to investigate damage of this type involving radiation at these low levels. Also, it is possible that these fish do not live long enough for effects of damage to become manifest. Last, it is possible that recovery occurs faster than damage is induced. However, in this study consistent differences in blood parameters were found; specifically a decrease in packed cell volume, total protein, and red blood cell numbers, and an increase in numbers of white blood cells. These differences occurred in the two species with the highest radionuclide levels. This information and the fact that little is known about α and β radiation in comparison with γ radiation indicate, we believe, that laboratory studies are needed to further investigate and verify the effects of this type of radiation in fish.

ACKNOWLEDGMENTS

This work was supported by the Saskatchewan Health Research Board, Environment Canada, and the Saskatchewan Research Council. Our thanks to Barry Paquin, Dan Richert, Ross Barclay, and Gerry Sibbick for fieldwork and Katherine Meeres for help with statistical analysis.

REFERENCES

Anderson, J.B. Tsivoglon, E.C., and Shearer, S.D. (1963). Effects of uranium mill wastes on biological fauna of the Animas River (Colordao–New Mexico). In V. Schultz and A.W. Klement (Eds.), *Radioecology*. Reinhold, New York, p. 373.

Barnhart, R.A. (1969). Effects of certain variables on hematological characteristics of rainbow trout. *Trans. Am. Fish. Soc.* **98**: 411–418.

Beck, L.S. (1969). *Uranium Deposits of the Athabasca Region*. NTS Area 74N, 74O, 74P, Report 126, Saskatchewan Mineral Resources.

Blaylock, B.G. (1969). The fecundity of a *Gambusia affis* population exposed to chronic environmental radiation. *Rad. Res.* **37**: 108.

Blaylock, B.G., and Trabalka, J.R. (1978). Evaluating the effects of ionizing radiation on aquatic organisms. *Ad. Rad. Biol.* **7**: 103.

Bonham, K., Donaldson, L.R., Foster, R.F., Welander, A.D., and Seymour, A.H. (1948). The effect of x-rays on mortality, weight, length and counts of erythrocytes and hematopoietic cells in fingerling chinook salmon, *Oncorhynchus tshawytscha* Walbaum. *Growth* **12**: 107–121.

Cairns, M.A., and Christian, A.R. (1978). Effects of hemorrhagic stress on several blood parameters in adult rainbow trout (*Salmo gairdneri*). *Trans. Am. Fish. Soc.* **107**: 334–340.

Casarett, L.J. (1980). *Casarett and Doull's Toxicology: The Basic Science of Poisons*. MacMillan, Toronto.

Christensen, G.M., Fiandt, J.T., and Poeschl, B.A. (1978). Cells, proteins and certain physical-chemical properties of brook trout (*Salvelinus fontinalis*) blood. *J. Fish Biol.* **12**: 51–60.

Cosgrove, G.E., and Blaylock, B.G. (1972). Acute and chronic irradiation effects in mosquito fish at 15° or 25°C. In D.J. Nelson (Ed.), *Proceedings of the Third National Symposium on Radioecology*. USAEC, Oak Ridge, TN.

Dawson, A.B. (1935). The hemopoietic response in the catfish *Ameiurus nebulosus* to chronic lead poisoning. *Biol. Bull.* **68**: 335–346.

Dean, J.R., Chiu, N., Neame, P., and Bland, C.J. (1982). Background levels of naturally occurring radionuclides in the environment of a uranium mining area of northern Saskatchewan, Canada. Nat. Radiat. Environ. Proc. Spec. Symp. Nat. Radiat. Environ 2d, pp. 67–73.

Demayo, A., Taylor, M.C., and Reeder, S.W. (1980). *Guidelines for Surface Water Quality*, Vol. 1, *Inorganic Chemical Substances. Lead*. Environment Canada, Ottawa.

Dick, P.T., and Dixon, D.G. (1985). Changes in circulating blood cell levels of rainbow trout, *Salmo gairdneri* Richardson, following acute and chronic exposure to copper. *J. Fish Biol.* **26**: 475–481.

Eisenbud, M. (1973). *Environmental Radioactivity*. New York: Academic.

Eldorado Nuclear Ltd. (1978). *Environmental Overview Assessment for the Dubyna 31-Zone Uranium Production Program*. Eldorado Nuclear Ltd., Eldorado, Sask.

Hardig, J., and Hoglund, L.B. (1984). Seasonal variation in blood components of reared Baltic salmon, *Salmo salar* L. *J. Fish Biol.* **24**: 565–579.

HASL (1972). Health and Safety Laboratory. *Procedures Manual.* New York: USAEC.

Hille, S. (1982). A literature review of the blood chemistry of rainbow trout, *Salmo gairdneri* Rich. *J. Fish Biol.* **20**: 535-569.

Hodson, P.V., Blunt, B.R., and Spry, D.J. (1978). Chronic toxicity of water-borne and dietary lead to rainbow trout (*Salmo gairdneri*) in Lake Ontario water. *Water Res.* **12**: 869–878.

Holtzman, R.B. (1969). Concentrations of the naturally occurring radionuclides ^{226}Ra, ^{210}Pb, and ^{210}Po in aquatic fauna. In Symp. Radioecology, Proc. 2d Nat. Symp. on Radioecology, CONF-670503. NTIS, Springfield, VA.

Joshi, D.B., and Tandon, R.S. (1977). The correlation of body size and some hematological values of the freshwater fish, I. *Clarias batrachus* (L.). *J. Anim. Morphol. Physiol.* **24**: 339–343.

Justyn, J., and Lusk, S. (1976). Evaluation of natural radionuclide contamination of fishes in streams affected by uranium ore mining and milling. *Zoologicke Listy* **25**: 265.

Lane, H.C., Weaver, J.W., Benson, J.A., and Nicholas, H.A. (1982). Some age related changes of rainbow trout, *Salmo gairdneri* Rich., peripheral erythocytes separated by velocity sedimentation at unit gravity. *J. Fish Biol.* **21**: 1–13.

Lehmann, J., and Stürenberg, F.-J. (1974). Haematologisch–serologische substratuntersuchungen an der Regeubo geuforelle (*Salmo gairdneri* Richardson). *Gewässer und Abwässer* **53/54**: 114–132.

Lucas, H.F., Edgington, D.N., and Colby, P.J. (1970). Concentrations of trace elements in Great Lakes fishes. *J. Fish. Res. Board Can.* **27**: 677.

McCarthy, D.H., Stevenson, J.P., and Roberts, M.S. (1975). Some blood parameters of the rainbow trout (*Salmo gairdneri* Richardson), II. The Shasta variety. *J. Fish Biol.* **7**: 215–219.

McLeay, D.J. (1973). Effects of 12-hr. and 25-day exposure to kraft pulpmill effluent on the blood and tissues of juvenile Coho salmon (*Oncorhynchus kisutch*). *J. Fish. Res. Board Can.* **30**: 395–400.

Morin, K.A., Cherry, J.A., Lim, T.P., and Vivyurka, A.J. (1982). Contaminant migration in a sand aquifer near an inactive uranium tailings impoundment, Elliot Lake, Ontario, Canada. *Geotech, J.* **19**: 49–62.

Patel, B., and Patel, S. (1979). Techniques for detecting radiation effects on biochemical and physiological systems in aquatic organisms. In *Methodology aorAssessing Impacts of Radioactivity on Aquatic Systems*. IAEA technical reports series, No. 190, Vienna, pp. 237–264.

Raabe, O.G. (1984). Comparison of the carcinogenicity of radium and bone-seeking actinides. *Health Phys.* **46**: 1241–1258.

Roberts, R.J. (1978). *Fish Pathology*. Bailliere Tindall, London.

Ruggles, R.G., and Rowley, W.J. (1978). A study of water pollution in the vicinity of the Eldorado Nuclear Limited Beaverlodge Operation. EPS 5-NW-78-10. Environment Canada, Ottawa.

Shleifer, G.S., and Shekanova, I.A. (1980). The effect of ionizing radiation on the immunophysiological condition of fish. *In* A.I. Ileuko (Ed.), *Problems and tasks of the radioecology of animals*. Nauka, Moscow.

Swanson, S.M. (1982). Levels and effects of radionuclides in aquatic fauna of the Beaverlodge area (Saskatchewan). SRC C-806-5-E-82. Sask. Res. Council, Saskatoon, Sask.

Swanson, S.M. (1983). Levels of ^{226}Ra, ^{210}Pb and $^{total}U$ in fish near a Saskatchewan uranium mine and mill. *Health Phys.* **45**: 67–80.

Swanson, S.M. (1985). Food-chain transfer of U-series radionuclides in a northern Saskatchewan aquatic system. *Health Phys.* **49**: 747–770.

Ulrikson, G.U. (1973). Radiation effects on serum proteins, hematocrits, electrophoretic patterns and protein components in the bullgill (*Lepomis macrochirus*). In D.J. Nelson (Ed.), *Radionuclides in ecosystems, Proc. Third Nat. Symp. Radioecol.* USAEC CONF-710501-P2, pp. 1100–1105.

Wedemeyer, G.A., and Nelson, N.C. (1975). Statistical methods for estimating normal blood chemistry ranges and variance in rainbow trout (*Salmo gairdneri*) Shasta strain. *J. Fish. Res. Board Can.* **32**: 551–554.

Whicker, F.W., and Schultz, V. (1982). *Radioecology: Nuclear Energy and the Environment*. CRC Press, Boca Raton, FL.

5

COLORIMETRIC DETERMINATION OF CYANIDE IN AQUATIC SYSTEMS

M.C. Mehra and A. Arseneau

Department of Chemistry and Biochemistry, Université de Moncton, Moncton, New Brunswick, Canada E1A 3E9

1. Introduction
2. Experimental
 2.1. Reagents and equipment
 2.2. Procedure
 2.3. Distillation of CN^- from a complex matrix
3. Results and discussion
4. Summary
 References

1. INTRODUCTION

Cyanide compounds are used in mineral processing, metal cleaning, and electroplating operations in industry. Organic cyanides are used in the manufacture of textiles, plastics, and agricultural and research chemicals. Though industrially important, these compounds are toxic to all kinds of living beings, and are regarded as metabolic inhibitors and chemical asphyxiants that prevent the utilization of oxygen supply by the tissues. Cyanide combines preferentially with ferricytochrome oxidase, and inhibition of cytochrome oxidase produces cytotoxic anoxia, a condition characterized by interference with cell metabolism in the presence of adequate supply of blood and oxygen (Casarett and Doull, 1975; Manahan, 1979). Waste effluents from mining and milling industries, cake ovens in steel plants, and metal cleaning and electroplating operations are potential sources of cyanide release into the aquatic environment. Concentrations ranging from 0.3 to 60 mg/L have been observed in such effluents (Canadian Council of Resource and Environmental Ministers, 1987). For these reasons, waste effluents discharged from these industries are carefully monitored. At present, tolerance limits of 0.1 ppm for oxidizable cyanide and 3.0 ppm for total cyanide are recommended for metal-finishing liquid effluents (Environment Canada, 1977), while a limit of 0.2 ppm is recommended for potable waters (Health and Welfare Canada, 1978).

The analytical procedures for cyanide (CN^-) determination generally involve its volatilization from the sample and entrapment in an alkaline solution. The final analysis is completed by titrimetry (Harzdorf and Dorn, 1977; Burger, 1985; Besada et al., 1983); electrometry (Conrad, 1971; Lynch, 1984); gas chromatography (Nota and Importa, 1979); colorimetry (Snell, 1981); or atomic absorption spectrometry (Matseuda, 1983; Varilikitois and Straits, 1984; Xu et al, 1984; Haj-Hussain et al., 1986). A recent review on CN^- detection and determination summarizes the current analytical methods (Singh et al., 1986).

Because of their simplicity, selectivity, and cost-effectiveness, colorimetric procedures are still popular. In a vast majority of cases this approach is based on quantitative determination of the absorbance supression of a metal chelate by CN^-, which remains proportional to the CN^- concentration in the sample of interest. The chelates of silver (Dagnall et al., 1968; Zhu et al., 1983), nickel (Wei et al., 1981), copper (Wei et al., 1984), palladium (Balanco and Maspoch, 1984), and mercury (Humphrey and Hinze, 1971; Hinze and Humphrey, 1973; Verma et al., 1979) have been successfully employed in this context for CN^- determination in surface and waste waters.

This chapter describes the use of a new chromogenic reagent for CN^- determination. The mercuric complex of the reagent 3-(2-pyridyl)-5,6-bis(5-(2-furyldisulfonic acid))-1,2,4-triazine disodium salt, commercially named ferene, absorbs strongly at 382 nm. The decrease in absorbance of the mercuric chelate in the presence of CN^- is quantifiable even in the sub-ppm range. The experimental conditions have been evaluated for CN^- determination in surface and simulated waste waters.

2. EXPERIMENTAL

2.1. Reagents and Equipment

Ferene reagent, 0.01 M. Dissolve 0.4943 g of the reagent in 1000 mL deionized distilled water to obtain 0.01 M solution. The reagent is prepared fresh every 2 or 3 days and may be procured from Diagnostic Chemicals Inc., Charlottetown, PEI, Canada.

Mercury solution, 1000 ppm. Dissolve 0.1354 g of pure mercuric chloride in 100 mL deionized distilled water and maintain at pH 2.0 acidified with HCl. Prepare 100- and 10-ppm solutions by dilution when required.

Buffer solutions. Prepare by mixing appropriate volumes of acetic acid (2 M) and sodium acetate (2 M) in the 4–5.5 pH range.

Cyanide solution, 1000 ppm. Prepare by dissolving 0.1884 g of solid sodium cyanide in 100 mL of deionized distilled water. Prepare standards at 100, 10, and 1 ppm by dilution when needed.

Record system absorbance using a double-base precision spectrophotometer. In the present study a Varian Cary-118 recording spectrophotometer sensitive to 0.0001 AUS was employed.

2.2. Procedure

To a 50-mL standard flask add 1.3 mL acetic acid, 3.5 mL sodium acetate, 2 mL mercury (100 ppm), and 3.5 mL ferene solution, and dilute to about 35 mL with deionized distilled water. After thorough mixing for 1 min, add the CN^- solution, not exceeding 100 μg, to complete the volume. Prepare another set similarly without CN^- to serve as system reference. Prepare a blank solution without mercury and cyanide. Measure the absorbance of the reference and sample solutions against the blank system in a 1-cm matched glass cell after 30 min at 382 nm. Obtain by subtraction the net absorbance decrease (ΔA) for the cyanide solution. Repeat for several CN^- concentrations and prepare a calibration curve.

2.3. Distillation of CN^- from a Complex Matrix

When the CN^- concentration in a sample is low and interferences from reactive species are suspected, separation of the CN^- concentration from the matrix is highly recommended. Put 500 mL of the sample solution ($CN^- < 40$ μg) in a double-necked distillation flask; add 50 mL H_2SO_4 (1 : 1), 20 mL $MgCl_2$ (50%), and 1–2 g solid mercurous sulfate. Fix a long (20–30 cm) tubing (2.5 mm i.d.) covered with cotton, for air intake, in one opening, and in the other adjust a water-cooled condenser at a slight upward inclination. Project the other end of the condenser to the base of a 25-mL conical suction flask containing an alkaline receiving solution. Distill for about 2 h at a rate of 100–200 bubbles/min, while maintaining mild air suction in the receiving flask. Receive the distillate in 2.5 mL of 0.1 M NaOH, neutralize to pH 5.0 with the acetic acid, and analyze the CN^- content by taking a suitable aliquot or the entire distillate as described above by bringing the final volume to 50 mL with deionized distilled water.

3. RESULTS AND DISCUSSION

The reaction of mercuric ion with the reagent ferene forms a yellow-colored chelate whose characteristics have been investigated in detail. The complex, though negatively charged, absorbs at 382 nm and has its maximum stability at pH 5.0 ± 0.05. The molar absorptivity of the chelate at 9.62×10^3 l mol^{-1} cm^{-1} permits quantitative determination of Hg(II) in the 1- to 10-ppm range (Arseneau, 1986). The color intensity of the chelate, however, is reduced in the presence of CN^- due to ligand exchange reaction which generates the colorless $Hg(CN_2)$:

$$[Hg(ferene)_2]^{-x} + 2\,CN^- \rightleftharpoons Hg(CN)_2 + 2[ferene]^{-x}$$

The absorbance decreases linearly and quantitatively for CN^- in the 0.2- to 1.0-ppm range at a constant Hg(II) concentration. Six replicate analyses for 0.2 ppm CN^- at 4 ppm Hg(II) show an average absorbance decrease of 0.0366 units with a coefficient of variation of 6% or less, while at 1 ppm CN^- the absorbance decrease is 0.1896 units with a coefficient of variation of 1% or less. Detection of CN^- at still lower concentrations is possible, but the coefficient of variation increases (e.g., 19% at 0.05 ppm CN^-). These data suggest that Hg–ferene chelate can be useful in the micro analysis of CN^- in the aqueous systems.

For drinking water and other effluents associated with low CN^- concentration (<0.2 ppm), the distillation procedure can be suitably modified such that the CN^- concentration in the final distillate may fall within the linear concentration range.

When CN^- analysis is performed in the presence of cations from transition metals at a concentration exceeding two- to threefold excess over the Hg(II) concentration, either the mercuric chelate formation is inhibited or inconsistent data are produced. The common anions also interfere at elevated concentrations (100–200 ppm). These interferences limit the use of this procedure for CN^- analysis in contaminated systems such as industrial effluents; therefore, separation of CN^- from the matrix becomes necessary, and is achieved by distillation wherein CN^- is displaced as volatile HCN, which is trapped in an alkaline solution. The experimental conditions are so adjusted that distillation of other reactive species is supressed or eliminated, while the binding of CN^- with heavy metals is prevented. It is known that an acidic condition favors HCN formation, while mercury and magnesium salts accelerate cyanometallate decomposition (Royer et al., 1973; Burger, 1985). The presence of mercury ions also prevents volatilization of sulfur compounds through the formation of insoluble mercury sulfides. In this study, the addition of mercurous sulfate in the distillation system has been preferred to eliminate such interferences and also to supress the chloride influence in the matrix due to the formation of partially soluble mercurous chloride.

Waste water samples prepared in the laboratory and spiked with known CN^-

concentrations were analyzed by this method, incorporating the distillation step. The data show that quantitative CN^- recovery occurs even at 0.02 ppm (Table 5.1). This is encouraging since it is an order of magnitude lower than the tolerance limit of 0.2 ppm CN^- for potable waters.

The heavy metals are known to form stable cyanometallates. The success of this procedure depends on their total decomposition in the distillation step such that CN^- is completely liberated from the system. To verify this, a known concentration of the cyanometallates of Cd, Fe, Cr, Zn, Ni, W, Co, and Ru was directly distilled, the CN^- released was analyzed by the proposed method, and the analytical data were compared to the theoretical value. The experimental data in Table 5.2 show that except for Co, W, and Ru, complete decomposition of the cyanometallates occurs in all cases. Obviously, in the presence of the first three elements, more drastic conditions are required for CN^- analysis. It may be assumed, however, that these elements do not occur in surface or environmental waters at significant concentrations to block the free cyanide in a system.

The surface waters of environmental concern spiked with a known amount of CN^- were also analyzed by the procedure developed in this study. In all cases, satisfactory results were recorded, as seen in the experimental data in Table 5.3. Analysis of CN^- in a complex matrix such as seawater is noteworthy, which augments the usefulness of the analytical procedure developed.

Table 5.1 CN^- Recovery from Pure and Saline Aqueous Samples by Distillation: Volume 500 mL

Conc. (μg) CN^- Spiked	No. of Distillations	Average % Recovery
	Aqueous	
50	6	101.1
35	6	102.1
25	10	100.3
15	10	99.5
10	10	102.5
	Anionic[a]	
15	8	102
	Cationic[b]	
15	8	99.6

[a] 200 μg each: S^{2-}, NO_3, SO_4^{2-}, Cl^-, SO_3^{2-}, and PO_4^{3-} as sodium or potassium salts.
[b] 200 μg each: Fe^{3+}, Cd^{2+}, Ni^{2+}, Cu^{2+}, and Zn^{2+} as chloride or nitrate salts.

Table 5.2 CN^- Recovery from Cyanometallates by Acid Distillation

Complex	CN^- Equivalent (μg)	No. of Determinations	% Recovery
$K_2[Cd(CN)_4]$	50	10	102
$K_3[Fe(CN)_6]$	50	10	101
$K_4[Fe(CN)_6]$	50	10	102
$K_2[Zn(CN)_4]$	50	10	98.6
$K_3[Cr(CN)_6]$	50	10	102
$K_2[Cu(CN)_4]$	50	10	98.8
$K_2[Ni(CN)_4]$	50	10	99.2
$K_4[Ru(CN)_6]$	50	10	86.4
$K_3[Co(CN)_6]$	50	10	19.7
$K_4[W(CN)_8]$	50	10	20.2

The reagent ferene has been employed recently for the trace analysis of iron (Mehra et al., 1985), chromium (Mehra et al., 1984), palladium (Mehra et al., 1986), and ruthenium (Arseneau et al., 1986) in aqueous samples. Current investigations further extend the use of this versatile reagent in the determination of toxic contaminants.

4. SUMMARY

A novel chromogenic reagent 3-(2-pyridyl)-5,6-bis(5-(2-furyl disulfonate)-1,2,4-triazine disodium salt, commonly known as ferene, has been employed in the trace analysis of CN^- in diverse aquatic samples. The absorbance of its chelate with mercury(II) is supressed in the presence of CN^- and this occurs quantitatively in the 0.2- to 1.0-ppm range. Distillation of CN^- from a complex matrix eliminates most of the interferences due to other reactive species and offers a preconcentration such that CN^- quantification at 0.02 ppm is realizable. The method has been applied for CN^- determination in surface waters and contaminated waters.

Table 5.3 CN^- Analysis in Spiked Surface Waters

Sample	No. of Distillations	μg CN^- Spiked	μg CN^- Recovered
River water	6	25	24.9 ± 0.9
Potable city water	6	25	25.2 ± 0.6
Well water	6	25	25.3 ± 1.0
Seawater	6	25	24.9 ± 0.7

ACKNOWLEDGMENTS

The authors are grateful to the National Sciences and Engineering Research Council of Canada and the Université de Moncton for the financial support provided in the realization of this research.

REFERENCES

Arseneau, A. (1986). M.Sc. Thesis, Université de Moncton, Moncton, NB, Canada.

Arseneau, A., Mehra, M.C., and Campanella, L. (1986). Spectrophotometric determination of ruthenium with a novel chromogen—5-5′ (3-(2-pyridyl)-1,2,4-triazine-5,6-diyl)-bis-2-ferene sulfonate. *Rassegna chimica (Italy)* **5**: 269–274.

Balanco, M., and Mospoch, S. (1984). Determination of CN^- by a highly sensitive indirect spectrophotometric method. *Talanta* **31**: 85–87.

Besada, A., Gawarigous, Y.A., Flatoos, B.N., and El-Sahat, M.F. (1983). Micro and sub micro determination of CN^- and SO with N-bromosuccinimide. *Mikrochim. Acta* **3**: 197–201.

Burger, N. (1985). Determination of iron and cyanide in cyanoferrate complexes. *Talanta* **32**: 49–50.

Canadian Council of Resource and Environmental Ministers (1987). Canadian Water Quality Guidelines, Task force report, Inland Water Directorate, Ottawa.

Casarett, L.J., and Doiull, J. (1975). *Toxicology—The Basic Science of Poisons*. New York: Macmillan, pp. 157, 2023–204.

Conrad, F.J. (1971). Potentiometric titration of cyanide and chloride with the use of a silver ion selective electrode. *Talanta* **18**: 952–955.

Dagnall, R.M., El-Ghamry, M.T., and West, T.M. (1968). Analytical applications of ternary complexes. V. Indirect spectrophotometric determination of cyanide. *Talanta* **15**: 107–110.

Environment Canada (1977). Metal finishing liquid effluent guidelines: Regulations, code and protocols. Report EPS1-WP-77-05.

Haj-Hussain, A.T., Christian, G.D., and Ruzicka, J. (1986). Determination of cyanide by atomic absorption using flow injection conversion method. *Anal. Chem*. **58**: 38–42.

Harzdorf, C., and Dorn, L. (1977). Voltametric titration of cyanide with mercury(II) chloride: Application to determination of cyanide in water and waste water. *Fres. Z. Anal. Chem*. **284**: 189–192.

Health and Welfare Canada (1978). *Guidelines for Drinking Water Quality*. Supply and Services Canada, Hull.

Hinze, W., and Humphrey, R.E. (1973). Mercury(II) iodate as an analytical reagent: Spectrophotometric determination of certain anions by an amplification procedure employing the linear starch–iodine system. *Anal. Chem*. **45**: 385–388.

Humphrey, R.E., and Hinze, W. (1971). Spectrophotometric determination of cyanide, sulphite, and sulphite with mercuric chloranilate. *Anal. Chem*. **43**: 1100–1102.

Lynch, T.P. (1984). Determination of free cyanide in mineral leachates. *Analyst* **109**: 421–423

Manahan, S. (1979). *Environmental Chemistry*, 3d ed. Grant, Boston, p. 455.

Matseuda, T. (1983). Indirect atomic absorption method for anions with thiourea CU(I) complex—Determination of CN^-. *Buniseki Kagaku* **32**: 373–377 (cf. *Anal. Abst*. 1 H 29–1984).

Mehra, M.C., Francoeur, B., and Satake, M. (1984). Indirect spectrophotometric determination of chromium in aqueous samples with a novel chromogen—Ferene. *Mikrochim. Acta* **3**: 61–68.

Mehra, M.C., Francoeur, B., and Katyal, M. (1985). Spectrophotometric determination of iron in environmental and industrial samples with a new chromogenic reagent—Ferene. *J. Bangladesh Acad. Sci.* **9**: 47–54.

Mehra, M.C., Arseneau, A., and Katyal, M. (1986). Spectrophotometric determination of palladium with a novel reagent—5-5′ (3-(2-pyridyl)-1,2,4-triazine-5,6-diyl)-bis-2-furone sulfonic acid disodium salt. *Orient. J. Chem.* **12**: 1–5.

Nota, G., and Importa, C. (1979). Determination of CN^- in cake oven waters. *Water Res.* **13**: 177–179.

Royer, J.L., Twichell, J.E., and Muir, S.M. (1973). Automated colorimetric method for total cyanide in water and waste water. *Anal. Lett.* **6**: 619–627.

Singh, H.B., Wasi, A., and Mehra, M.C. (1986). Detection and determination of cyanide–a review. *Intern. J. Environ. Anal. Chem.* **26**: 115–126.

Snell, F.D. (1981). Photometric and fluorometric methods of analysis–Nonmetals. New York: Wiley, pp. 652–679.

Varilikitois, G.S., and Straits, J.A. (1984). Determination of CN^- by atomic absorption and spectrophotometry through Cu–2-2′ dipyridyl-2-pyridyl hydrazone complex formation. *Microchem. J.* **29**: 209–218.

Verma, Y.S., Singh, I., Garg, B.S., and Singh, R.P. (1979). Spectrophotometric determination of cyanide ions through ligand exchange reaction in solution. *Mikrochim. Acta* **35**: 445–451.

Wei, F.S., Liu, Y.Q., Yin, F., and Shen, N.K. (1981). Determination of cyanide by an indirect spectrophotometric method using 5-Br-PADAP (2-(5-bromopyridylazo)-5-dimethylamino-phenol). *Talanta* **28**: 694–696.

Wei, F.S., Han, B., and Sham, N. (1984). Copper-cation 2B-triton X-100 system to spectrophotometric determination of micro amounts of CN^- in waste waters. *Analyst* **109**: 167–169.

Xu, B.X., Xu, T.M., and Fang, Y.m.(1984). Indirect determination of CN^- by atomic absorption. *Talanta* **31**: 141–143.

Zhu, Y.R., Wei, F.S., and Yiu, F. (1983). Indirect spectrophotometric determination of cyanide by means of colar reaction of silver with cation 2B in the presence of triton X-100. *Talanta* **30**: 795–799.

6

NAFION DIALYSIS PROCEDURE FOR SPECIATION OF METAL CATIONS

Charles L. Bourque

Department of Chemistry and Biochemistry, Université de Moncton, Moncton, New Brunswick, Canada

Robert D. Guy

Trace Analysis Research Centre, Department of Chemistry, Dalhousie University, Halifax, Nova Scotia, Canada

1. Introduction
- 1.1. Experimental speciation techniques
  - 1.1.1. Speciation of free metal ions
  - 1.1.2. Speciation methods involving a kinetic parameter
  - 1.1.3. Speciation methods involving a size parameter
- 1.2. Speciation and bioassay studies
- 1.3. Nafion dialysis

2. Experimental
- 2.1. Nafion dialysis speciation
- 2.2. Analysis of dialyzate

3. Results and discussion
- 3.1. Sorption characteristics of Nafion membranes
- 3.2. Permselectivity of Nafion membranes
- 3.3. Simple dialysis studies
- 3.4. Multication dialysis studies
- 3.5. Studies involving natural ligands

4. Summary

References

1. INTRODUCTION

Trace metal speciation in natural waters can be divided into three classes of problems: (1) the detection and determination of specific organometallic species; (2) the determination of the oxidation state of a metal ion; and (3) the differentiation of free versus bound metal ion in either complexing media or in the presence of colloids. The first two types of problems are amenable to solution by either gas or liquid chromatographic procedures or by electrochemical methods. The third type is considerably more complex, because the bound species are often poorly defined or in a labile equilibrium with the free metal ion.

1.1. Experimental Speciation Techniques

1.1.1. Speciation of Free Metal Ions

The free versus bound metal ion speciation methods can be divided into three groups. The first group is based on determination of the free metal ion. Potentiometry using ion-selective electrodes is suitable for the determination of Cu^{2+}, Pb^{2+}, Cd^{2+}, Ca^{2+}, and Mg^{2+} in water matrices of well-defined composition (Page and VanLoon, 1978). The ionic strength of samples and standards must be matched and the concentration of free metal ion should be greater than 1.0×10^{-6}M in unbuffered media. Heijne and VanderLinden (1978) have reported that aminopolycarboxylic acid ligands (i.e., EDTA and NTA) can interfere with the cupric ion-selective electrode. Cantwell et al. (1982) have used equilibrium ion exchange coupled with AAS detection to determine free nickel ion in sewage. The method assumes that only the free cation interacts with the resin and is applicable to any cation of interest. The method is also unique because it effects a preconcentration of the free ion during the speciation.

1.1.2. Speciation Methods Involving a Kinetic Parameter

The second group of speciation methods is based on a kinetic parameter. In these methods, the fraction measured includes free metal ion as well as metal ion present in "labile" complexes and able to dissociate in the time frame of the experiment. Thus, this fraction will depend at least in part upon the operational parameters of the method involved. The first of these methods, anodic stripping voltammetry, has been widely used as a speciation tool. Davison (1978) has provided a lucid description relating the terms "labile" and "nonlabile" to the Nernst diffusion layer, maximum substitution rate for complexation, and conditional stability constants for metal complexes. The second kinetic method characterized by Figura and McDuffie (1979) is based on passage of the sample through a short Chelex 100 column. The method determines any free metal ion and complexes that can dissociate and interact with the resin during a residence time of 4 to 6 s.

1.1.3. Speciation Methods Involving a Size Parameter

The third group of methods is based on a separation of species by size. The most common method uses equilibrium dialysis with small-pore cellulose membranes. Truitt and Weber (1981) have used equilibrium dialysis to characterize the complexation capacity of fulvic acid solutions and natural waters. Benes and Steinnes (1974) have used *in situ* dialysis to monitor metal ions in lake waters. The advantages of equilibrium dialysis are, first, the ability to optimize sample to receiver volumes to minimize shifts in labile equilibria and, second, the ease of interpretation of the results. The difficulties are potentially long equilibration times (greater than 24 h for the smaller membrane pore sizes) and shifts in equilibria induced by adsorption of metal ions onto the dialysis membrane. Ultrafiltration, a second type of size speciation, was used by Guy and Chakrabarti (1975) to characterize trace metal interactions with humic materials. Adsorption onto the membrane was a problem at a very low metal concentrations, but the study did indicate a correlation between the results obtained using the ultrafiltration method and those obtained using anodic stripping voltammetry.

1.2. Speciation and Bioassay Studies

One of the potential applications of free versus bound metal ion speciation is the determination of species toxic to aquatic biota. Experiments comparing bioassays and computer-calculated speciation of copper suggest that the free aquated cupric ion is the toxic species to algae (Guy and Kean, 1980) and to the microscopic crustacean *Daphnia Magna* (Andrew et al., 1977). The cupric ion, for example, exhibits toxicity to algae at a free metal ion concentration of 10^{-8}M. The kinetic methods and size-separation methods described above do not provide speciation information that is selective only to the free cation. In addition, potentiometry using ion selective electrodes does not have the sensitivity to provide speciation at the very low levels required for toxicity studies.

1.3. Nafion Dialysis

Cellulose dialysis membranes contain residual anionic sites that retard the dialysis of anionic species (Guy and Chakrabarti, 1975). The charge density is too low, however, to provide complete exclusion of anions. Dupont Company–USA produces a family of ion exchange membranes (e.g., Nafion 811X) that have an inert fluorocarbon backbone with sulfonic acid exchange sites bonded to it. One conceptual model of this membrane is a hydrophobic matrix with islands of hydrated exchange groups connected by channels in the matrix (Yeager and Kipling, 1979).

If one uses such an ion exchange membrane to separate the sample solution from a receiver electrolyte solution, one can then envision a speciation procedure based upon the following mechanism. An ion exchange equilibrium is established between the free metal cation and membrane sites on the sample side of the membrane:

$$2NaX_{(M,S)} + Cu^{2+}_{(S)} \rightleftharpoons CuX_{2(M,S)} + 2Na^{+}_{(S)} \tag{1}$$

The metal ions diffuse through the connecting channels and redistribute to establish ion exchange equilibrium with all the available sites within the membrane:

$$CuX_{2(M,S)} \rightleftharpoons CuX_{(M,R)} \tag{2}$$

and an ion exchange equilibrium is finally established with the receiver solution:

$$CuX_{(M,R)} + 2Na^{+}_{(R)} \rightleftharpoons Cu^{2+}_{(R)} + 2NaX_{(M,R)} \tag{3}$$

In the above equations M, R, and S represent membrane, receiver side, and sample side, respectively. If the electrolyte concentration is the same in both the sample and the receiver solutions, then Donnan preconcentration can be avoided. In addition, under experimental conditions where sample solution volumes are significantly larger receiver solution volumes, equilibrium shifts due to dilution are minimized and at equilibrium the free metal concentrations are the same on both sides of the membrane. Subsequently, the receiver solution may be analyzed by any appropriate analytical technique, yielding speciation information pertinent to water quality criteria.

Cox and Twardowski (1980) have illustrated the advantages of using the tubular membrane Nafion 811X for rapid preconcentration of trace metal ions by Donnan dialysis. These same advantages hold true for speciation studies, namely a large sample volume to receiver volume ratio and an increase in ion transport across the membrane that results from pumping the receiver solution through the tubular membrane. This chapter characterizes the application of Nafion 811X tubular membrane to trace metal speciation. The advantage of this method is that it combines the selectivity of speciation by Donnan exclusion of anions with size speciation resulting from the small diameter channels in the hydrophobic matrix.

2. EXPERIMENTAL

2.1. Nafion Dialysis Speciation

The Nafion speciation procedure used a tubular Nafion 811X cation exchange membrane (Dupont, Wilmington, DE). A 2.0-m length of tubing was tied in a 3.0-cm-diameter coil with Teflon tape. One end of the tubing was push-fit into

0.0315-in.(0.80-mm). Tygon tubing and the other end was placed into the receiver solution (Fig. 6.1). The Tygon tubing was part of a Masterflex pump equipped with a model No. 7013 pump head (Cole Parmer, Chicago, IL). The receiver solution was pumped through the tubular membrane at a rate of 3.0 mL/min. The sample volume of 400 mL was contained in a Nalgene beaker and the 25.00 mL of receiver solution was stored in a volumetric flask. The ionic strength of sample and receiver solutions was adjusted to 0.30 with $NaNO_3$. The membrane coil was placed in the magnetically stirred sample solution with one end in the receiver flask, and the flow circuit was completed by placing the remaining end of the Tygon tubing in the receiver flask. The receiver solution flowed in through the Nafion tubing, the coil, the Nafion–Tygon tubing connection, the pump, and then to the receiver. This arrangement ensured that the Nafion tubing–Tygon tubing connection would not "pop out" accidentally. The Masterflex pump was capable of accepting six pump heads, hence six samples were run simultaneously.

The Nafion membrane was treated between experiments to ensure uniform membrane characteristics, to minimize cross-contamination, and to prevent the membrane from altering the sample pH. Metal contamination was removed by placing the Nafion coils in 2 M HNO_3 and pumping the acid through the coils for 15 min. The coils were emptied, rinsed with distilled water, and placed in 2 M NaOH for 15 min to convert the membrane to the sodium form. The coils were again emptied, rinsed with distilled water, and placed in 0.30 M $NaNO_3$ containing acetic acid buffer at the same pH as the sample. This ensured that the membrane would not alter the pH of a poorly buffered sample solution.

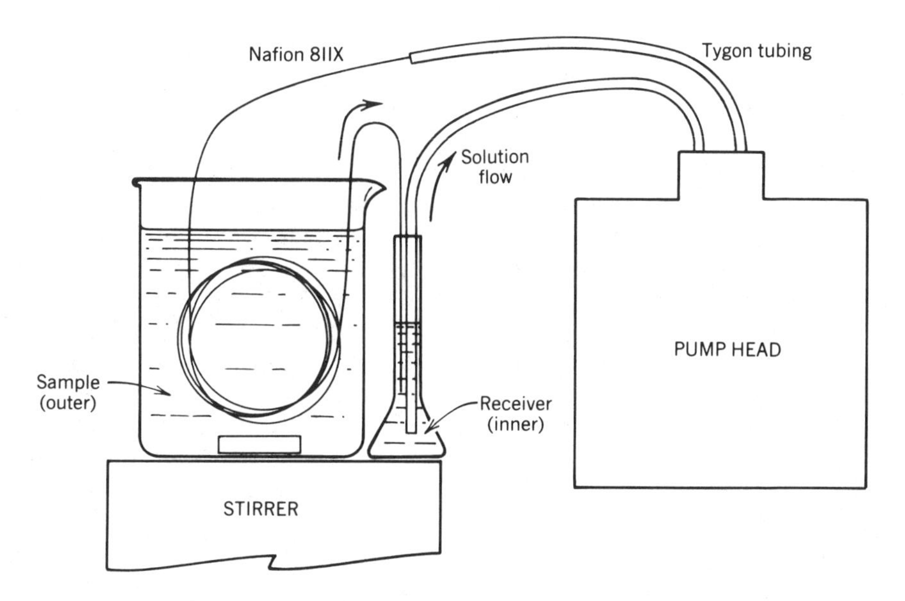

Figure 6.1. Speciation using Nafion 811X tubular membranes.

2.2. Analysis of Dialyzate

Trace metal ion analysis in the receiver solution was done by differential pulse polarography for Cu, Pb, Zn, and Cd using a PAR Model 174A Polarographic Analyzer equipped with a Model 1746 Drop Timer (Princeton Applied Research, Princeton, NJ). Multication receiver solutions with Cu, Zn, Ni, Pb, Co, and Cd were analyzed by ion exchange chromatography. A 20×0.42-cm cation exchange column packed with Aminex A-9 resin (Biorad, Richmond, CA) in sodium form was used with a 2.20 M tartrate eluent at pH 3.6. The eluent was mixed postcolumn with a 200 μM PAR (4–(2–pyridylazo)resorcinol) solution in a pH 10 ammonia buffer at a ration of 1 mL eluent to 0.5 mL detection reagent, and was coupled directly to a simple photometer consisting of a 520-nm light source (tungsten lamp and interference filter) and phototransistor. The ion exchange/photometer system was capable of separation and detection of the six metal ions at the 0.25 μM level in 16 min with a 1.0-mL sample injection. Replicate injection of standards in 0.30 M $NaNO_3$ gave relative standard deviations of 3.8% for Cu, 0.5% for Zn and Ni, 1.2% for Pb and Cd, and 0.8% for Co. The copper value is high because of a poor peak shape.

3. RESULTS AND DISCUSSION

3.1. Sorption Characteristics of Nafion Membranes

A problem with any membrane separation technique is the adsorption of the species of interest onto the membrane. This sorption produces an inaccurate speciation by lowering the free metal ion concentration and shifting any labile equilibria present in the sample. The ion exchange membrane can sorb significant amounts of free cation using the ion exchange reactions 1 and 3. The equivalent weight of Nafion 811X is 1100 and the 2-m length of tubing weighs about 1.0 g. This suggests that there are, on the average, 9.1×10^{-4} mol of exchange sites capable of binding the free ion. The amount of an ion sorbed will depend on the exchange constant and the concentration of the competing ion. A mass balance experiment investigating the effect of ionic strength on the amount of cupric ion sorbed gave the following results (the sorbed metal was determined by acid leaching the membrane): 53% sorbed with 0.05 M $NaNO_3$, 35% with 0.075 M $NaNO_3$, 25% with 0.10 M $NaNO_3$, 8.9% with 0.20 M $NaNO_3$, 5.2% with 0.30 M $NaNO_3$, and 2.2% with 0.50 M $NaNO_3$. A second experiment for cadmium using 0.10 M $NaNO_3$ electrolyte showed that the amount of metal sorbed increased linearly over the concentration range 0.5 to 50 μM. Similar sorption behavior was noted for the other ions and preliminary experiments indicated that the sorption was onto the Nafion membrane and not onto the glassware or Tygon tubing. The behavior noted in these experiments is characteristic of ion exchange with a large excess of sites over the potential amount of cation to be sorbed. No indication of a secondary strong adsorption site was noted for the metals studied over the concentration range 0.25 to 25 μM in 0.30 and 0.50 M $NaNO_3$. Secondary adsorption sites have been noted for similar membranes by Blaedel and Niemann (1975). To minimize sorption shifts of binding equilibria, the ionic strength of all samples was adjusted to 0.30 using $NaNO_3$.

3.2. Permselectivity of Nafion Membranes

The permselectivity of the Nafion membrane was characterized using a series of test samples. The first sample was 10 μM Cd^{2+} at pH 5.0, adjusted with 0.010 M acetate buffer. The second sample was 10 μM Cd^{2+} in the presence of 100 μM nitrilotriacetic acid (NTA) at pH 5.0, also adjusted with 0.010 M acetate buffer. The third sample was 20 μM ϕ_3Sn^+ cation in pH 5.0, adjusted with 0.010 M acetate buffer. These samples represented a free cation, an anionic complex, and a large organocation, respectively. The fourth and fifth samples were 2-methyl-5-nitroimidazole and 10 μM Pb^{2+} in 0.010 M phthalic acid at pH 5.0. All samples were maintained at 0.30 M $NaNO_3$.

The kinetics of dialysis of these systems are shown in Fig. 6.2. The free cadium ion attains dialysis equilibrium in 120 min while the anionic species was completely excluded. The triphenyltin cation was also completely excluded, which suggests that the species must be size excluded. The neutral nitroimidazole compound dialyzes slowly and dialysis was not complete even after 4 h. The lead ion in phthalic acid attained dialysis equilibrium after 120 min with 28.5 ±1.0% of the lead dialyzable. Equilibrium calculations using literature constants (corrected to 0.30 ionic strength) show a species distribution of 9.1% Pb^{2+}, 90.6% $[Pb(phthalate)]^{\circ}$, and 0.4% $[Pb(phthalate)_2]^{2-}$.

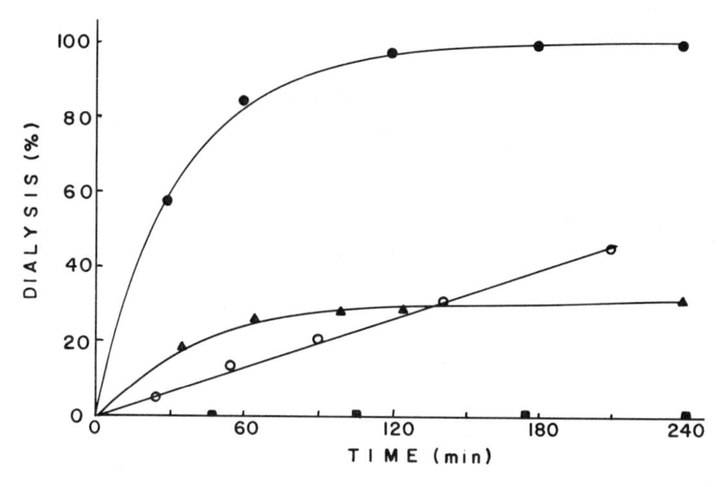

Figure 6.2. Permselectivity of the Nafion 811X membrane at pH 5.0 in 0.30 M $NaNO_3$. ●, Cd^{2+}; ▲, 10 μM Pb^{2+} in 0.01 M phthalic acid; ○, 2-methyl-5-nitroimidazole; ■, ϕ_3Sn^+ and 10 μM Cd^{2+} in 100 μM NTA.

Our results suggest that if only the free lead was dialyzable the conditional stability constant for lead phthalate should be about 250 instead of the value of 1000 reported in the literature (Ringbom, 1979). If one compares the dialysis behavior of the neutral nitroimidazole species and the neutral lead phthalate complex, one may conclude that the latter was size excluded. The Nafion dialysis procedure thus appears to be selective for free aquated cations and possibly small neutral species, excluding anionic species such as Cd–NTA^-. A tentative speciation procedure to differentiate between cations and neutral species would be to compare the results at 120 and 240 min. An increase in receiver concentration at 240 min would indicate the presence of a small neutral dialyzable species and the two results should allow one to estimate the contribution due to this small neutral species. All studies reported in this paper are limited to speciation experiments at pH values less than 6.5; hence no information can be provided on the permeability of simple cationic complexes (i.e., $CuOH^+$) or ion pairs ($CuCO_3$).

3.3. Simple Dialysis Studies

Simple dialysis studies using pH 4.5 buffer and 0.30 M $NaNO_3$ gave linear calibration curves for the six metal cations over the range 0.25 to 20 μM. The relative standard deviation of six replicate speciation experiments for copper, zinc, and nickel (5 μM) in 0.010 M pH 5 acetate buffer and 0.30 M $NaNO_3$ was 4.5, 0.9, and 0.9%, respectively. Replicate experiments for the speciation of 10 μM cadmium in 5.0 mM citric acid at pH 5.0 gave a dialyzable cadmium concentration of 3.78 μM with a relative standard deviation of 3.3% for experiments done on three different days. This suggests that the Nafion

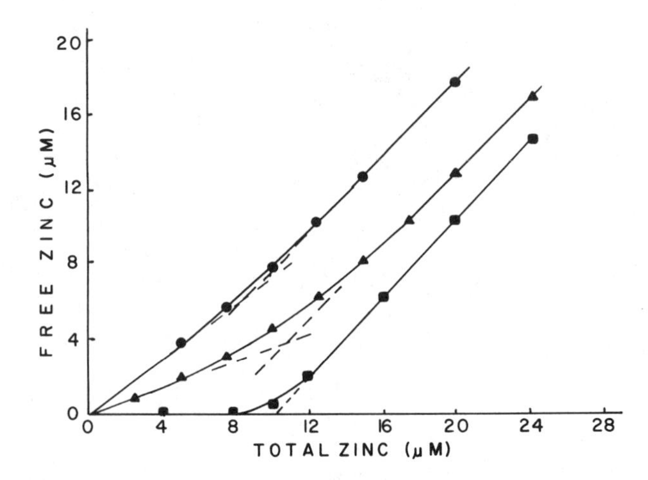

Figure 6.3. Binding capacity titration curves at pH 5.0 using zinc ion as the titrant. ■, EDTA; ▲, NTA; and ●, EGTA present at a concentration of 10 μM.

membrane can provide a speciation that is reproducible to within 5% at the micromolar level of free metal concentration.

The Nafion membrane speciation is suitable for determining the free aquated metal ion in a complexing media. Two speciation procedures commonly use the free ion as a monitor of changes in water chemistry. The first type of experiment is a binding capacity experiment in which the ligands present in the sample are titrated with a metal cation. The titration is usually done with a strongly complexed metal such as copper or nickel, but the curves in Fig. 6.3 use zinc as a titrant to illustrate the role of the conditional formation constant on the shape of the binding capacity curve. For each titration the concentration of ligand was 10 μM and the free zinc ion was determined by Nafion dialysis for 180 min.

The experimental data were used in the expression

$$K = \frac{[M_B]}{[M_F][L_F]} = \frac{[C_T] - [M_F]}{[M_F]([L_T] - [C_T] + [M_F])} \tag{4}$$

where K, C_T, M_F, M_B, and L_T are the conditional stability constant, total metal added, metal free, metal bound, and binding capacity, respectively. This expression assumes no secondary complexing agents and the formation of 1 : 1 complexes only. The data for the ethyleneglycol bis(2-aminoethylether)tetraacetic acid (EGTA) and nitrilotriacetic acid (NTA) titrations, i.e., C_T and M_F, were fit by a nonlinear least-squares program (Duggleby, 1981) to give the conditional binding constant and the potential binding capacity. The zinc–NTA titration gave a value of $2.16 \times 10^5 \pm 1.8 \times 10^4$ for the conditional constant and 1.05×10^{-5}M $\pm 5 \times 10^{-7}$M for the binding capacity. The zinc–EGTA data gave $4.71 \times 10^4 \pm 2.5 \times 10^4$ for the conditional constant and 7.6×10^{-6}M $\pm 3.1 \times 10^{-6}$M for the potential binding capacity. The greater uncertainty in the Zn–EGTA data reflects the greater relative error in the amount of zinc bound (calculated as the difference between the total concentration of zinc added and the Nafion dialyzable zinc).

3.4. Multication Dialysis Studies

The simple one ligand–one metal titration shown in Fig. 6.3 is an ideal titration. In real natural waters, however, there exists a more complex system, which contains competing ions that are already bound to the ligands. One has, therefore, a mixture of binding sites with binding constants that may vary over orders of magnitude. The binding capacity titration should use a multication analysis to determine the relative changes in free metal ion concentrations on the addition of the titrant metal ion. Figure 6.4 illustrates the potential of multication analysis after Nafion speciation for two simple model systems. Curve a in the diagram is the same titration curve for 10 μM NTA given in Fig. 6.3 but curve b presents the results of a 10 μM NTA titration in the presence of 10 μM nickel ion. Curve b in the inset is the Nafion dialyzable

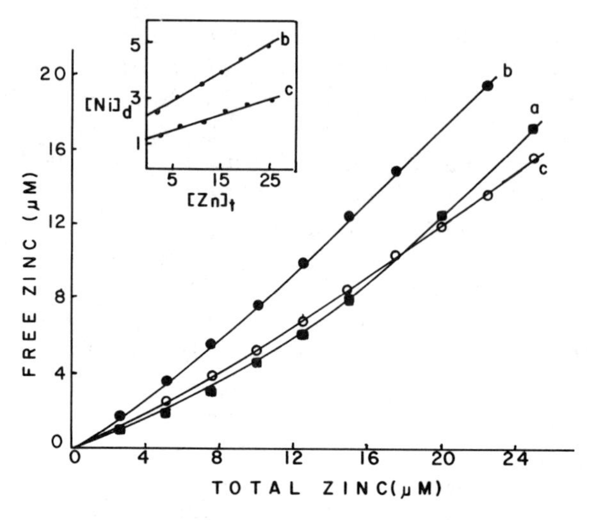

Figure 6.4. Binding capacity titrations of mixed systems at pH 5.0 using zinc ion as the titrant. ■, 10 μM NTA; ●, 10 μM NTA + 10 μM Ni^{2+}; ○, 10 μM NTA + 10 μM Ni^{2+} + 100 μM citric acid.

Dialyzable nickel as a function of total zinc added.

nickel during the titration. Nickel has a larger stability constant with NTA than zinc; hence the 2.5-fold excess zinc concentration displaces only 37.5% of the initially bound nickel. Curve c represents the same titration of 10 μM NTA and 10 μM Ni^{2+} with zinc ion but with an added 100 μM of citric acid as a secondary complexing agent. One now has a complicated equilibrium in which the two metal ions compete for the two types of binding sites. The Nafion speciation can yield information only on the concentration of the free metal ion and provides no information about the distribution of nickel and zinc between the NTA and citric acid ligands. The Nafion speciation can, however, provide some information about how the ligand mixture couples the chemistry of the two metal ions together. The two curves in the inset illustrate the changes in the free nickel concentration as a function of the total added zinc concentration. Within experimental error the free nickel concentration increases linearly with the total zinc concentration. The slopes of the two lines ($d[\mathrm{Ni}]_F/d[\mathrm{Zn}]_T$) can be used to calculate an interaction intensity parameter, as described by Morel et al. (1973). The interaction intensity parameter can be used to estimate the dependence of one metal's speciation in the presence of a second metal. The interaction intensity parameters were 0.060 and 0.122 with and without citric acid, respectively. The presence of an excess amount of weak binding ligands lowers the interactions between the two metal ions. One potential application of the Nafion speciation method coupled with a multication analysis during a binding capacity experiment is to obtain plots of interaction intensity parameters to characterize metal competition in multiligand systems.

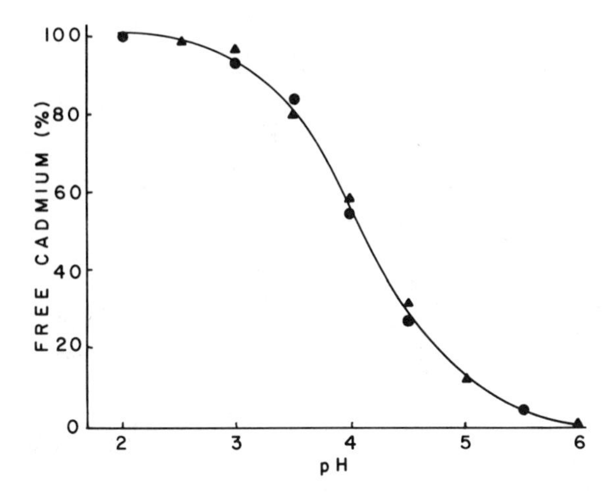

Figure 6.5. Speciation of 10 μM cadmium in 100 μM NTA as a function of solution pH. The curve through the points is the calculated speciation using literature equilibrium constants corrected for ionic strength. Speciation by differential pulse polarography (●) and Nafion dialysis (▲).

The second common speciation procedure to characterize the water chemistry of a sample is the determination of the concentration of free metal ion after adjusting the sample pH. Figure 6.5 presents the cadmium speciation in the presence of 100 μM NTA as a function of sample pH. The curve is drawn through three sets of data: the Nafion speciation results, direct determination of Cd^{2+} in sample by differential pulse polarography (one can distinguish between Cd^{2+} and $Cd{-}NTA^{-}$ polarographically), and the calculated speciation using stability and protonation constants corrected to an ionic strength of 0.30. Good agreement was obtained between the three sets of results. A multication pH speciation is given in Fig. 6.6 for six metals (10 μM each) in 5 mM citric acid. The large excess of ligand over metal ion masks any competition between metals for binding sites. The relative amount of the free metal at any pH reflects the conditional stability constand for the metal–citric acid species. The experimental Nafion speciation suggests that the conditional constants should decrease in the order:

$$Cu > Ni > Pb > Co, Zn > Cd$$

A computer-calculated speciation using constants from Ringbom (1979) gives the order Pb > Cu > Ni > Co, Zn > Cd, whereas the calculated speciation using Martell and Smith's (1977) selected stability constants gives Cu > Ni > Co > Zn > Pb > Cd. Both tabulations of stability constants appear to have selected a poor lead value. Our data at pH 5, for example, give a conditional stability constant of $4.2 \pm 0.7 \times 10^3$, whereas the literature values were 1.0×10^4 (Ringbom, 1979) and 1.7×10^3 (Martell and Smith, 1977). The dependence of the

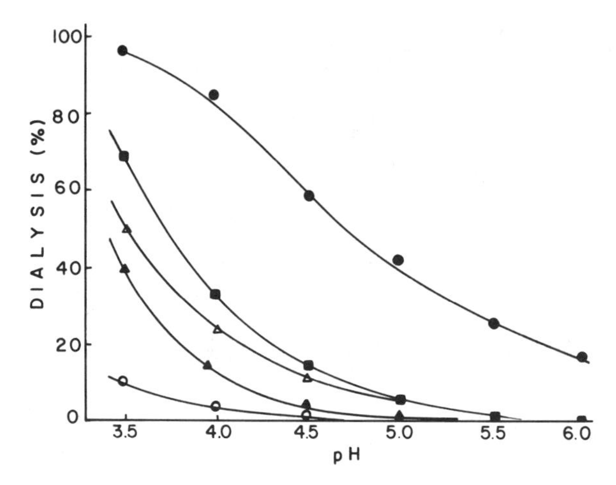

Figure 6.6. Multication speciation of 10 μM (each) of copper (○), lead (△), cadmium (●), cobalt (■), zinc (■), and nickel (▲) in 5 mM citric acid.

calculated speciation on the selected stability constants indicates that caution must be applied when using calculated speciation results for evaluation of metal toxicity to aquatic biota in multimetal systems. This dependence will be more critical when the potential binding capacity to metal concentrations ratio is small and metal competition is important. The Nafion speciation with multication analysis should prove useful in basic studies of metal ion interactions with aquatic biota.

3.5. Studies Involving Natural Ligands

The Nafion speciation requires a high sample ionic strength for two reasons: to prevent metal sorption by ion exchange and to minimize Donnan preconcentration. This requirement may limit the application of the method to strongly binding systems. Humic colloids are complex multisite binding substances that can interact with transition metal ions by a combination of inner sphere complexation using bidentate salicylic acid type groups and ion exchange. Figure 6.7 presents the adsorption isotherm obtained for cupric ion onto Aldrich humic acid and fulvic acid using Nafion speciation. The isotherm indicates that at the humic colloid has a maximum capacity for copper of 0.30 mmol/g acid. The speciation of 10 μM cupric ion in 20 ppm humic acid, buffered to pH 5.0 and containing 0.05, 0.10, or 0.30 M $NaNO_3$, can be used to illustrate the complex role that ionic strength can play in Nafion speciation. The data are given in Table 6.1. The total metal in solution was determined by an acid wash of the membrane. If one were to assume incorrectly that the membrane does not sorb metal ion and just uses the receiver copper concentration, one would conclude that ion exchange plays an important role in humic

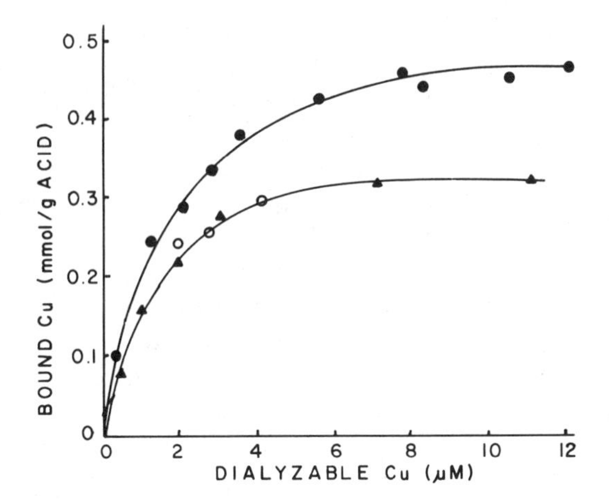

Figure 6.7. Copper speciation in the presence of humic colloids at pH 5 in 0.30 M $NaNO_3$. ▲, Aldrich humic acid; ●, soil fulvic acid; ○, copper adsorption at lower ionic strengths (see text).

acid binding. The equilibrium distribution in 0.05 and 0.10 M $NaNO_3$ is a result of the competition between Nafion exchange sites and humic acid binding sites for free cupric ion. The effective total metal concentrations at 0.05 and 0.10 M $NaNO_3$ are 6.7 and 8.1 μM, respectively, the remainder being bound by the Nafion membrane. The three experimental points have been included in the isotherm given in Fig. 6.7. The data suggest that the humic acid/metal ion ratio employed in these studies ion exchange is not important in the binding process over the range 0.05 to 0.30 M Na^+.

It has been reported by Blaedel and Haupert (1966) that Donnan exclusion breaks down at higher ionic strengths (i.e., >0.2). However, our results for anionic species such as Cd–NTA (Fig. 6.2 and 6.5), $Zn\text{–}NTA^-$ and $Zn\text{–}EDTA^{2-}$ (Fig. 6.3), and $M\text{–}citrate^-$ (Fig. 6.6) indicate that these species do not dialyze under our experimental conditions. This incongruity may possibly be due to exclusion of such anionic species based on a combination of size and charge parameters. Similar anionic species have been reported to dialyze

Table 6.1 Ionic Strength Effect on Metal Speciation

I	Initial Cu Total (μM)	Total CU in Solution (μM)	Nafion Dialyzable (μM)	mmol Cu/g HA[a]
0.05	10	6.7	1.92	0.24 ± 0.01
0.10	10	8.1	3.00	0.26 ± 0.01
0.30	10	10.0	3.96	0.30 ± 0.01

[a] Average of three experiments.

through cellulose membranes, albeit at a lower rate than the free metal ion (Guy and Chakrabarti, 1976). The total exclusion of such anionic complexes by Nafion speciation may possibly yield a dialyzate better resembling that which is bioavailable.

The Nafion 811X tubular membrane has been found to be selective to free aquated metal ions, excluding anionic complexes such as metal–polycarboxylates and large cationic species. The membrane has been shown to be useful for studies in acidic waters if a salt is added to increase ionic strength and thereby minimize cation interaction with ion exchange sites. Dialysis separations using 400-mL samples, 25-mL receiver solution, and 0.30 M $NaNO_3$ ionic strength adjuster will shift equilibria by approximately 6% (sample dilution) and 5% (ion exchange sorption). These shifts in equilibria will be in excess of shifts induced in ion exchange binding to colloids by adding the $NaNO_3$ salt.

4. SUMMARY

A tubular cation exchange membrane (Nafion 811X, Dupont) was used to distinguish between free metal cations and bound metals in model freshwater systems. The sample ionic strength must be augmented to 0.30 to prevent Donnan preconcentration in the receiver and to minimize metal sorption onto the membrane. The membrane was found to be permeable to small free cations but impermeable to anions, large cations (i.e., triphenyl tin), and neutral complexes (i.e., lead phthalate). Dialysis equilibrium is attained in 120 min and a multication speciation is possible using a suitable metal detector. The Nafion dialysis procedure was suitable for metal cation levels above 0.20 μM and sample pH below 6.5. The studies using model systems suggest that the procedure would be suitable for binding capacity measurements.

ACKNOWLEDGMENTS

This research was funded by a grant from the Natural Sciences and Engineering Research Council of Canada. C. Bourque acknowledges a research scholarship from the W. E. Sumner Foundation.

REFERENCES

Andrew, R.W., Biesinger, K.E., and Glass, G.E. (1977). Effects of inorganic complexing on the toxicity of copper to Daphnia Magna. *Water Res.* **11**: 309–315.

Benes, P., and Steinnes, E. (1974). In situ dialysis for the determination of the state of trace elements in natural waters. *Water Res.* **8**: 947–953.

Blaedel, W.J., and Haupert, T.J. (1966). Exchange equilibrium through ion exchange membranes. *Anal. Chem.* **38**: 1305–1308.

Blaedel, W.J., and Niemann, R.A. (1975). Application of ion exchange membranes to sampling and enrichment: Interference of metal ion binding groups. *Anal. Chem* **47**: 1455–1457.

Cantwell, F.F., Nielsen, J.S., and Hrudey, S.E. (1982). Free nickel ion concentration in sewage by an ion exchange column-equilibrium method. *Anal. Chem*. **54**: 1498–1503.

Cox, J.A., and Twardowski, Z. (1980). Tubular flow Donnan dialysis. *Anal. Chem*. **52**: 1503–1505.

Davison, W. (1978). Defining the electroanalytically measured species in a natural water sample. *J. Electroanal. Chem*. **87**: 395–404.

Duggleby, R.G. (1981). A nonlinear regression program for small computers. *Anal. Biochem*. **110**: 9–13.

Figura, P., and McDuffie, B. (1979). Use of Chelex resin for determination of labile trace metal fractions in aqueous ligand media and comparison of the method with anodic stripping voltammetry. *Anal. Chem*. **51**: 120–125.

Guy, R.D., and Chakrabarti, C.L. (1975). Analytical techniques for speciation of trace metals. Proc. Int. Conf. Heavy Metals in the Environment, Univ. of Toronto, Vol. 1, pp. 275–294.

Guy, R.D., and Chakrabarti, C.L. (1976). Studies of metal–organic interactions in model systems pertaining to natural waters. *Can. J. Chem*. **54**: 2600–2611.

Guy, R.D., and Kean, A.R. (1980). Algae as a chemical speciation monitor. A comparison of algal growth and computer calculated speciation. *Water Res*. **14**: 891–899.

Heijne, G.J.M., and VanderLinden, W.E. (1978). The formation of mixed copper sulfide–silver sulfide membranes for copper(II)-selective electrodes. *Anal. Chim. Acta* **96**: 13–22.

Martell, A.E., and Smith, R.M. (1977). *Critical Stability Constants*, Plenum, New York.

Morel, F., McDuff, R.E., and Morgan, J.J. (1973). Interactions and chemostasis in aquatic chemical systems: Role of pH, pE, solubility and complexation. In P.C. Singer (Ed.), *Trace Metals and Metal–Organic Interactions in Natural Waters*, Ann Arbor Science, Ann Arbor, MI, pp. 157–200.

Page, J.A., and VanLoon, G.W. (1978). A report on the application of electroanalytical methods to the study of trace metals in the marine environment. Report 5, NRCC No. 16924, Marine Analytical Chemistry Standards Program, National Research Council of Canada.

Ringbom, A. (1979). *Complexation in Analytical Chemistry*. Kreiger, Huntington, NY, pp. 320–328.

Truitt, R.E., and Weber, J.H. (1981). Determination of complexing capacity of fulvic acid for copper(II) and cadmium(II) by dialysis titration. *Anal. Chem*. **53**: 337–342.

Yeager, H.L., and Kipling, B. (1979). Ionic diffusion and ion clustering in a perfluorosulfonate ion-exchange membrane. *J. Phys. Chem*. **83**: 1836–1839.

7

METALLOTHIONEIN MESSENGER RNA: Potential Molecular Indicator of Metal Exposure

K.M. Chan

Marine Sciences Research Laboratory, Memorial University of Newfoundland, St. John's, Newfoundland, Canada

W.S. Davidson

Department of Biochemistry, Memorial University of Newfoundland, St. John's, Newfoundland, Canada

G.L. Fletcher

Marine Sciences Research Laboratory, Memorial University of Newfoundland, St. John's, Newfoundland, Canada

1. Introduction
- 1.1. Winter flounder
- 1.2. Metallothioneins: Structures and possible functions
- 1.3. Fish metallothioneins
- 1.4. Metallothionein in the winter flounder
- 1.5. Other low molecular weight metal-binding proteins
- 1.6. Aim of this study

2. Materials and methods
- 2.1. Winter flounder
- 2.2. Metallothionein induction
- 2.3. Metallothionein purification

2.4. Isolation of hepatic mRNA
2.4.1. Total RNA isolation
2.4.2. Isolation of Poly A RNA
2.4.3. Sucrose gradient centrifugation
2.5. Cell free translation
2.6. Analysis of translation products
2.7. cDNA synthesis
3. Results
3.1. Hepatic metallothionein in the winter flounder
3.2. Purification of 9-S messenger RNA
3.2.1. RNA isolation
3.2.2. *In vitro* translation
3.2.3. Polyacrylamide gel electrophoresis
3.2.4. Sucrose density gradients
3.3. Complementary DNA to 9-S messenger RNA
4. Discussion
5. Summary
References

---

1. INTRODUCTION

1.1. Winter Flounder

Winter flounder can be found throughout the year in inshore areas of the northeastern coast of North America (Leim and Scott, 1966). They inhabit soft, muddy to sand–gravel bottoms at depths of 1–40 m and tolerate wide ranges of temperature, salinity, and oxygen concentrations (Klein-MacPhee, 1978). Their close proximity to the shore makes them prone to exposure to a wide range to water-borne contaminants of land, river, and oceanic origin. In addition, their habit of burrowing deep into sediments and ingesting mud and sand while feeding makes them vulnerable to contaminants that may not be present in the water (Fletcher, 1975; Fletcher et al., 1981, 1982). The winter flounder's inshore habitat, its availability, and its relative ease of maintainance under laboratory conditions have made it a useful and valuable East Coast marine species for fish biochemistry, physiology, and toxicology, and for the biological monitoring of coastal environmental quality. These attributes have made the winter flounder one of the most studied marine fish, resulting in a large bank of information that is available to all who need an excellent fish model.

We have been studying various aspects of Zn^{2+} metabolism in the winter flounder in an attempt to understand how this essential trace element is regulated in body tissues, and to determine what happens to the regulatory mechanisms when they are challenged by excess Zn^{2+} or other trace elements (Fletcher and King, 1978; Fletcher and Fletcher, 1978, 1980; Shears and Fletcher, 1979, 1983). More recently we have turned our attention to metallothionein (MT) and its possible role in trace element metabolism and detoxification (Shears and Fletcher, 1984, 1985).

1.2. Metallothioneins: Structures and Possible Functions

MTs are metal-binding proteins that characteristicly have a high cysteine content (30%) but do not contain aromatic amino acids or histidine residues. Their molecular weights are in the range 6000 to 10,000 and they usually consist of about 60 amino acids (Kagi and Nordberg, 1979). A highly conserved central segment of 10 amino acid residues has been found in MTs from yeast, crab, sea urchin, and mammals (Nemer et al., 1985). Two domains of clusters of Cys–X–Cys sequences allow MTs to bind six or seven atoms on Zn^{2+} or Cd^{2+} per molecule of protein using the thiol groups as ligands (for reviews, see Karin and Richard, 1984; Hamer, 1986). The extensive homology among the amino acid sequences of MTs isolated from these diverse species indicates that there has been a conservation of heavy metal-binding function throughout evolution.

MTs are induced in many organisms and cultured cells after administration of heavy metal ions. It has been proposed that MTs are important in regulating metal ions during mammalian development, are active in copper and zinc metabolism, are involved in cadmium detoxification, and are induced as part of a stress response (Brady, 1982; Cherian and Nordberg, 1983; Karin, 1985; Hamer, 1986). However the physiological functions of MTs are still unclear and further studies are required. MTs are intracellular proteins and they may be responsible for maintaining the correct balance of trace metals within cells so that metal-dependent enzymic reactions may occur.

1.3. Fish Metallothioneins

In fish, preexposure to sublethal levels of heavy metals results in the induction of MTs and can result in an acclimation to a potentially toxic level of these metals in the water (McCarter and Roch, 1984; Klaverkamp et al., 1984). MTs have been characterized from various fish, including carp, salmon, trout, plaice, sculpin, eel, and winter flounder (Klaverkamp et al., 1984; Shears and Fletcher, 1985). It is not uncommon to observe more than one isoform of MT in a particular species. For example, according to amino acid composition data, two isoforms were identified in the liver of skipjack tuna (Takeda and Shimuzu, 1982) and rainbow trout (Bonham and Gedamu, 1984; Olsson and Haux, 1985), and in the kidney and hepatopancreas of carp (Kito et al., 1982,

1984). However, only one MT was identified in plaice (Overnell and Coombs, 1979) and in the winter flounder (Shears and Fletcher, 1985) (Table 7.1).

1.4. Metallothionein in the Winter Flounder

The winter flounder produces a single MT after an injection of zinc chloride or cadmium chloride. This protein has been purified by conventional techniques involving gel filtration and ion-exchange chromatography (Shears and Fletcher, 1984, 1985). The minimum molecular weight for the MT from winter flounder is 7129. This value is based on its amino acid composition, which is shown in Table 7.1. Table 7.1 also shows the amino acid compositions of some other fish MTs and the MTs of mouse and horse.

Available evidence suggests that marine fish attain Zn^{2+} from food rather than by direct accumulation from the water (Hoss, 1964; Renfro et al., 1975; Pentreath, 1973a,b). Thus, the gastrointestinal tract should be an important site for the regulation of this essential trace metal. Studies on the winter flounder have shown that the amount of Zn^{2+} absorbed, or capable of being absorbed, varies with dietary composition, body Zn^{2+} levels, and season (Fletcher and King, 1978; Shears and Fletcher, 1983). Thus, there is a considerable capacity for homeostatic mechanisms to operate at the level of the digentive tract.

Experiments designed to look for a relation between MT and body Zn^{2+} regulation have not met with success. Shears and Fletcher (1983, 1984) and Shears (1983) demonstrated that the induction of intestinal mucosa and liver MT by parenteral administration of Zn^{2+} (25% body level) did not alter the rate of Zn^{2+} uptake or excretion by the intestine of the winter flounder. Moreover, the rate of Zn^{2+} loss (biological half time) from the Zn^{2+} treated fish did not differ from normal (control) flounder (Shears, 1983).

1.5. Other Low Molecular Weight Metal-Binding Proteins

Although it is generally accepted that MTs play a role in Zn^{2+} and Cu^{2+} homeostasis, there is no evidence that MT is normally present in winter flounder tissues. Chromatographic methods identical tho those used to purify MT from Zn^{2+}- or Cd^{2+}-treated flounder yielded low molecular weight metal-binding proteins with a cysteine content much lower than that of MT. For example, low molecular weight proteins isolated from mucosal and liver cytosols had only 2.3 and 13% cysteine, respectively (Shears and Fletcher, 1984, 1985). The liver low molecular weight cysteine-containing protein is present throughout the year. In contrast, the low molecular weight MT-like protein in the intestinal mucosa has been observed only in trace amounts during the last half of the year, particularly when the flounder have ended their seasonal period of feeding (Shears and Fletcher, 1984; Fletcher and King, 1978). The functions of these low molecular weight proteins are unknown.

Non-MT low molecular weight metal-binding proteins in the liver of

Table 7.1 Amino Acid Composition (Residues per Molecule) of Hepatic Metallothionein Isolated from Cadmium-Injected Fish (1–4), mouse (5), and Normal Horse (6)

Amino Acid	(1) Winter Flounder[a]	(2) Plaice[b]	(3) Carp[c]		(4) Rainbow Trout		(5) Mouse[e]		(6) Horse[f]	
			MT_1	MT_2	MT_1	MT_2	MT_1	MT_2	MT_1	MT_2
Asx	6	6	5	6	6	8	4	4	3	3
Thr	7	6	4	5	4	4	5	1	3	1
Ser	6	6	7	6	8	10	9	10	8	8
Pro	4	4	3	3	2	2	2	2	3	2
Glx	3	3	2	2	3	2	1	3	2	3
Gly	6	7	6	7	6	6	5	4	7	5
Ala	2	2	3	3	3	3	5	6	5	7
Cys	18	19	20	20	20	17	20	20	20	20
Val	1	1	1	1	1	1	2	1	1	3
Met	1	2	1	1	1	1	1	1	1	1
Ile	—	—	—	—	—	—	—	1	—	—
Leu	—	1	—	—	—	—	—	—	—	—
Try	—	—	—	—	—	—	—	—	—	—
Phe	—	—	—	—	—	—	—	—	—	—
Lys	6	6	7	6	6	6	7	8	6	7
His	—	—	—	—	—	—	—	—	—	—
Arg	—	—	—	—	—	—	—	—	2	1

Source: Modified from Shears and Fletcher (1985).

Note: MT, metallothionein.

[a] Shears and Fletcher (1985).

[b] Calculated by Shears and Fletcher (1985) as nearest whole unit number of residues based on 19 half cystine residues per molecule (Overnell and Coombs, 1979).

[c] Expressed to nearest integer (Kito et al., 1982).

[d] Olsson and Haux (1985).

[e] Sequence analysis from Huang et al. (1979).

[f] Sequence analysis from Kojima et al. (1979).

rainbow trout are also documented (Pierson, 1985a,b). Thomas et al. (1985) found that Cd^{2+} accumulated in two of these non-MT low molecular weight metal-binding proteins in rainbow trout exposed to cadmium salts. However, if the fish had been preexposed to Zn^{2+}, the Cd^{2+} became bound to MT in addition to the other proteins (Thomas et al., 1983a,b). The interrelation between MT and non-MT metal-binding proteins is not clear.

1.6. Aim of This Study

One of the major difficulties in coming to firm conclusions about the role of MT in the metabolism of these trace metals stems from our inability to positively identify and quantify MT in normal and metal exposed fish. To correct this we have taken a molecular biological approach by attempting to develop a complementary DNA (cDNA) probe that will enable us to measure MT messenger RNA (mRNA) levels with a high degree of specificity and precision. As a first step we report the isolation from Cd^{2+} treated flounder of hepatic mRNA that translates to yield MT. Some of the results reported in this communication have been published (Chan et al., 1987). The general applicability of a cDNA probe to measure MT mRNA in winter flounder tissues is elaborated and discussed.

2. MATERIALS AND METHODS

2.1. Winter Flounder

Winter flounder (*Pseudopleuronectes americanus*) were caught during August by scuba divers in Conception Bay, Newfoundland. The flounder were maintained in laboratory aquaria (250–500 L) supplied with continuously flowing seawater under ambient temperature (average = 10.8°C) and photoperiod (Fletcher, 1977).

2.2. Metallothionein Induction

Solutions of cadmium chloride were prepared in 1.1% sodium chloride such that the final concentration of Cd^{2+} was 2 or 4 mg/mL. The induction protocol for the preparation of MT was similar to the schedule described by Shears and Fletcher (1985): Day 1, 0.2 mg Cd^{2+}/kg body w; Day 3, 0.6 mg/kg; Day 7, 2.0 mg/kg. The fish were killed on Day 14. The induction protocol for the isolation of MT mRNA was similar to that of Bonham and Gedamu (1984): Day 1, 0.4 mg/kg: Day 2, 0.6 mg/kg; Day 3, 1.0 mg/kg; Day 4, 2.0 mg/kg. The fish were killed 24 h after the final injection. Flounder of both sexes were used for controls (1.1% saline injections) and cadmium chloride treatment.

2.3. Metallothionein Purification

Hepatic MT was purified from Cd^{2+}-treated winter flounder using a combination of the procedures outlined by Shears and Fletcher (1984, 1985). Livers were pooled and homogenized in 0.1 M ammonium bicarbonate, pH 8.5, containing 2 mM 2-mercaptoethanol. A cytosolic extract, prepared by ultracentrifugation (65,000 g for 1 h), was heat-treated at 70°C for 1 min and then centrifuged as before to remove heat-labile high molecular weight proteins. The resulting supernatant was treated with ammonium sulfate and the proteins precipitating between 40 and 100% saturation were collected. The precipitated proteins were dissolved in 0.1 M ammonium bicarbonate, pH 8.5, containing 2 mM mercaptoethanol. This fraction was desalted by dialysis before having $^{109}Cd^{2+}$ added to it and being subjected to gel filtration on Sephadex G75. The low molecular weight $^{109}Cd^{2+}$-binding proteins were pooled and applied to an ion-exchange column of Whatman DE52. Flounder MT was eluted using a linear salt gradient. That this protein was indeed MT was confirmed from analysis of its amino acid composition (Table 7.1), molecular weight, and spectral properties.

2.4. Isolation of Hepatic mRNA

2.4.1. Total RNA Isolation

Total RNA was prepared from flounder liver according to the procedure of Davies and Hew (1980). Ten grams of liver tissue pooled from control or cadmium chloride-treated fish was homogenized on ice in a mixture containing 100 mL phenol (saturated with 0.1 M Tris–HCl, pH 8.0) and 100 mL water containing 0.05% diethylpyrocarbonate (DEP), 0.5% sodium dodecylsulfate (SDS), 25 mg/mL polyvinylsulfate, and 35 μg/mL spermine. The homogenate was centrifuged at 8000 *g* for 10 min at 4°C and then the supernatant was extracted with equal volumes of chloroform until no white interphase was visible. The RNA was precipitated from the aqueous phase by adjusting the sodium chloride concentration to 0.2 M and adding two volumes of redistilled ethanol (–20°C). The solution was mixed and stored overnight at –70°C. The RNA was precipitated by centrifugation at 10,000 *g* for 35 min at 4°C. The pellet was washed twice with 3 M sodium acetate, pH 6.0 (150 mL each time), redissolved in water, and centrifuged at 100,000 *g* for 45 min at 4°C to precipitate glycogen. The RNA in the supernatant was precipitated in the presence of sodium chloride and ethanol as described above. The RNA was washed twice with 70% ethanol, lypholized, and dissolved in a small volume of water. Proteinase K digestion was carried out by incubating the RNA in a solution containing 0.5 M sodium chloride, 0.5 mg/mL proteinase K, 0.5% SDS, 5 mM EDTA (ethylenediaminetetraacetic acid), and 10 mM Tris–HCl, pH 7.4, for 1 h at 37°C. After incubation, the solution was extracted with an equal volume of phenol/chloroform (1 : 1). The aqueous phase was further extracted twice with an equal volume of chloroform/isoamylalcohol (49 : 1). The RNA was precipitated as described above, washed with 70% ethanol lypholized, and stored frozen in DEP water at –70°C. All glassware and water used in the purification steps wer treated with 0.05% DEP and then autoclaved.

2.4.2. Isolation of Poly A RNA

Polyadenylated messenger RNA (poly A RNA) was isolated by affinity chromatography on oligo-(dT)-cellulose following the precedure of Aviv and Leder (1972). Total RNA was incubated with oligo-(dT)-cellulose for 3 h at 37°C in a high-salt buffer (0.2 M sodium chloride, 10 mM Tris–HCl, pH 7.4, 1 mM EDTA, 0.1% SDS). The unbound RNA (poly A minus RNA) was washed from the column with this buffer. Poly A RNA was eluted from the oligo-(dT)-

cellulose with a low salt buffer (10 mM Tris–HCl, pH 7.4, 1 mM EDTA, 0.1% SDS) at 65°C. Poly A RNA and poly A minus RNA were precipitated and stored as described above. The integrity of the RNA samples was determined by agarose gel electrophoresis in the presence of methylmercury (Bailey and Davidson, 1976).

2.4.3. Sucrose Gradient Centrifugation

Poly A RNA (100–200 μg) in water was heated at 65°C for 5 min and then quickly chilled on ice before being loaded on a linear 5 to 30% sucrose gradient made up in a buffer consisting of 0.1 M sodium chloride, 1 mM EDTA, 10 mM Tris–HCl, pH 7.5. Centrifugation was for 15 h in a Beckman SW41 rotor at 250,000 *g*. Thirty fractions (400 μL each) were collected. The RNA in each fraction was precipitated as above and redissolved in 10 μL of water containing the ribonuclease inhibitor RNasein. Poly A minus RNA and globin mRNA were used as sedimentation markers.

2.5. Cell Free Translation

In vitro cell translations of RNA were carried out using [^{35}S]cysteine and a nuclease-treated rabbit reticulocyte lysate translation kit from Amersham. The optimal potassium ion concentration was 80 mM. Reactions were allowed to proceed for 120 min at 32°C or for 90 min at 37°C. The incorporation of [^{35}S]cysteine into trichloroacetic acid-precipitable material was measured according to the procedure of Pelham and Jackson (1976) and was linear under these conditions.

2.6. Analysis of Translation Products

Translation products were carboxymethylated by incubation in the dark for 1 h at 37°C in 0.2 M iodoacetic acid titrated to pH 8.6 with Tris base. Polyacrylamide slab gel electrophoresis followed the procedure of Laemmli (1970). Nondenaturing polyacrylamide gel electrophoresis used the Laemmli system without any SDS present. Gels were stained for protein with 0.1% Coomassie blue R250 in a solution of isopropanol/acetic acid/water (25:10:65) containing 0.1% cupric acetate. Excess stain was removed by soaking the gels in methanol/acetic acid/water (30:10:60) then in methanol/acetic acid/water (5:7:88). If protein staining was not required, the gel was soaked in trichloroacetic acid/acetic acetic acid/methanol/water (20:10:30:40) for 1–2 h. For fluorography, the gels were treated with Enhance (New England Nuclear) or Amplify (Amersham) according to the manufacturer's instructions, soaked in 5% glycerol, and then dried under vaccuum at 60°C for 2 h followed by 2 h at room temperature. Kodak XAR-5 or AGFA CURIX RP-1 film was used and exposures were carried out at −70°C.

2.7. cDNA Synthesis

Double-stranded cDNA was prepared according to the method of Gubler and Hoffman (1983). Synthesis of the first strand was carried out in a reaction volume of 50 μL containing 5 μg poly A RNA, 100 mM Tris–HCl, pH 8.3, 10 mM magnesium chloride, 140 mM potassium chloride, 1 mM deoxynucleotide triphosphates, 5 μg oligo-$(dT)_{12-18}$ primer, 40 units RNasein, 30 μCi α-[^{32}P]dATP (3000 Ci/mmol, Amersham), and 50 units of reverse transcriptase (Life Sciences) for 30 min at 43°C. The reaction was terminated by the addition of EDTA to a final concentration of 20 mM, a phenol–chloroform extraction, and then precipitation of the nucleic acid in the presence of ammonium acetate and ethanol. For synthesis of the second strand the nucleic acid was redissolved in a final volume of 100 μL containing 20 mM Tris–HCl, pH 7.5, 5 mM magnesium chloride, 10 mM ammonium sulfate, 100 mM potassium chloride, 0.5 mM NAD, 50 μM deoxynucleotide triphosphates, 30 μCi α-[^{32}P]dATP, 0.8 units RNase H, 23 units DNA polymerase 1, and 1 unit of DNA ligase. The reaction was allowed to proceed at 12°C for 1 h and then at 22°C for another hour before being terminated by the addition of EDTA to 20 mM and SDS to 0.4%. After a phenol–chloroform extraction, the aqueous phase was applied to a column of Sephadex G50 in 10 mM Tris–HCl, 1 mM EDTA, pH 8.0. The double-stranded cDNA eluted in the void volume and was precipitated with ammonium acetate and ethanol. The double-stranded cDNA was analyzed by electrophoresis in a 6% polyacrylamide gel in a Tris borate buffer containing 7 M urea (Maniatis et al., 1982).

3. RESULTS

3.1. Hepatic Metallothionein in the Winter Flounder

Heat-treated and ammonium sulfate-fractionated liver cytosol obtained from cadmium chloride-treated flounder was chromatographed on Sephadex G75 (Fig. 7.1). As most of the high molecular weight metal-binding proteins had been removed by the heat treatment and salt-fractionation steps, the major peak of $^{109}Cd^{2+}$ binding activity was observed in the position of low molecular weight proteins (range 10,000 to 14,000). This peak was characterized by its high absorbance at 250 nm (metalthiolates) and corresponding low absorbance at 280 nm (lack of aromatic amino acids). Anion exchange chromatography on DE52 was used to purify this low molecular weight fraction further and this resulted in a pure preparation of cadmium MT (Shears and Fletcher, 1984, 1985). This preparation was used as the winter flounder MT standard in the polyacrylamide gel electrophoresis experiments.

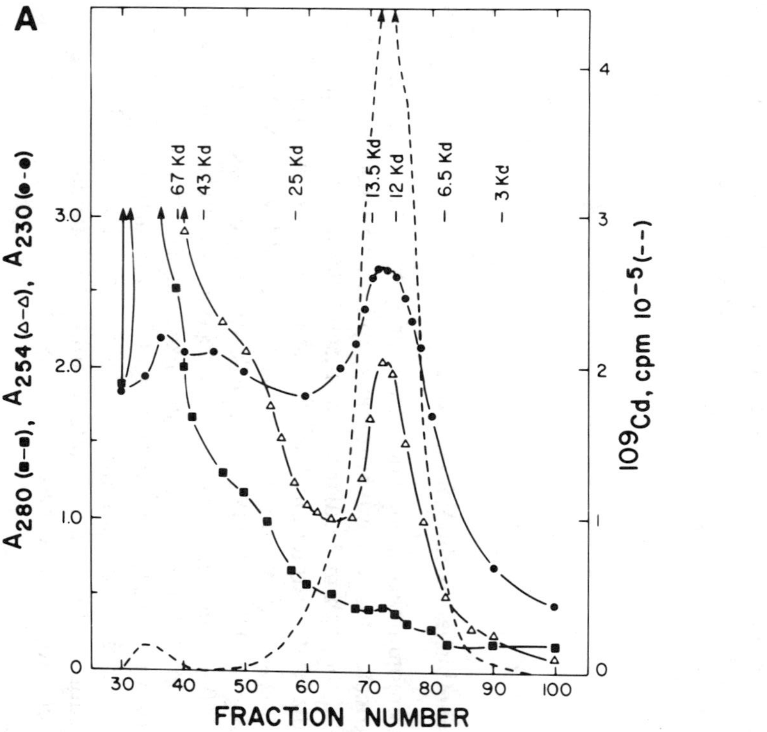

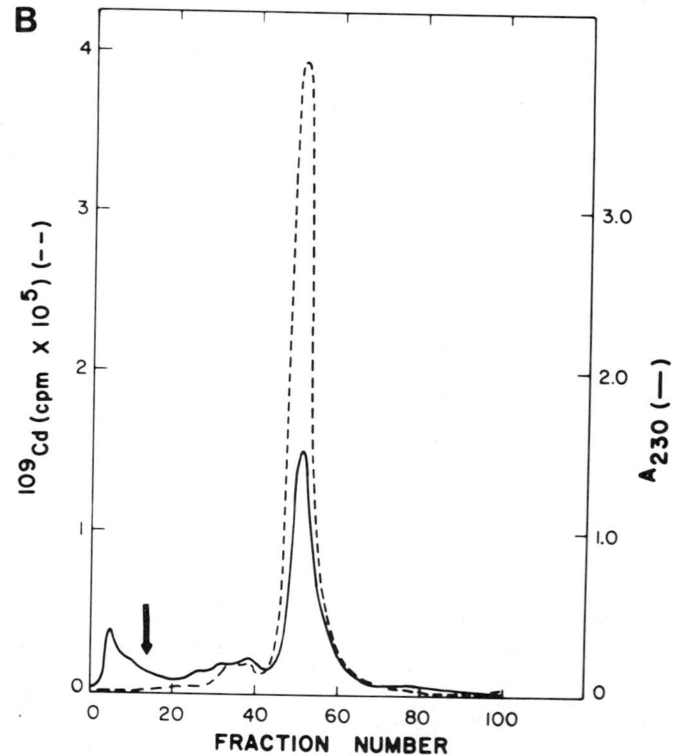

Figure 7.1. Gel filtration profiles of heat-treated and ammonium sulfate-fractionated cytosol obtained from cadmium chloride-injected winter flounder. (*a*) Sephadex G75 fractionation shows a peak of MT, which is indicated by the high absorbance at 230, the low relative absorbance at 280, and the associated radioactivity ($^{109}Cd^{2+}$). Incubation with the radioisotope was carried out at 4°C for 30 min (25 μCi per 20 mL cytosol from 25 g liver) prior to loading the sample. The molecular weight markers were in order of elution: bovine serum albumin, ovalbumin, chymotrypsinogen A, ribonuclease A, cytochrome *c*, apotinin, and insulin B. The column size was 2.5 × 90 cm, fraction size 4 mL, and the buffer was 0.1 M ammonium bicarbonate, pH 8.5, containing 2 mM mercaptoethanol. (*b*) Ion-exchange (Whatman DE52) chromatographic profile of MT obtained from the Sephadex G75 column. The column size was 1 × 15 cm, fraction size 4 mL, and the buffer was 0.05 to 0.3 M Tris/HCl, pH 8.6, linear gradient. The arrow indicates the beginning of the gradient.

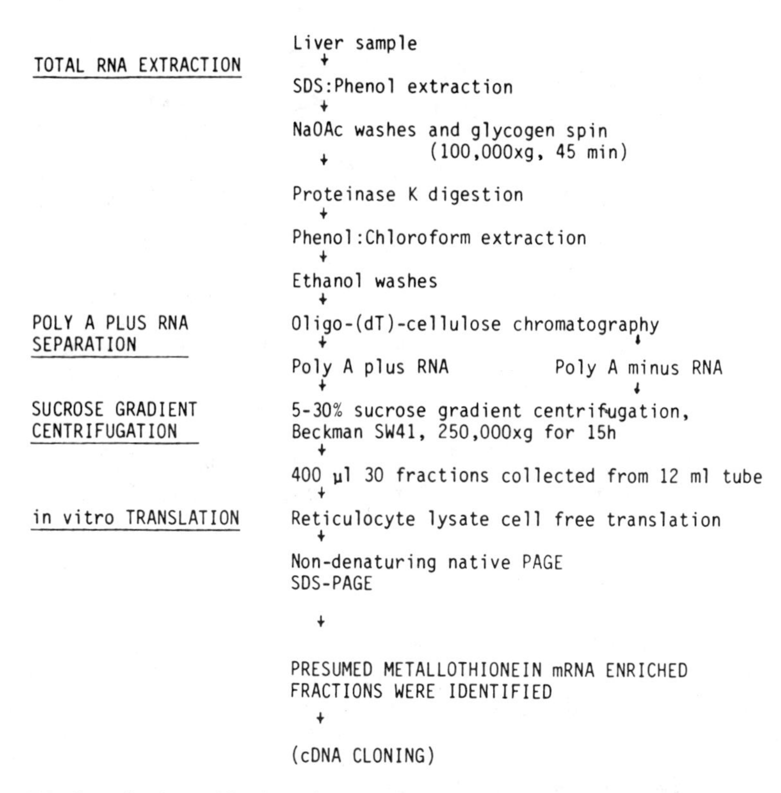

Figure 7.2. Steps in the purification of MT-enriched mRNA fractions from cadmium chloride-injected winter flounder. Saline-injected fish were processed in the same manner as controls.

3.2. Purification of 9-S Messenger RNA

3.2.1. RNA Isolation

The steps in the isolation of an enriched fraction of mRNA coding for MT are outlined in Fig. 7.2. Both cadmium chloride- and saline-injected flounder were studied. The yield of total RNA varied according to season and the purification procedure used. However, it was routinely possible to obtain 2–3 mg of RNA per gram of liver. Two to 4% of the total RNA was poly A RNA. Cadmium chloride injections did not appear to affect the yield of total RNA or poly A RNA.

3.2.2. In Vitro Translation

As shown in Fig. 7.3, the incorporation of [^{35}S]cysteine into protein in an in vitro translation system was essentially the same if poly A RNA from cadmium chloride-treated or saline-injected flounder was used. In both cases the system

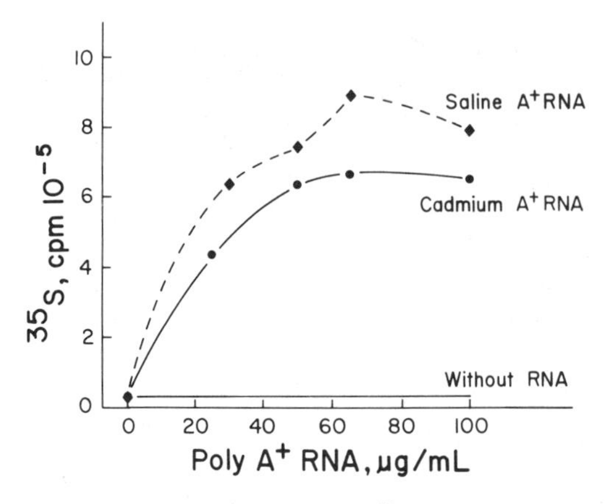

Figure 7.3. Influence of poly A RNA concentration on [^{35}S]cysteine incorporation into protein in a rabbit reticulocyte lysate cell free translation system. Each value is the average of two reactions.

became saturated with RNA at a concentration of approximately 70 μg/mL. These results indicate that despite the induction of MT mRNA, there was no detectable increase in the total amount of [^{35}S]cysteine incorporated into cell free translation products of Cd^{2+}-treated fish.

3.2.3. Polyacrylamide Gel Electrophoresis

MTs are characterized by their high cysteine content. Iodoacetic acid carboxymethylates cysteine residues and for every cysteine that is modified the protein gains an additional negative charge. Therefore, carboxymethylated MT has a very fast migration rate in native polyacrylamide gel eletrophoresis because of its small size and large negative charge (Anderson and Weser, 1978; Karin and Herschman, 1980; Koizumi et al., 1982, 1985). Figure 7.4*a* shows the results of a fluorographic analysis of the carboxymethylated [^{35}S]cysteine-labeled cell free translation products of poly A RNA from cadmium chloride- and saline-treated flounder liver after electrophoresis in a native 15% polyacrylamide gel. Only a few proteins entered the gel and MT was easily recognized as it migrated close to the dye front. No band corresponding to MT was observed in the translation products of saline-treated flounder.

Polyacrylamide gel electrophoresis in the presence of SDS separates proteins primarily according to size. Several low molecular weight cysteine-containing proteins were found as cell free translation products in addition to MT. MT could not be resolved from these other low molecular weight cysteine-containing proteins unless the cell free translation products were carboxymethylated prior to electrophoresis (Fig. 7.4*b*). As was the case in Fig. 7.4*a*, no MT band could be observed among the cell free translation products of the control flounder.

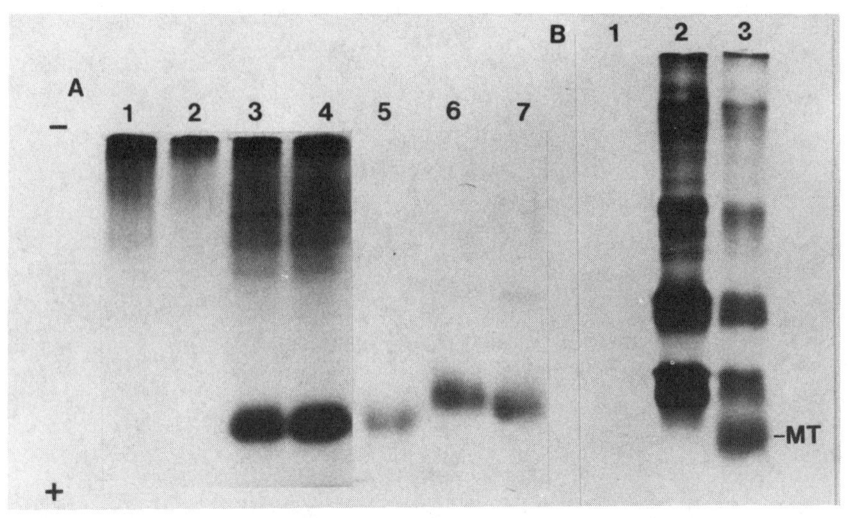

Figure 7.4. Fluorographic analysis of cell free translation products (carboxymethylated) directed by poly A RNA. (*a*) Native polyacrylamide (15%) gel electrophoresis. Hepatic poly A RNA from saline-treated flounder (lanes 1 and 2) and cadmium-treated flounder (lanes 3 and 4). Lanes 2 and 4 had 15 μg Cd^{2+} in the translation reaction mixes. Lanes 5, 6, and 7 are flounder MT, rabbit MT 1, and rabbit MT 2, respectively, stained with Coomassie blue. (*b*) 15 to 20% linear gradient SDS polyacrylamide gel electrophoresis. Lane 1, without RNA added; lane 2, same sample as lane 2 in (*a*); lane 3, same sample as lane 4 in (*a*). MT indicates where the standard flounder MT migrates in this system. (Modified from Chan *et al*., 1987).

3.2.4. Sucrose Density Gradients

RNA was fractionated on 5 to 30% sucrose density gradients (Fig. 7.5*a*). mRNA-encoding MT was enriched in the 8- to 10-S fractions of poly A RNA from cadmium chloride-treated flounder liver. There was no evidence of any MT production in poly A RNA from saline-treated flounder liver (Fig. 7.5*b*). Other low molecular weight cysteine-containing proteins were also present in these RNA fractions (Fig. 7.6), but these were common to the cadmium chloride-treated and the control samples.

3.3. Complementary DNA to 9-S Messenger RNA

Double-stranded complementary DNA was prepared from total poly A RNA and the 9-S fraction of poly A RNA from cadmium chloride-treated flounder liver. Analysis of these products by electrophoresis under denaturing conditions in a DNA sequencing gel revealed that the 9-S poly A RNA had produced fragments that ranged in size from 500 to 600 nucleotides (Fig. 7.7). In neither case was there a definite band that stood out above the background. The double-stranded DNA derived from the 9-S fraction is now being used to prepare a MT cDNA clone.

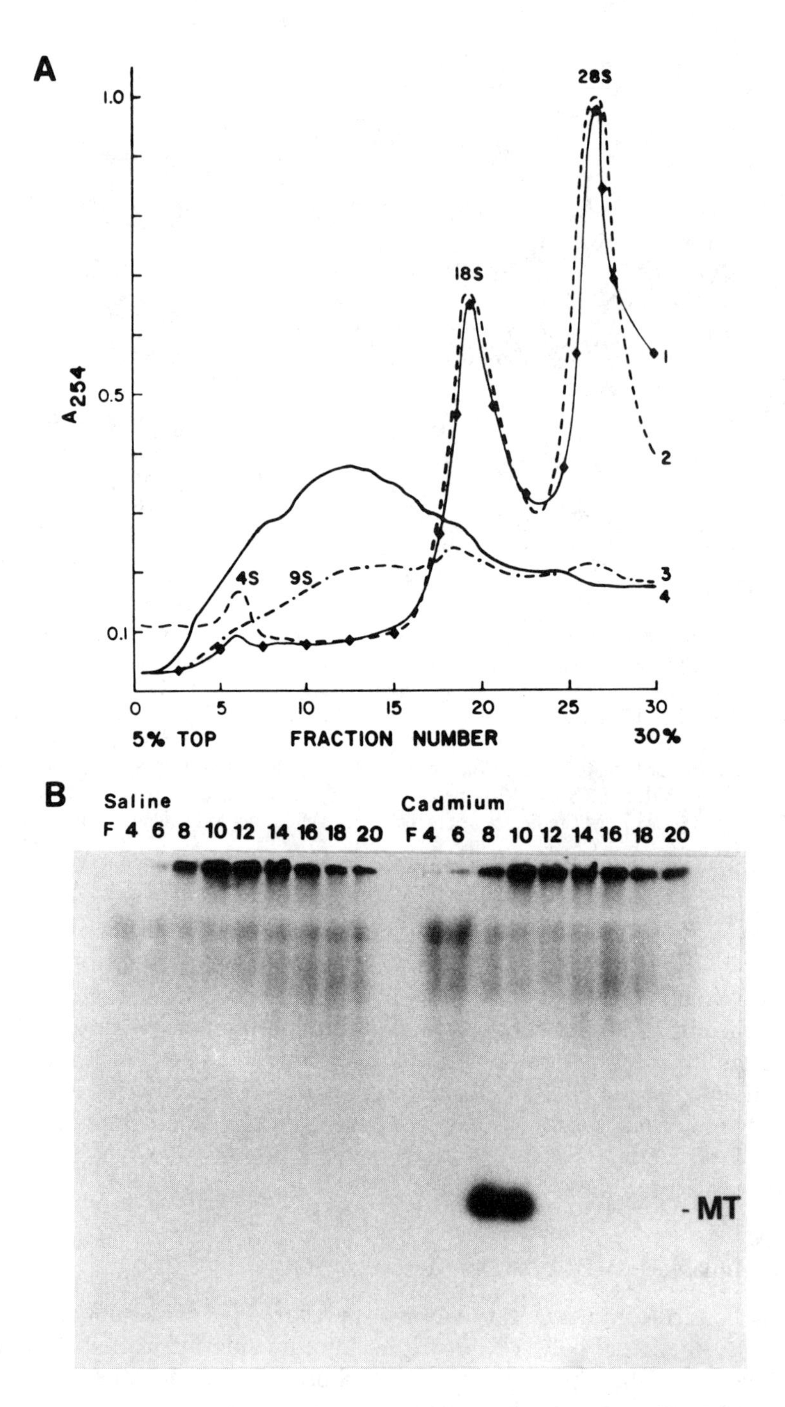

Figure 7.5. (*a*) Separation of RNA by centrifugation in a linear 5 to 30% sucrose density gradient. Line 1, poly A minus hepatic RNA (200 μg) from saline-treated fish; line 2, poly A minus hepatic RNA (200 μg) from cadmium chloride-treated founder; line 3, poly A hepatic RNA (150 μg) from cadmium chloride-treated flounder; line 4, poly A hepatic RNA (250 μg) from saline-treated flounder. The size markers for 4 S and 9 S were yeast transfer RNA and globin RNA. (*b*) Cell free translation products directed from RNA isolated from individual fractions were carboxymethylated and electrophoresed on a 20% polyacrylamide nondenaturing gel and analyzed by flourography. Fraction numbers refer to the gradients shown in (*a*). (Modified from Chan *et al*., 1987).

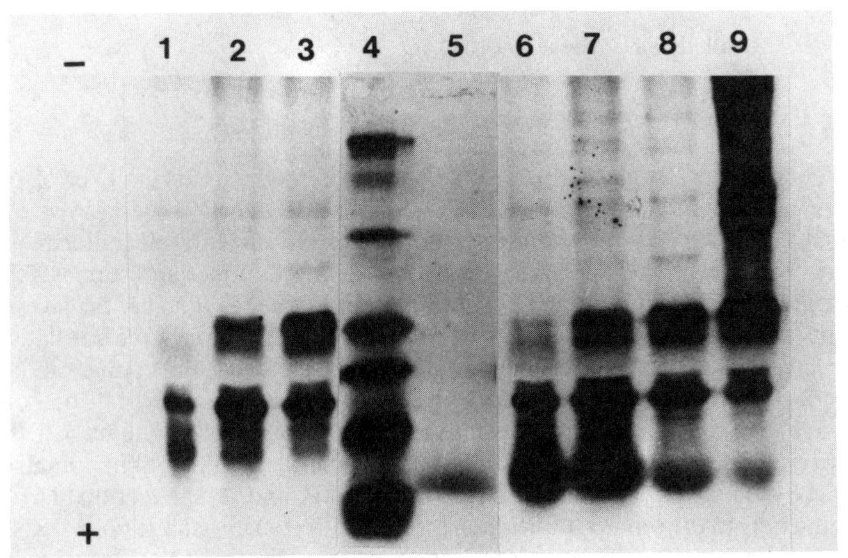

Figure 7.6. Fifteen to 20% SDS polyacrylamide gel electrophoresis of cell free translation products directed by sucrose gradient fractions 8, 9, and 10 of saline-treated controls (lanes 1 to 3) and fractions 8, 9, and 10 of cadmium chloride-treated flounder (lanes 6 to 8). Lane 4 is Coomassie blue-stained protein molecular weight markers and lane 5 is flounder MT. All proteins were carboxymethylated. (Modified from Chan *et al.*, 1987).

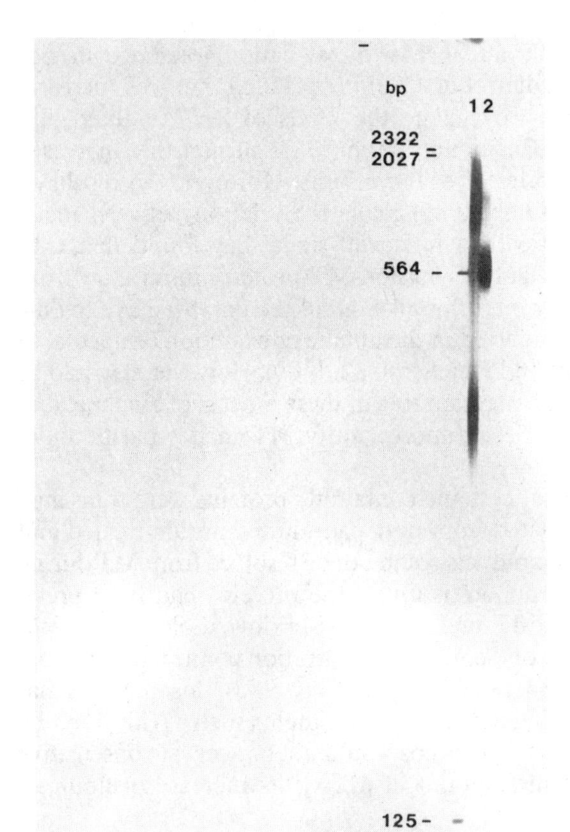

Figure 7.7. Autoradiograph of a 6% polyacrylamide gel electrophoresis in the presence of 7 M urea of ^{32}P-labeled cDNA made to poly A RNA (lane 1) and the 9-S fraction (2) of hepatic RNA from cadmium chloride-treated flounder. The size standards are ^{32}P-labeled λ DNA digested with *Hin*dIII.

4. DISCUSSION

The present investigtion confirmed that MT is produced in the liver of the winter flounder in response to cadmium chloride injection. Moreover, our results indicate that the administration of cadmium chloride induces the accumulation of MT mRNA in the liver of winter flounder and supports the hypothesis that MT genes in fish, as in mammals, are regulated at the level of transcription. The conclusion that MT was a product of the cell free translation directed by winter flounder poly A RNA was based on (1) the comigration of the carboxymethylated derivative with similarly treated flounder MT on both native and denaturing polyacrylamide gel electrophoresis systems, (2) the observation that the size of the MT mRNA was similar to those of mammalian MTs (Anderson and Weser, 1978; Ohi et al., 1981), and (3) the fact that MT is found only in the cell free translation products directed by hepatic poly A RNA from cadmium chloride-treated flounder.

In view of the biochemical properties of MT and the mechanisms known to control its biosynthesis, it seems reasonable that MT plays an important role in trace metal metabolism and detoxification. Evidence for this hypothesis comes from studies on salmonids where hepatic MT levels correlate well with the amount of heavy metal contamination of the water (Roch et al., 1982; Roch and McCarter, 1984a,b; McCarter and Roch, 1984). Other studies have examined the relation between MT and heavy metal cation tolerance in fish (Klaverkamp et al., 1984). Bradley et al. (1985) concluded that MT increases the ability of acclimated fish to regulate the levels of Zn^{2+} in their gills. McCarter et al. (1982) and McCarter and Roch (1984) suggest that increased metal tolerance is related to rates of MT synthesis. However, the results of studies by Thomas et al. (1983a,b, 1985) make the relation between metal regulation and MT more difficult to interpret, since they found that Cd^{2+} accumulated in two low molecular weight non-MT proteins in rainbow trout exposed to cadmium salts. In winter flounder, all of the data we have to date argue that MT is not directly involved in the uptake or excretion of normal or excell body Zn^{2+} levels (Shears and Fletcher, 1983, 1984). However, it would be premature to conclude that MT plays no role in these aspects of zinc metabolism where we cannot readily detect and quantify MT in normal flounder tissues.

Several low molecular weight cysteine-containing proteins were translated from hepatic poly A RNA isolated from both cadmium chloride-treated and control winter flounder. These products could not be resolved from MT during SDS polyacrylamide gel electrophoresis unless the proteins had been previously carboxymethylated. As MT and the non-MT low molecular weight metal-binding proteins coelute on Sephadex gel filtration column chromatography (Shears and Fletcher, 1984, 1985; Thomas et al., 1985), this indicates that quantitative analysis of MT using indirect measurements such as Cd^{2+} binding capacity or the number of sulfhydryl groups would not be accurate or sensitive enough to measure the absolute amount of MT or its increase in flounder tissues. In view of the results of the present study, equating low molecular weight metal-binding proteins to MT should be done with caution.

One striking feature of the present study was the complete lack of evidence for the presence of MT mRNA in normal winter flounder. This suggests that MT is not synthesized until the fish is exposed to an excess of heavy metal. In addition, the lack of MT mRNA points to one reason why we were unable to isolate MT from livers of normal flounder (Shears and Fletcher, 1985). MT may not be present in normal flounder, and if it is, its levels are well below the detection limits of standard chemical procedures.

It is evident from the literature that MT and other low molecular weight metal-binding proteins play important roles in heavy metal regulation and detoxification in fish. The problem lies in determining exactly what role MT does play, and how MT interacts with the MT-like proteins. Some of the confusion about the role of MT in fish originates from the methodology used to identify and quantify MT, and from the tendency for investigators to equate low molecular weight MT-like proteins to MT itself.

Standard chromatographic procedures (such as gel filtration) are useful in determining the relative amounts of heavy metals bound to low molecular weight proteins (MT-like). These methods are inexpensive and straightforward, and they can be adapted to process large number of samples, a common requirement of environmental toxicologists. However, such methods suffer from being nonspecific, there being invariably a number of proteins eluting from the column in the 10,000- to 15,000-Dal range.

Polarographic techniques for determining the number of sulfhydryl groups are more specific for MT (Olafson and Sim, 1979), particularly if they are combined with column chromatography. However, the occurrence of low molecular weight cysteine-containing, but non-MT proteins in fish warns up that polarographic methods can also be nonspecific (Pierson, 1985b). This is particularly true for winter flounder where a relatively high cysteine-containing (13%) metal binding protein can be found in the liver throughout the year (Shears and Fletcher, 1985).

If we are to understand fully the role of MT in heavy metal metabolism in fish, sensitive and specific methods must be employed for its determination. Two highly specific and sensitive methods could be developed: The first is a radioimmunoassay such as those used in mammals, which utilizes antibodies to purified MT (Garvey, 1984), and the second is based on a cDNA probe specific to the MT mRNA. For the present we have chosen to develop the cDNA probe to measure the induction of MT mRNA. Such a probe will enable use to determine, for example, the levels of heavy metals in water, sediments, and food necessary to induce MT mRNA production in any flounder tissue. The fact that there appears to be no MT mRNA in normal flounder liver indicates that this technique could be useful in the field as a molecular indicator of environmental quality. Although a cDNA probe to MT mRNA could be a powerful tool, it is realized that a radioimmunoassay will eventually be developed so that the protein as well as the mRNA can be measured.

In spite of the fact that the initial development of a cDNA probe to measure MT mRNA is technically difficult and costly, once developed the screening of large numbers of samples using dot-blot hybridization procedures is relatively routine (Fourney et al., 1984). In addition, once a cDNA probe is available it can be used on winter flounder throughout its geographical range. It may also be equally as useful in all other pleuronectids, regardless of their geographical location.

5. SUMMARY

Total messenger RNA (mRNA) was purified from liver of winter flounder treated with a series of intraperitoneal injection of $CdCl_2$. The mRNA translated in a cell free translation system to yield metallothionein. Several other low molecular weight cysteine-labeled proteins were also found in the cell free translation products of cadmium- and saline-treated fish. Further analysis of the total mRNA using sucrose density gradient centrifugation revealed that metallothionein mRNA was greatest in the 8–10 S fractions. Cloning of cDNA from the metallothionein mRNA will provide a probe that will be useful for quantitative analysis of the rate of metallothionein synthesis in fish subjected to metal exposure.

ACKNOWLEDGMENTS

We thank M. Shears and R. Fourney for help with MT isolation and RNA extractions. The technical assistance of M. King and A. Cadigan is also appreciated. A. Burness and A. Vaisius kindly provided advice and loaned us equipment. This research was supported by grants from the Natural Sciences and Engineering Research Council of Canada to W.S.D. and G.L.F. This is Marine Sciences Research Laboratory Contribution No. 688.

REFERENCES

Anderson, R.D., and Weser, U. (1978). Partial purification, characterization and translation *in vitro* of rat liver metallothionein messenger ribonucleic acid. *Biochem. J.* **175**: 841–852.

Aviv, J., and Leder, P. (1972). Purification of biologically active globin mRNA by chromatography on oligothymidylic acid cellulose. *Proc. Nat. Acad. Sci. USA* **69**: 1409–1412.

Bailey, J.M., and Davidson, N. (1976). Methylmercury as a reversible denaturing agent for agarose gel electrophoresis. *Anal. Biochem.* **70**: 75–85.

Bonham, K., and Gedamu, L. (1984). Induction of metallothionein and metallothionein mRNA in rainbow trout liver following cadmium treatment. *Biosci. Rep.* **4**: 633–642.

Bradley, R.W., DuQuesnay, C., and Sprague, J.B. (1985). Acclimation of rainbow trout, *Salmo gairdneri* Richardson, to zinc: Kinetics and mechanism of enhanced tolerance to induction. *J. Fish. Biol.* **27**: 367–379.

Brady, F.O. (1982). The physiological function of metallothionein. *Trends Biochem. Sci.* **7**: 143–145.

Chan, K.M., Davidson, W.S., and Fletcher, G.L. (1987). Hepatic metallothionein RNA in the winter flounder (*Pseudopleuronectes americanus*). *Can. J. Zool.*, **65**: 472–480.

Cherian, M.G., and Nordberg, M. (1983). Cellular adaptation in metal toxicology and metallothionein. *Toxicology* **28**: 1–15.

Davies, P.L., and Hew, C.L. (1980). Isolation and characterization of the antifreeze protein messanger RNA from winter flounder. *J. Biol. Chem.* **255**: 8729–8734.

Fletcher, G.L. (1975). The effects of capture "stress" and storage of whole blood on the red blood cells, plasma proteins, glucose and electrolytes of the winter flounder (*Pseudopleuronectes americanus*). *Can. J. Zool* **53**: 197–206.

Fletcher, G.L. (1977). Circannual cycles of blood plasma freezing point and Na^+ and Cl^- concentrations in Newfoundland winter flounder (*Pseudopleuronectes americanus*): Correlation with water temperature and photoperiod. *Can. J. Zool.* **55**: 789–795.

Fletcher, P.E., and Fletcher, G.L. (1978). The binding of zinc to the plasma of winter flounder (*Pseudopleuronectes americanus*): affinity and specificity. *Can. J. Zool.* **56**: 114–120.

Fletcher, P.E., and Fletcher, G.L. (1980). Zinc and copper binding proteins in the plasma of winter flounder (*Pseudopleuronectes americanus*). *Can. J. Zool.* **58**: 609–613.

Fletcher, G.L., and King, M.J. (1978). Seasonal dynamics of Cu^{2+}, Zn^{2+}, Cd^{2+} and Mg^{2+} in gonads and livers of winter flounder (*Pseudopleuronectes americanus*): Evidence for summer storage of Zn^{2+} for winter gonad development in females. *Can. J. Zool.* **56**: 284–290.

Fletcher, G.L., Kiceniuk, J.W., and Williams, V.P. (1981). Effects of oiled sediments on mortality feeding and growth of winter flounder (*Pseudopleuronectes americanus*). *Mar. Ecol. Prog. Ser.* **4**: 91–96.

Fletcher, G.L., King, M.J., Kiceniuk, J.W., and Addison, R.F. (1982). Liver hypertrophy in winter flounder following exposure to experimentally oiled sediments. *Comp. Biochem. Physiol.* **73C**: 457–462.

Fourney, R.M., Fletcher, G.L., and Hew, C.L. (1984). The effects of long day length on liver mRNA in the winter flounder (*Pseudopleuronectes americanus*). *Can. J. Zool.* **62**: 1456–1460.

Garvey, J.S. (1984). Metallothionein: structure/antigenicity and detection/quantitation in normal physiological fluids. *Environ. Health Perspec.* **54**: 117–127.

Gubler, U., and Hoffman, B.J. (1983). A simple and very efficient method for generating cDNA libraries. *Gene* **25**: 263–269.

Hamer, D.H. (1986). Metallothionein. *Annu. Rev. Biochem.* **55**: 913–951.

Hoss, D.E. (1964). Accumulation of zinc-65 by flounder of the genus *Paralichthys*. *Trans. Am. Fish. Soc.* **93**: 364–368.

Huang, I.Y., Tsunoo, H., Kimura, M., Nakashima, H., and Yoshida, A. (1979). Primary structure of mouse liver metallothionein-I and II. In J.H.R. Kagi and M. Nordberg (Eds.), *Metallothionein. Proceedings of the 1st International Meeting on Metallothionein and Other Low Molecular Weight Metal-Binding Proteins, Zurich, July 17–22, 1978*. Birkhauser Verlag, Basel, Switzerland, pp. 169–172.

Kagi, J.H.R., and Nordberg, M. (Eds.). (1979). *Metallothionein. Proceedings of the 1st International Meeting of Metallothionein and Other Low Molecular Weight Metal-Binding Proteins, Zurich, July 17–22, 1978*. Birkhauser Verlag, Basel, Switzerland.

Karin, M. (1985). Metallothioneins: Proteins in search of function. *Cell* **41**: 9–10.

Karin, M., and Herschman, H.R. (1980). Characterization of the metallothioneins induced in HeLa cells by dexamethasone and zinc. *Eur. J. Biochem.* **107**: 395–401.

Karin, M., and Richards, R.I. (1984). The human metallothionein gene family structure and expression. *Environ. Health Perspect.* **54**: 111–115.

Kito, H., Ose, Y., Mizuhira, Z., Sato, T., Ishikawa, T., and Tazawa, T. (1982). Separation and purification of (Cd, Cu, Zn)- metallothionein in carp hepato-pancreas. *Comp. Biochem. Physiol.* **73C**: 121–127.

Kito, H., Ose, Y., Hayashi, K., Yonezawa, S., Sato, T., Ishikawa, T., and Nagase, H. (1984). Some properties of metallothionein from hepato-pancreas and kidney in carp (*Cyprinus carpio*). *Eisei Kagaku* **30**: 119–125.

Klaverkamp, J.F., MacDonald, W.A., Dunca, D.A., and Wagemann, R. (1984). Metallothionein and acclimation to heavy metals in fish—A review. In *Contaminant Effects in Fisheries*. V.W. Cairns, P.V. Hodson, and J.O. Nraigu (Eds.), Wiley, New York, pp. 99–113.

Klein-MacPhee, G. (1978). Synopsis of biological data of the winter flounder, *Pseudopleuronectes americanus* (Walbaum). NOAA Technical Report, NMFS Circ. No. 414, pp. 1–43.

Koisumi, S., Otaki, N., and Kimura, M. (1982). Estimation of thionein synthesis in cultured cells by slab gel electrophoresis. *Ind. Health* **20**: 101–108.

Koisumi, S., Otaki, N., and Kimura, M. (1985). Evidence for more than two metallothionein isoforms in primates. *J. Biol. Chem.* **260**: 3672–3675.

Kojima, Y., Berger, C., and Kagi, J.H.R. (1979). The amino acid sequence of equine metallothioneins. In J.H.R. Kagi and M. Nordberg (Eds.), *Metallothionein. Proceedings of the 1st International Meeting on Metallothionein and Other Low Molecular Weight Metal-Binding Proteins, Zurich, July 17–22, 1978*. Birkhauser Verlag, Basel, Switzerland, pp. 153–161.

Laemmli, U.K. (1970). Cleavage of structural proteins during the assembly of the head of bacteriophage T4. *Nature (London)* **227**: 680–685.

Leim, A.H., and Scott, W.B. (1966). Fishes of the Atlantic coast of Canada. *Bull. Fish. Res. Bd. Can.* **155**.

Maniatis, T., Fritsch, E.F., and Sambrook, J. (1982). *Molecular Cloning*. Cold Spring Harbour Lab. Publications.

McCarter, J.A., and Roch, M. (1984). Chronic exposure of coho salmon to sublethal concentration of copper, III. Kinetics of metabolism of metallothionein. *Comp. Biochem. Physiol.* **77C**: 83–87.

McCarter, J.A., Matheson, A.T., Roch, M., Olafson, R.W., and Buckley, J.T. (1982). Chronic exposure of coho salmon to sublethal concentrations of copper, II. Distibution of copper between high- and low-molecular-weight proteins in the liver cytosol and the possible role of metallothionein in detoxification. *Comp. Biochem. Physiol.* **72C**: 21–26.

Nemer, M., Wilkinson, D.G., Travaglini, E.C., Stornberg, E.J., and Butt, T.R. (1985). Sea urchin metallothionein sequence—Key to an evolutionary diversity. *Proc. Nat. Acad. Sci. USA* **82**: 4992–4994.

Ohi, S., Gardenosa, G., Pine, R., and Huang, P.C. (1981). Cadmium induced accumulation of metallothionein messenger RNA in rat liver. *J. Biol. Chem.* **256**: 2180–2184.

Olafson, R.W., and Sim, R.G. (1979). An electrochemical approach to quantitation and characterization of metallothioneins. *Anal. Biochem.* **100**: 343–351.

Olsson, P.E., and Haux, C. (1985). Rainbow trout metallothionein. *Inorgan. Chim. Acta* **107**: 67–71.

Overnell, J., and Coombs, T.L. (1979). Purification and properties of plaice metallothionein, a cadmium-binding protein from the liver of the plaice (*Pleuronectes platessa*). *Biochem. J.* **183**: 277–283.

Pelham, H.R.B., and Jackson, R.J. (1976). An efficient RNA-dependent translation system from reticulocyte lysates. *Eur. J. Biochem.* **67**: 247–251.

Pentreath, R.J. (1973a). The roles of food and water in the accumulation of radionucleotides by marine teleost and elasmobranch fish. Proc. Symposium on Radioactive Contamination of the Marine Environment. IAEA, Vienna, July 10–14, 1972, Seattle, Washington, pp. 421–436.

Pentreath, R.J. (1973b). The accumulation and retention of ^{65}Zn and ^{54}Mn by the plaice (*Pleuronectes platessa L.*). *J. Exp. Mar. Biol. Ecol.* **12**: 1–18.

Pierson, K.B. (1985a). Isolation and partial characterization of a non-thionein, zinc-bonding protein from the liver of rainbow trout (*Salmo gairdneri*). *Comp. Biochem. Physiol.* **80C**: 299–304.

Pierson, K.B. (1985b). Occurrence and synthesis of a non-thionein zinc-binding protein in the rainbow trout (*Salmo gairdneri*). *Comp. Biochem. Physiol.* **81C**: 71–75.

Renfro, W.C., Fowler, S.W., Hegraud, M., and La Rosa, J. (1975). Relative importance of food and water in long term $zinc^{65}$ accumulation by marine biota. *J. Fish. Res. Bd. Can.* **32**: 1339–1345.

Roch, M., and McCarter, J.A. (1984a). Hepatic metallothionein production and resistance to heavy metals by rainbow trout *Salmo gairdneri*, I. Exposed to an artificial mixture of zinc, copper, and cadmium. *Comp. Biochem. Physiol.* **77C**: 71–75.

Roch, M., and McCarter, J.A. (1984b). Hepatic metallothionein production and resistance to heavy metals by rainbow trout *Salmo gairdneri*, II. Held in a series of contaminated lakes. *Comp. Biochem. Physiol.* **77C**: 77–82.

Roch, M., McCarter, J.A., Matheson, A.T., Clark, M.J.R., and Olafson, R.W. (1982). Hepatic metallothionein in rainbow trout (*Salmo gairdneri*) as an indicator of metal pollution in the Campbell River system. *Can. J. Fish. Aquat. Sci.* **39**: 1596–1601.

Shears, M.A. (1983) Zinc metabolism in winter flounder (*Pseudopleuronectes americanus*). Ph.D. thesis, Memorial University of Newfoundland, St. John's, Nfld., Canada.

Shears, M.A., and Fletcher, G.L. (1979). The binding of zinc to the soluble proteins of intestinal mucosa in winter flounder (*Pseudopleuronectes americanus*). *Comp. Biochem. Physiol.* **64B**: 297–299.

Shears, M.A., and Fletcher, G.L. (1983). Regulation of Zn^{2+} uptake from the gastrointestinal tract of a marine teleost, the winter flounder (*Pseudopleuronectes americanus*). *Can. J. Fish. Aquat. Sci.* **40** (Suppl. 2): 197–205.

Shears, M.A., and Fletcher, G.L. (1984). The relationship between metallothionein and intestinal zinc absorption in the winter flounder (*Pseudopleuronectes americanus*). *Can. J. Zool.* **63**: 2211–2220.

Shears, M.A., and Fletcher, G.L. (1985). Hepatic metallothionein in the winter flounder (*Pseudopleuronectes americanus*). *Can. J. Zool.* **63**: 1602–1609.

Takeda, H., and Shimizu, C. (1982). Purification of metallothionein from the liver of skipjack and its properties. *Bull. Jpn. Soc. Sci. Fish.* **48**: 717–723.

Thomas, D.G., Solbe, J.F. del G., Kay, J., and Cryer, A. (1983a). Environmental cadmium is not sequestered by metallothionein in rainbow trout. *Biochem. Biophys. Res. Commun.* **110**: 584–592.

Thomas, D.G., Cryer, A., Solbe, J.F. del G., and Kay, J. (1983b). A comparison of the accumulation and protein binding of environmental cadmium in the gills, kidney and liver of the rainbow trout (*Salmo gairdneri*). *Comp. Biochem. Physiol.* **76C**: 241–246.

Thomas, D.G., Brown, M.W., Shurben, D., Solbe, J.F. del G., Cryer, A., and Kay, J. (1985). A comparison of the sequestration of cadmium and zinc in the tissues of rainbow trout (*Salmo gairdneri*) following exposure to the metals singly or in combination. *Comp. Biochem. Physiol.* **82C**: 55–62.

8

AVOIDANCE TESTS WITH A PLATING INDUSTRIAL EFFLUENT

J. Hadjinicolaou and L.D. Spraggs

Department of Civil Engineering and Applied Mechanics, McGill University, Montreal, Canada

1. **Introduction**
2. **Materials and methods**
   2.1. Avoidance apparatus
   2.2. Avoidance testing
   2.3. Data analysis and interpretation
3. **Results and discussion**
   3.1. Longitudinal distribution
   3.2. Vertical distribution
   3.3. Lateral distribution
4. **Conclusions**
   References

1. INTRODUCTION

This chapter is about the acute avoidance behavior of rainbow trout to sublethal concentrations of industrial effluent.

A toxicant may stress fish population in ways other than acute mortality, but only recently has substantial attention been given to sublethal effects. Field studies have shown that toxic solutions may render large portions of an environment undesirable to fish (Birtwell, 1977). The behavior of fish in these areas is often modified; any behavior patterns that allow fish to escape or to avoid deleterious conditions would contribute to their survival and well being (Cherry et al., 1982).

Among behavioral tests, preference–avoidance reactions are particularly important and practical because they can detect adverse effects of pollutants at sublethal concentrations. These tests are of relatively short duration (normally a matter of hours), and their outcome is measured by the alteration of behavior, rather than by effects on particular organs. Avoidance curves and thresholds might provide valuable insight to the behavior of fish in toxic solutions that cannot be obtained from static or chronic bioassays.

Three basic types of apparatuses have been used to analyze chemical avoidance by fish. In 1913, Shelford and Allee (1913) first tried shallow concentration gradients. Wells (1915), Ishio (1969), and Birtwell (1977) used and modified this technique. Jones (1947) introduced the steep gradient method. Finally, the fluviarium system was introduced by Hoglund (1951). Other pioneers, notably Sprague (1964), Kleerekoper and Morgensen (1963), Scherer and Novak (1973), and Westlake and Lubinski (1976), have used modifications of the three methods.

The system outlined in this chapter was designed to take advantage of the success of previous researchers—to use steep gradient design rather than shallow or fluviarium system, to have the ability to store results for later reexamination, to use actual effluent instead of only purified chemical components—and at the same time to provide possible new advantages—to have vertical, longitudinal, and lateral analysis of the organisms position; to increase the number of test organisms; and to use a large tank which eliminates end effects.

2. MATERIALS AND METHODS

2.1. Avoidance Apparatus

The entire design of the new flow-through system (Spraggs et al., 1982; Hadjinicolaou and Spraggs, 1981, 1982) can be divided into five basic components:

1. The water treatment facilities
2. The holding facilities
3. The temperature control system
4. The channel
5. The data acquisition facilities.

The system is depicted schematically in Figs. 8.1 and 8.2. A numbered key below each diagram indicates its constituent parts.

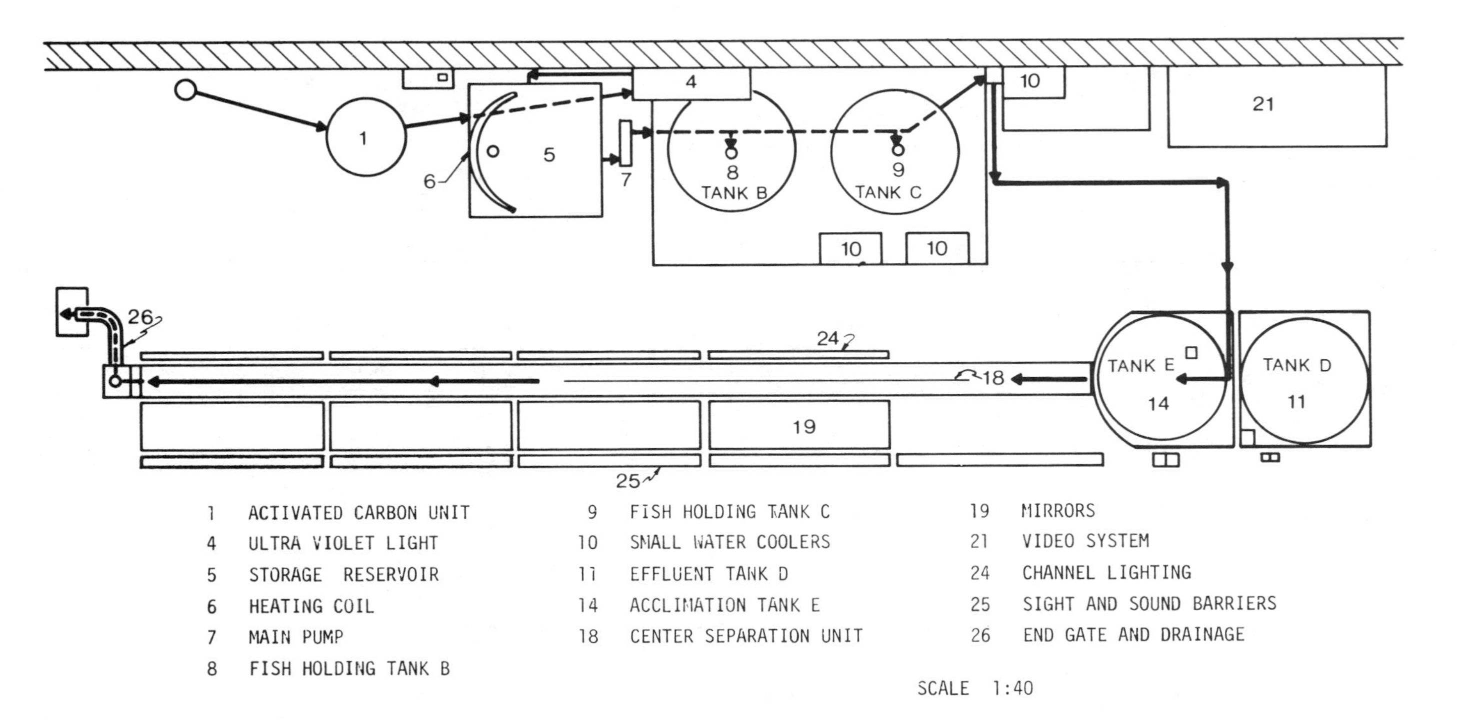

Figure 8.1. Flow diagram and plan view of the avoidance apparatus.

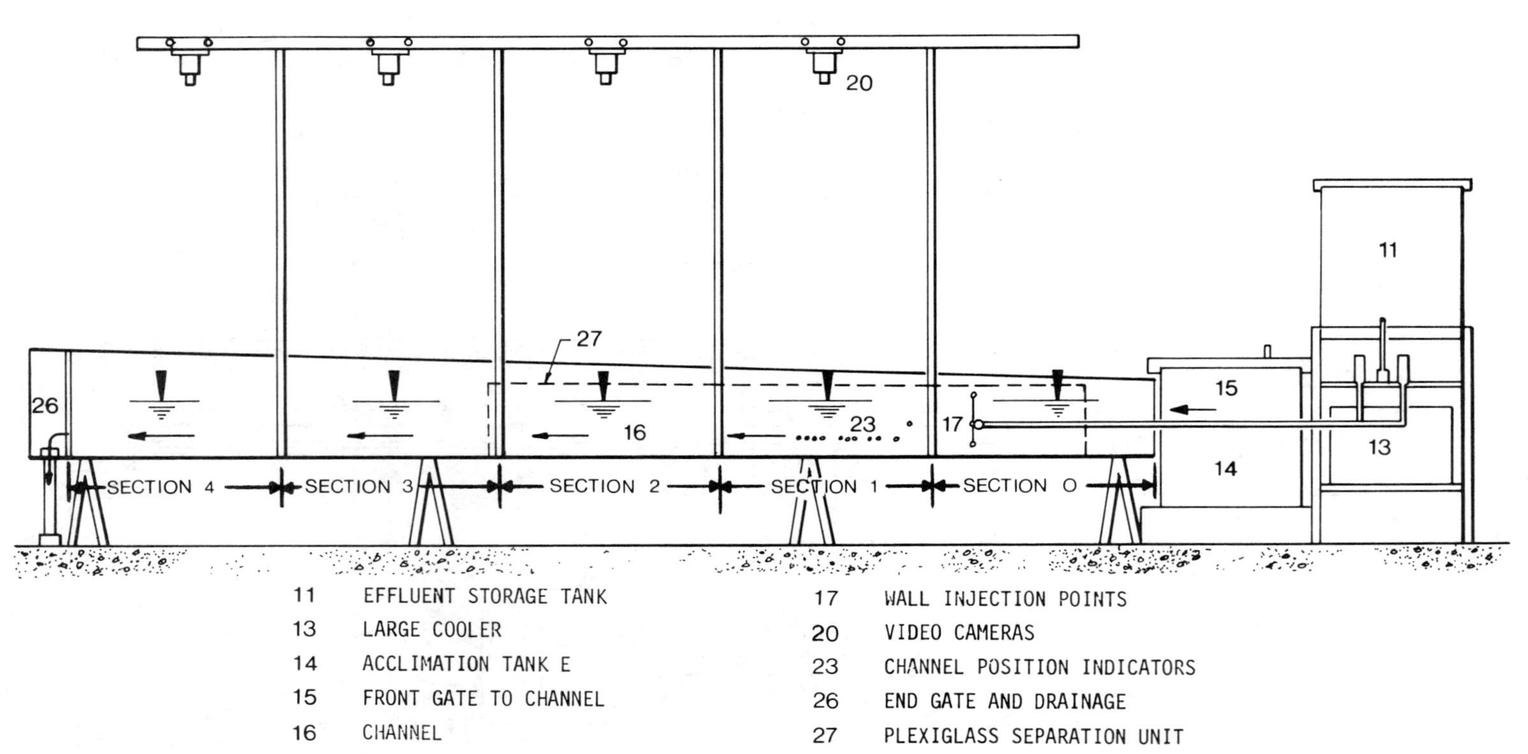

Figure 8.2. Side view of the avoidance apparatus.

Water for the system was taken from the domestic supply of the City of Montreal. Prior to actual use, it was treated to render it suitable for fish survival. The water treatment facilities consisted of an activated carbon filtration unit and ultraviolet lights to kill bacteria. The maximum flow of treated water was 1.26 L/s. Water analyses were done monthly to evaluate the water quality before its entrance into the holding facilities. The averages of measured values were pH 6.3 ± 0.3; temperature, 15 ± 1°C; sodium, 14.2 ± 0.8 ppm; calcium, 23.5 ± 2 ppm; chromium, <0.002 ppm; nickel, <0.002 ppm; iron, <0.002 ppm; total and fecal coliforms, 0.

The holding facilities consisted of the storage reservoir for the treated water (Tank A, 1890 L), the fish tanks (Tanks B, C, and E, 1134 L), and Tank D, a storage reservoir for industrial effluents. Flowmeters regulated the amount of water and air that flowed into the tanks and the channel. Tank D was mounted on a 2-m-high steel support structure and the industrial effluent was pumped through one of two flowmeters to give a constant flow of effluent into the channel. On one side of Tank E, a 0.4 × 0.3-m hole was made 0.25 m from the top of the tank to allow the fish to move from Tank E to the channel through the front gate.

A submerged heating unit in Tank A and an automatic temperature control unit provided a constant temperature of 15°C in the system. In addition, cooling units were used to cool the water to 15°C in the summer. Ten waterproof temperature probes were strategically located throughout the system (channel, holding tanks, etc.) and connected to a manually operated multichannel thermometer for rapid recording of the temperature at any point in the system. The maximum temperature differences between different sections of the channel were within 1°C.

The experimental channel was 10 m long and 0.305 m wide, and had water depth of 0.35 m. It was divided into five sections of 2 m each. The first section was constructed of plexiglass to allow holes for jets and diffusers, while the other four sections were made of glass. In the first section of the channel, 1.7 m from the front gate and 0.3 m from the second section, a provision was made for checking for the injection of the effluents and dyes from the effluent Tank D. The flow of the channel was laminar, with a rate of 0.37 L/s. The injection system used in this study was a wall momentum jet 10 cm from the bottom.

An 8-mm-thick plexiglas barrier with holding devices was placed in the center of the channel and extended from 1.35 m upstream of the first section through to the third section (4.65 m long). This separation unit rose 0.1 m above the water level and was introduced to give the fish an option to choose either the polluted side (polluted sections 1, 2, 3, and 4) or the nonpolluted side (section 0, or nonpolluted sections 1, 2, 3, and 4) of the channel. The fish could not go over or under the separation unit, only around it, from sections 0 and 3.

Because it was desirable to have a longitudinal, vertical, and lateral analysis of the disposition of the fish in the channel, mirrors were placed along one side of the channel. The mirrors were placed at an angle of 45° and, when photographed by video cameras from above, showed the longitudinal projection in the direct view and the vertical projection in the mirror. Consequently, any avoidance in either the lateral, longitudinal, or vertical direction was recorded. Direct longitudinal coverage was provided by four overlapping video cameras, mounted on moveable trolleys 3 m above the channel and along its length (Fig. 8.2). Each video camera covered approximately 2.2 m of the channel, resulting in total coverage of the area where the fish were able to swim. The video cameras recorded (1) an overhead view, which provided observation of the longitudinal and lateral distribution of fish within the section, and (2) a side view through the 45° mirrors, which provided information about the vertical distribution of fish within the section.

Simultaneously, four long-playing video recorders were used to record the video signals for later analysis. The recorders stored the events of every experiment, both vertically and longitudinally. Pictures from the four video cameras were displayed on a separate video monitor for each section of the channel.

The image in each monitor was composed of two parts: the upper part, which corresponds to the vertical distribution (depths (A) 0–0.11 m, (B) 0.11–0.22 m, (C) 0.22–0.33 m) of the particular section, and the bottom part, which illustrated the longitudinal distribution (sections (1) 2–4 m, (2) 4–6 m, (3) 6–8 m, (4) 8–10 m) and the lateral distribution (Sides A and B) within the particular section (Fig. 8.3).

The bottom and one side of the entire channel were divided into 0.101 × 0.101-m squares with black stripes. Each of these squares was color-coded horizontally and vertically to aid in determining the position of the fish in the channel. Each section had 18 squares.

Because the light reaching the iris of the cameras was not adequate and because the pictures on the monitors from the mirrors were not clear, fluorescent lights 1.2 m long, were placed along the length of the channel about 5 cm distance from one side. These lights had a 12-h photo period, with lamp strength of 2 × 34 W and were covered by a plastic shield.

Large portable modular wall panels were located along one side of the channel to minimize both noise and visual disturbances while experiments were in progress.

A plexiglas box with a perforated end gate was built at the downstream end of the channel. The perforated end gate was required to allow for flow, which approximated parallel flow in a cross-section of a river while at the same time holding the fish within the channel. The diameter of the holes (4 mm) and the length of slots (7.5 cm) in the end gate prevented small fish from passing through. The end gate was removable for rapid drainage and easy cleaning of the channel. Cleaning was performed daily and had to be completed 8 h before beginning an experiment, by flushing three times with water containing a low (1 : 60) Versa-Clean liquid concentrate dilution.

2.2. Avoidance Testing

The species used in the study was rainbow trout (*Salmo gairdneri*, Rich.) from two Quebec hatcheries, originally 2–3.5 g in weight (5–7 cm in length), with a

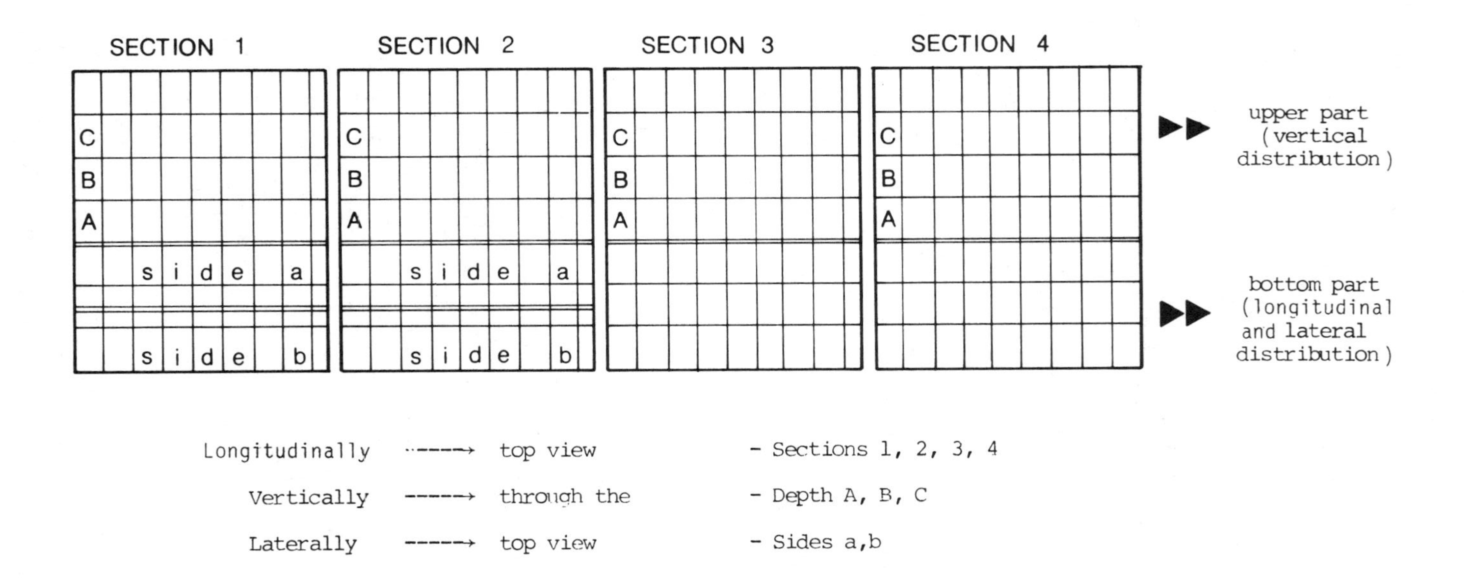

Figure 8.3. View on the TV monitors.

growth rate of approximately 2 cm per month. The fish food was Trout Food-Atlantide, size 3. There were no mortalities during holding or experimentation except once, caused by a power failure. The fish were kept in Tanks B and C until they became 10–11 cm in length before being transferred to the channel acclimation unit (Tank E), where they were acclimated to living in either the channel or Tank E.

Fish feeding was done at the same time each day at five specific locations in the channel to avoid grouping.

No experiments began until 3 h after feeding. Fish had to be counted prior to the commencement of any experiment. The count took 30 min and was done by comparing the number of fish on five still frames of a continuous 10-min recording, together with a corresponding visual count of the number of fish in section 0.

The temperature of the effluent and the channel were checked every half hour. The effluent was discharged for 15 min before actual recordings began. A switch box was connected to the four cameras. By turning the switch on, the experiment was simultaneously recorded on the four cassettes recorders for 2 h continuously. Any changes in the ambient conditions had to be recorded and any resultant behavior changes likewise noted. The experiment was supervised during its entirety, including changes in personnel. After 2 h, the experiment was terminated by turning the switch off.

2.3. Data Analysis and Interpretation

Following any experiment, a series of procedures were followed to ensure that subsequent experiments were not biased. Cassettes were rewound and temporarily labeled. Fish that had been exposed were removed and destroyed. No fish that had been exposed to the effluents were placed back in the acclimation Tank E or reused. No fish that had been moved to the experimental tank were returned to the stock tanks. The channel was drained and refilled, and fish were allowed to enter the channel for acclimation for subsequent experiments.

Because 100 fish were used in each experiment, their position in the channel had to be noted at selected time intervals. This analysis required an initial visual count followed by an analysis of the position of the fish. To determine the relative avoidance of the fish, the longitudinal, vertical, and lateral position of the fish was determined at each counting interval (5 min).

The data were transferred from the counting sheets to computer files. The statistical analysis consisted of preliminary calculations (mean value, standardization, range, and percentage of the number of fish per section); time series analysis to evaluate the stability of the response with time on the sides A and B of the lateral distribution; an analysis of variance to evaluate the likeness of having significantly different values among the percentage of fish present in each section and at each concentration; and Duncan's tests to analyze which value on a group of data differed from the rest, after it was found from ANOVA that the values were likely to be significantly different. To ensure that the results were not biased, the tests were conducted with effluent being inserted in alternate sides of the channel for alternate tests.

To obtain the longitudinal distribution curve of a particular experiment, the total number of fish per section was transformed into a percentage plotted against the four longitudinal sections.

To obtain the vertical distribution curve for a particular experiment, the total number of fish per depth was transformed into a percentage plotted against the three depths.

In the lateral distribution the main avoidance was defined as the mean percentage of fish in the nonpolluted side of the first two sections. Secondary avoidance was defined as the mean percentage of fish in the nonpolluted side of the third and fourth sections, due to the difficulty in deciding about the orientation of fish that were observed between the polluted and the nonpolluted sides of these sections. In these calculations the first four readings were excluded because the first 20 min were considered to be an adaptation period in the channel. The main avoidance after the elimination of the bias factors (water jet effect) represented the avoidance value of the experiment and was used for the production of the avoidance curve. The secondary avoidance was used for additional information.

To obtain the avoidance curve, the time spent in clear water as a percentage of total time was plotted against the effluent dilution concentration. Generally the avoidance curves for each experiment are represented by a single value, which is the percentage of total experimental time spent by the fish in clear water. Avoidance reactions have usually been portrayed in this fashion, and only rarely have attempts been made to use other parameters. However, to provide more insight of the meaning of this single value it was deemed worthwhile to elaborate on two further parameters for the graphical representation of experimental avoidance data: (1) the crossing time T of the avoidance values, and (2) the fluctuation σ_e around the avoidance value. Both were derived from the computer outputs of the time series analysis. The crossing time T defined the reaction time that elapsed until the fish first achieved the avoidance value for the particular experiment. The fluctuation σ_e was measured as the standard deviation of the avoidance value. When T was considered in conjunction with the σ_e, supplementary information about the experiment was produced by plotting time T versus the time spent by the fish in clear water as a percentage of the total time. This new representation is called herein "avoidance reaction representation" for each experiment.

The industrial effluent used was taken from the effluent of a representative metal-plating industry. Toxic substances in the effluent were chromium (20 mg/L), nickel (50 mg/L), and small amounts of iron (4.0 mg/L). The 96-h LC_{50} of the effluent was found to be an effluent–water mixture containing 60% effluent after analysis of three effluent samples at the Environment Canada laboratories at Longueuil, Quebec. Other properties of the effluent measured were pH 6.8; total solids, 540 ppm; and grease and oils, <2.5 ppm.

Twenty-one tests were conducted to determine the effect of different concentrations of the industrial effluent on the avoidance behavior of the fish in the channel.

3. RESULTS AND DISCUSSION

3.1. Longitudinal Distribution

The longitudinal distribution of fish in the industrial effluent experiments shown in Fig. 8.4 indicated a massive shift of fish from sections 3 and 4 to section 2, at effluent concentrations of 0.005 of the 96-hr LC_{50}. At the same time there was a change from the polluted side to the nonpolluted side of the channel. The distribution appeared to change dramatically as the effluent concentration was increased. With higher concentrations, there was a tendency for the fish to move into the clean water behind the separator in section 2 of the channel. This behavior occurred because section 2 was one of the locations where the fish could avoid the effluent.

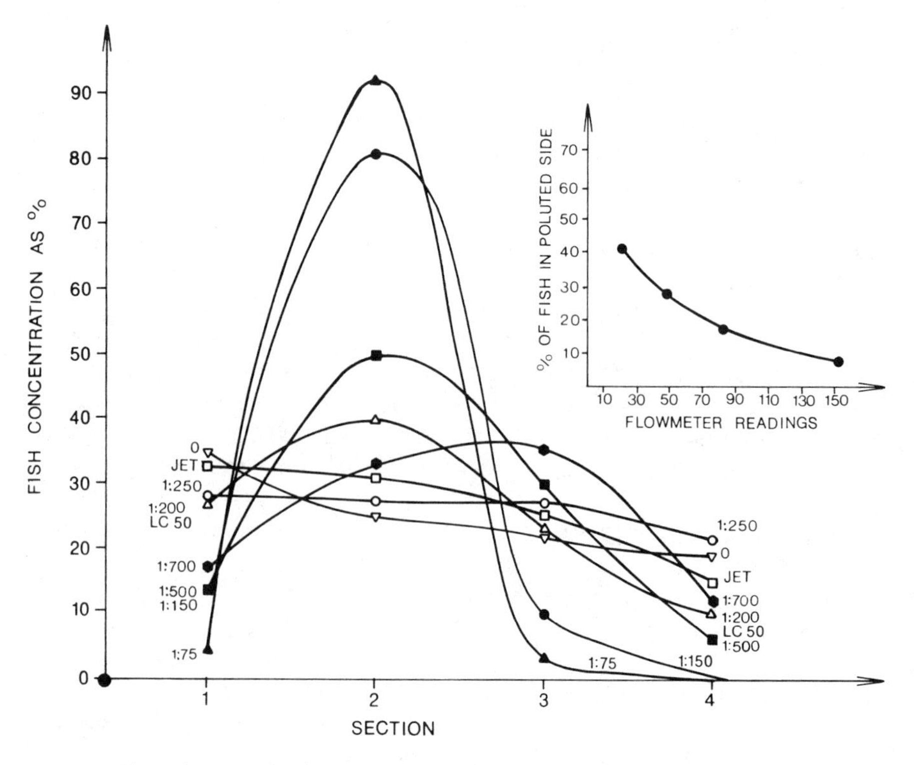

Figure 8.4. Longitudinal distribution of the industrial effluent experiments.

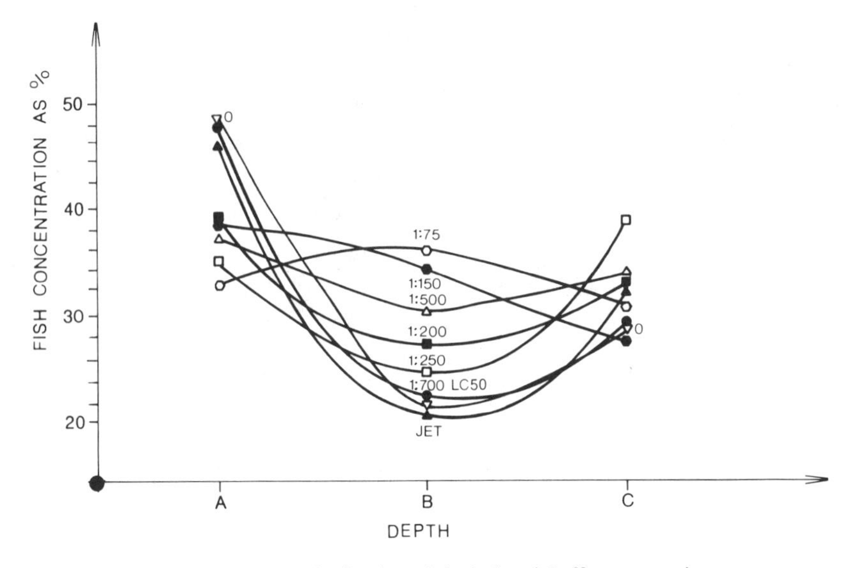

Figure 8.5. Vertical distribution of the industrial effluent experiments.

The ANOVA results for the longitudinal distribution of the effluent indicated that in sections 1, 3, and 4, the percentages of fish in unpolluted water for the different concentrations were not significantly different (i.e., they belonged to the same group, A). In section 2, which was significantly different from the other three sections, Duncan's tests revealed two groups (A and B), with the percentage of fish at concentrations of 0.013, 0.01, and 0.0067 of the 96-h LC_{50}.

The reference points in Fig. 8.5 for the evaluation of the effect of the different pollutants to the longitudinal distribution throughout the channel, were (1) the O concentration, indicating longitudinal distribution of the fish in unpolluted water and (2) the jet distribution, which takes into account the distribution of the fish in relation to an unpolluted water jet.

3.2. Vertical Distribution

The vertical distribution of fish in the industrial effluent experiments shown in Fig. 8.5 did not change significantly, indicating a movement of the fish to a similar elevation in the nonpolluted side. The ANOVA results indicate that there was a significant difference between the three depths of the vertical distribution of the effluent, while the results of Duncan's tests indicated no significant difference between the various concentrations at each depth. The test results shown in Fig. 8.5 seem to indicate a difference in the vertical distribution among the three depths, but it is unclear whether this was due only to the effluent. Part of the problem was the limited depth of the channel.

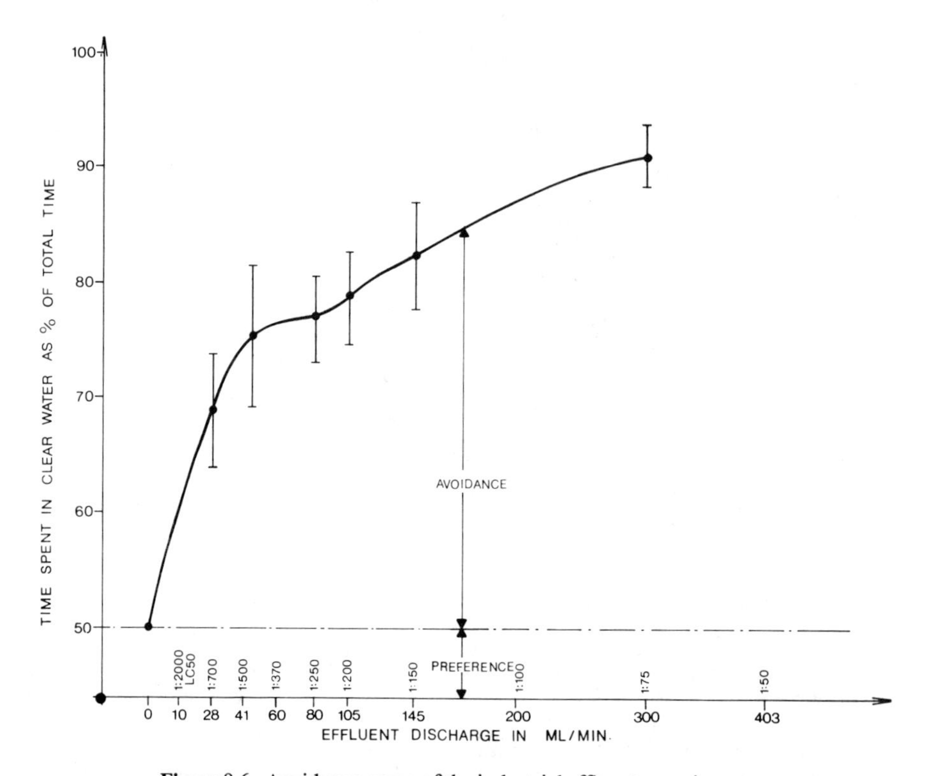

Figure 8.6. Avoidance curve of the industrial effluent experiments.

3.3. Lateral Distribution

From Fig. 8.6 it can be seen that avoidance of the industrial effluent began at very low concentrations and was already evident at a value of 1 : 700 LC_{50}. Extrapolation of the avoidance curve would indicate that avoidance can possibly begin as early as 1 : 3000 LC_{50}. The final avoidance was of the order of 90%, indicating that some of the fish did not avoid the effluent all the time, even at high concentrations. This could conceivably be explained by impairment of sensory organs in unresponsive fish.

Figure 8.7 presented the avoidance reaction representation (T, σ_e, overall avoidance, dilution ratio of LC_{50}) for the experiments of the actual industrial effluent conducted. It appears that T fluctuated in a range between 5 and 30 min (80% between 15 and 25 min) and the σ_e between 4 and 11% (80% between 5 and 10%).

The implication of the data portrayed in Fig. 8.7 is that avoidance reaction representation provided supplementary experimental avoidance information by evaluating the reaction time that elapsed before the avoidance value for a particular experiment was first achieved, and also the variability of the avoidance value for every experiment.

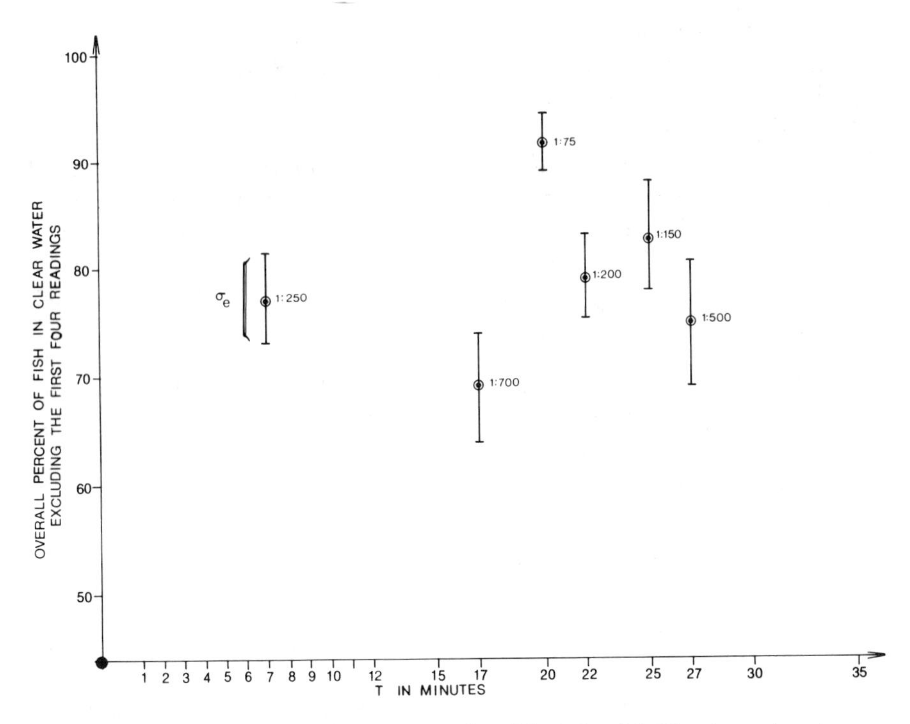

Figure 8.7. Avoidance reaction representation of the industrial effluent experiments.

4. CONCLUSIONS

1. An avoidance apparatus has been designed and implemented, using the advantages of previous avoidance systems and at the same time providing new advantages. The innovative characteristics of the design were the capacity for vertical, longitudinal, and lateral analysis of the position of test organisms; the use of a large number of test organisms; the ability to test different pollutants and to store results for later reexamination; and thc use of advanced video and analysis techniques for data interpretation.
2. The spectrum of plating industry effluent concentrations injected ranged between 1:3000 and 1:75 LC_{50} (well below the lethal threshold) and produced an avoidance reaction between 50 and 90%. This illustrated a significant potential hazard to aquatic organisms at sublethal concentrations that would be impossible to predict or estimate with the conventional static LC_{50}.
3. The sublethal threshold concentration for the avoidance reaction to the plating industry effluent was 1:3000 LC_{50}. The avoidance reaction reached 90% in dilution ratio of 1:75 LC_{50}.

4. The longitudinal and the vertical distribution of the position of the organisms tested was mainly related to the preference–avoidance pattern of the polluted and nonpolluted sections of the channel and partly related to the effluent characteristics.
5. The avoidance reaction representation was proposed as a new way of depicting supplementary experimental avoidance information related to the stability of these reactions with time.

ACKNOWLEDGMENTS

The assistance of Gilles LaRoche, Terry Rigby, Norman Birmingham, Christian Blaise, Robert Hutcheon, Joan Cornell, and Lesley-Ann Judge is gratefully appreciated.

REFERENCES

Birtwell, I.K. (1977). A field technique for studying the avoidance of fish to pollutants. *Proceedings of the 3rd Aquatic Toxicity Workshop, Halifax, N. S.*, Environmental Protection Service Technical Report No. EPS-5-AR-77-1, Halifax, Canada, pp. 69–86.

Cherry, D.S., and Cairns, J., Jr. (1982). Biological monitoring, Part V. Preference and avoidance studies. *Water Res.* **16**: 263–301.

Hadjinicolaou, J., and Spraggs, L.D. (1981). Methodology for assessing toxicity avoidance. Proceedings of the 8th Aquatic Toxicity Workshop in Guelph, Canada, *Can. Tech. Report of Fisheries and Aquatic Sciences*, No. 1151, pp. 68–82.

Hadjinicolaou, J., and Spraggs, L.D. (1982). Toxicity avoidance of toxic effluents. Proc. of the 9th Aquatic Toxicity Workshop in Edmonton, Alberta. *Can. Tech. Report of Fisheries and Aquatic Sciences*, No. 1163, p. 210.

Hoglund, L.B. (1951). A new method of studying the reactions of fishes in stable gradients of chemicals and other agents. *Oikos* **3**: 246–267.

Ishio, J. (1969). Behaviour of fish to exposed substances. In O. Tang (Ed.), *Advances in Water Pollution Research*, Vol. 1, *Proc. 2nd Int. Conf., Tokyo*. Pergamon, New York, pp. 19–40.

Jones, J.R.E. (1947). The reactions of *Pygosteus pugitius* to toxic solutions. *J. Exp. Biol.* **24**: 11–122.

Kleerekoper, H., and Morgensen, J. (1963). Role of olfaction in the orientation of *Petromyzon marinus*. I. Response to a single aurine in prey's body odour. *Physiol. Zool.* **36**: 347–360.

Scherer, E., and Novak, J. (1973). Apparatus for recording avoidance movements of fish. *J. Fish. Res. Board. Can.* **30**: 1594–1596.

Shelford, V.E., and Allee, W.C. (1913). The reactions of fishes to gradient of dissolved atmospheric gases. *J. Exp. Zool.* **14**: 107–166.

Spraggs, L.D., Gehr, R., and Hadjinicolaou, J. (1982). Polyelectrolyte toxicity tests by fish avoidance studies. *Water Sci. Technol.* **14**: 1564–1567.

Sprague, J.B. (1964). Avoidance of copper-zinc solutions by young salmon in the laboratory. *J. Water Pollut. Cont. Fed.* **36** (8): 990–1004.

Wells, M.M. (1915). Reactions and resistance of fishes in their natural environment to acidity, alkalinity and neutrality. *Bio. Bull.* **29**: 221–257.

Westlake, G.F., and Lubinski, K.S. (1976). A chamber to monitor locomotor behaviour of free swimming aquatic organisms exposed to simulated spills. *In* M.D. Rockville (Ed.), *Proc. of the National Conference on Control of Hazardous Materials Spills*, pp. 64–69.

9

DEVELOPMENT OF MULTIRESISTANCE PATTERNS IN THE BACTERIAL FLORA OF TROUT FOLLOWING AN ANTIBIOTIC THERAPY

Rachel Léger

Aquarium de Montréal, La Ronde, Ile Ste-Hélène, Montreal, Quebec, Canada

Réal Lallier

Faculty of Veterinary Medicine, University of Montreal, St-Hyacinthe, Quebec, Canada

1. **Introduction**
2. **Materials and methods**
   - 2.1. Treatment of trout
   - 2.2. Postmortem examination
   - 2.3. Antibiotic sensitivity tests
   - 2.4. Bacterial growth conditions
   - 2.5. Transfer of antibiotic resistance in vitro
3. **Results**
   - 3.1. Effects of SxT and RO5 on the intestinal flora of trout
   - 3.2. Antibiotic sensitivity tests
   - 3.3. Transfer of antibiotic resistance
4. **Discussion**
5. **Summary**

References

1. INTRODUCTION

Salmon aquaculture is very important in many countries. Unfortunately, to obtain a good financial profit, the fish have to be packed in order to utilize the installations at full capacity. This cramped environment favors a rapid spreading of infection, either by contagious fish, by stress, or through the water.

Species from fish ponds are therefore susceptible to diseases caused by different microbes, such as bacteria, virus, fungus, and protozoa. These diseases can reach the whole population of a fish tank or of many tanks supplied by the same water source, and cause important losses.

To fight bacterial infection, the most frequent disease in fish aquaculture, many antimicrobial agents are used as either a preventive or a cure. They are added to the feed, dissolved in water, or injected into the fish.

The repeated use of an antibiotic can incite the appearance of multiresistant bacterial strains. This fact has been observed in the majority of domestic and farm animals where the use of antibiotics in feed as a growth factor is standard (Bourque et al., 1980; Dubourguier et al., 1980; Smith, 1970). In hatcheries, though, the use of antibiotics as a preventive or growth factor is still infrequent.

Two synergic antimicrobials, SxT and, more recently, RO5–0037 are added to the feed as a treatment during furunculosis epidemics. However, very little is known about the effects of these agents on the microbial flora of the fish.

The objectives of this research were to evaluate the effects of SxT and RO5 on the intestinal flora of trout, verify the appearance of resistant strains, and determine if the resistance is transferable to bacteria of the same and different species.

2. MATERIALS AND METHODS

2.1. Treatment of Trout

Two experiments were performed with speckled trout, *Salvelinus fontinalis*, at the Faculty of Veterinary Medicine in St. Hyacinthe. The trout measured between 11 and 16 cm and had a mean weight of 30 g. The fish were purchased from a private hatchery, were disease free, and had never been treated with antibiotics.

For each of the experiments, the speckled trout were divided into two experimental groups of 30 fish each. Each group was maintained in a 300-L Plexiglas tank provided with aeration and a biological filter. The tanks were filled with totally fresh water and were in a closed-circuit system, although part of the tank water was changed when necessary. The dissolved oxygen level was calculated by the Winkler method (American Public Health Association, 1976) and remained above 10.0 mg/L. The pH, measured with a pH meter (Radiometer Copenhague), gave a value above 7.3 during all the experiments. The water temperature was between 14 and 17°C during the first experiment and between 9 and 16.5°C during the second experiment.

During the first experiment, the trout were fed Nutribec commercial feed. The trout received the feed at a rate of 2% of their body weight per day every morning. The drugs were coated onto the pelleted diet using 5% oil as a binder. The pellets, medicated with SxT, a drug containing a 5 : 1 ratio of sulfamethoxazole and trimethoprim, were prepared. Both the sulfamethoxazole and trimethoprim were furnished by Burroughs Wellcome Inc., Ville La Salle, Québec (see Appendixes A to D). 291.9 mg of sulfamethoxazole plus 58.1 mg of trimethoprim were suspended in 7.0 mL of sterile corn oil. The mixture was then poured on 140 g of pellets, which were stirred and allowed to dry for at least 24 h. The rate of feeding allowed the trout to receive 50 mg of SxT per kilogram of fish per day. The control group was fed with the same kind of feed coated with oil but without antibiotics; 200 g of feed mixed with 10 mL of corn oil was therefore prepared for the control group.

For the second experiment, another antimicrobial compound equivalent to RO5-0037, a potentiated sulfonamide containing sulfadimethoxine and ormetoprim at a 5 : 1 ratio, was used. Samples of the drugs were furnished by Hoffmann-LaRoche Inc., Nutley, New Jersey.* The medicated feed was prepared as above: 291.9 mg of sulfadimethoxine plus 58.1 mg of ormetoprim were mixed with 7.0 mL of corn oil and coated onto 140 g of pelleted diet. The trout, fed at a rate of 2% of their body weight per day every morning, received 50 mg of RO5 per kilogram of fish per day. For the control group, 600 g of feed was coated with 30 ml of corn oil and allowed to dry for at least 24 h.

During the two experiments, the treated groups were fed every day for 5 days with the feed medicated with their specific antimicrobials whereas the control groups received the feed coated with only corn oil. For 20 days following the treatments, all the trout received the same nonmedicated control diet. Three trout from each treated and control group were sampled and sacrificed every day before the morning feeding, and then every 5 days post-treatment until the end of the experiment.

2.2. Postmortem Examination

Aseptic surgery was employed to open the ventral surface of the fish. An intestinal segment about 4 cm in length from the anus was attached at both ends and cut free. The segment was then weighed and emptied of its content into a test tube filled with 5.0 mL of physiological buffered saline (PBS), pH 7.3. The intestinal membrane was also weighed to determine the weight of its contents (by subtracting its empty weight from its full weight).

The test tubes were agitated with a vortex mixer, and tenfold dilutions to 10^{-5} in PBS were done. 0.1 mL of each of the 10^{-2} to 10^{-5} dilutions was spread on triptic soy agar plates supplemented with 5% bovine blood. Following incubation at 22°C for 3 days, the bacterial numbers were assessed. Primary identifi-

*Since this research was completed, RO5-0037 has been cleared by the FDA for commercial use and is now marketed under the name Romet 30 by Hoffman-LaRoche Inc.

cation of the representative colonies was facilitated by examination of colonial morphology and pigmentation, as well as examination of the shape, arrangement, Gram staining characteristics, and motility of the cells. In addition the ability to produce oxidase and catalase was tested. Some 20 isolates of each dominant colony for each trout and each sampling day were picked and purified by inoculating them onto fresh brain heart infusion broth. Antibiotic resistance profiles of each isolate were determined.

2.3. Antibiotic Sensitivity Tests

There are no standard methods for determining antibiotic resistance of fish bacteria, these being bacteria that grow at low temperatures. We therefore modified the Bauer et al. (1966) technique. The Mueller-Hinton culture media were supplemented with 5% lysed horse blood and incubated at room temperature (22°C) and 30°C. Before the addition of the horse blood, the zones of inhibition around some antibiotic discs, such as the sulfonamides and trimethoprim, were confused. The antimicrobials we used act at the level of the synthesis of folic acids. Most of the interfering effects could be accounted for by the thymidine contents in the culture media. Thymidine permits a pathway for the synthesis of the folic acid and the inhibitory effect of sulfonamides and trimethoprim is greatly diminished (Bushby, 1969; Koch and Burchall, 1971). The horse blood cells contain the enzyme thymidine phosphorylase, which inhibits the action of thymidine in the culture media (Ferone et al., 1975). Agar supplemented with lysed horse blood allows complete inhibition of bacterial growth, giving a clear zone around the antibiotic discs.

The horse blood was lysed by adding 2.0 mL of a 10% saponin solution, sterilized by filtration, to 100 mL of defibrinated horse blood (Ferone et al., 1975). The blood is then added to the Mueller-Hinton preparation, just before pouring into the petri dish.

To verify the accuracy of the blood culture media, the quality control strain of *Escherichia coli* ATCC 25922 (Barry and Thornsberry, 1980) was tested for each new batch of agar. There were no significant changes in the diameter of the inhibitory zones following the addition of lysed horse blood. The zones were clearer and easier to read when lysed horse blood was added.

For the susceptibility testing, selected bacterial colonies inoculated in brain heart infusion broth were incubated at room temperature for 18–24 h. A 1 : 100 dilution in PBS was then prepared, and the Mueller-Hinton blood agar plates were inoculated with a cotton swab of the inoculum. Inoculated agar plates were allowed to stand undisturbed for 5 min before the antibiotic discs were added by an automatic disc distributor. The discs were ampicillin (Am) 10 μg; neomycin (N) 30 μg; tetracycline (Te) 30 μg; chloramphenicol (C) 30 μg; sulfamethoxazole–trimethoprim 19 : 1 (SxT) 25 μg; triple sulfa (SSS) 30 μg (Oxoid Canada Inc.). The RO5 discs were not available commercially, therefore the antimicrobial compounds were dissolved and added directly to the

Table 9.1 Preparation of Antibiotics

Antimicrobial[a] (AMB)	mg AMB per 20 mL	Solvent[b]	Dilution Liquid	Final Concentration (mg/mL)
Sulfadimethoxine	0.0834	NaOH 10%	H_2O	4.17
Ormetoprim	0.0166	DMF, lactic acid[c]	H_2O	0.83
Ampicillin	0.04	Phosphate buffer pH 8.0 0.1 M[d]	Phosphate buffer pH 6.0 0.1 M[d]	2
Neomycin	0.04	H_2O	H_2O	2
Tetracycline	0.04	H_2O	H_2O	2
Chloramphenicol	0.04	Methanol	H_2O	2
Sulfamethoxazole	0.0834	NaOH 10%	H_2O	4.17
Trimethoprim	0.0166	DMF, lactic acid[c]	H_2O	0.83
Triple sulfa	0.6	NaOH 10%	H_2O	30
Nalidixic acid	0.2	NaOH 10%	H_2O	10

Note: All antibiotics were filtered on a Milex-HA 0.45-μm filter unit.

Source: Barry (1976).

[a]The companies that furnished the antibiotics are listed in Appendix A.

[b]Only a few drops of the solvent are needed to dissolve the antibiotic powder. Complete to 20 mL with the dilution liquid.

[c]Refer to Appendix B for the method of preparation.

[d]Refer to Appendix C for the method of preparation.

[e]Refer to Appendix D for the components of triple sulfa.

Mueller-Hinton media (Table 9.1). The RO5 at 25 μg/mL was obtained by adding 5.0 mL of the 4.17-mg/mL concentration of sulfadimethoxine and 5.0 mL of 0.83-mg/mL concentration of ormetoprim to 1.0 L Mueller-Hinton solution.

The Bauer et al. (1966) agar dilution method was also modified by the incubation temperatures. The Mueller-Hinton plates, supplemented with horse blood and antibiotic discs or RO5, were incubated at room temperature and 30°C for 18–24 h. The zones of inhibition were then measured and the resistance or sensitivity was determined by referring to the chart of interpretation of zone of inhibition of Bauer et al. (1966). The chart was conceived for susceptibility testing at 37°C, but following some experiments with known sensitive and resistant *Aeromonas* strains, it was realized that the chart could be used to determine extreme zones at other temperatures. Therefore, the sensitive and resistant strains could be determined, whereas the intermediate zone data were not kept for compilation.

2.4. Bacterial Growth Conditions

Many bacterial strains were selected for their antibiotic resistance patterns. One group of bacterial strains was obtained from Dr. Réal Lallier's laboratory. The strains *Aeromonas hydrophila*, *A. sobria*, *A. salmonicida*, and *E. coli* (Table 9.2) were kept frozen at −80°C in bottles of 2.0 mL of BHI broth. Another group of bacterial strains was obtained from different experiments on trout. These strains were identified as *Aeromonas hydrophila* and *Pseudomonas* sp.

The strains from Dr. Lallier's lab were thawed and cultured on tryptic soy agar (TSA) plates supplemented with 5% bovine blood, and in BHI broth. Depending on the bacterial species, different temperatures and lengths of incubation were used. *A. salmonicida* was incubated at 18°C for 48 h; *A. hydrophila*, *A. sobria*, and *Pseudomonas* sp. were incubated at 30°C for 18–24 h; and *E. coli* was incubated at 37°C for 18–24 h.

The grown cultures were then subcultured on TSA slanted tube agar and incubated as mentioned above. The isolates were maintained by weekly transfer from the TSA agar maintained at 4°C for 4 months onto new TSA blood agar plates maintained at 4°C for one week.

The identification of isolates was verified using the following characteristics described in *Bergey's Manual of Systematic Bacteriology* (1984) and Lennett (1980): examination of colonial morphology and pigmentation, shape of bacteria, Gram staining characteristics, motility, and ability to produce oxidase and catalase. Moreover, the identification of the different types of dominant colonies was confirmed with the API 20E system.

2.5. Transfer of Antibiotic Resistance in Vitro

The bacterial strains used as donors and recipients are described in Table 9.2. Nalidixic acid at 100 μg/mL was used as the marker for the recipient strains. These were totally sensitive to all the antimicrobials tested or resistant to ampicillin only.

Mueller-Hinton agar plates supplemented with horse blood and antibiotics were prepared to verify the antibiotic susceptibility of the recipient and donor strains. The antibiotic powders were prepared as described in Table 9.1, at a concentration 100 times higher then needed. The antibiotic solution was diluted 100 times in the Mueller-Hinton medium before being poured in the petri dish. These supplemented Mueller-Hinton plates were only used for one week.

All donor and recipient strains were cultured on these media and in BHI broth and incubated at their optimal temperatures. Following incubation in BHI broth, one donor and one recipient strain were mixed together; 0.1 mL of each culture was inoculated in tryptic soy broth (TSB). Different donor–recipient mixtures were prepared. Each mixture, prepared in triplicate, was incubated at three temperatures: 18, 24, and 30°C, or 24, 30, and 37°C, for 24–48 h.

0.2 mL of each mating mixture was spread on Mueller-Hinton blood agar supplemented with one antibiotic and 100 μg/mL nalidixic acid. Colonies growing on these double-inhibitor supplemented medium after 24–48 h of incubation at different temperature were scored as presumptive conjugants. The conjugants were subcultured in BHI broth and incubated at the optional temperature of the recipient strain. Ten conjugants from each mating were picked and tested for their biochemical characteristics and antibiotic resistance.

3. RESULTS

3.1. Effect of SxT and RO5 on the Intestinal Flora of Trout

The intestinal flora of trout from the control group had a mean number of 10^5 to 10^9 bacterial per intestine (Fig. 9.1). Each point on the graph is the mean value of the number of bacteria per intestine of three or more fish. The gram-positive flora from the control and treated groups remained quite stable, ranging from an undetectable number of bacteria to 10^4 bacteria per intestine.

Notice in this graph and the following tables that the gram-negative and gram-positive bacteria are considered separately. This is because the gram-positive bacteria were not affected by the treatments.

The gram-negative flora were mainly represented by the *Aeromonas* and *Pseudomonas* species whereas the gram-positive bacteria were mainly of the *Bacillus* species. These results reflect well what is found in the literature. Colonies other than the dominant types were not identified any further than the morphology and Gram reaction.

The group treated with 50 mg of SxT per kg of fish per day showed a gram-negative flora dropping down to an undetectable number of bacteria per intestine (Fig. 9.1). After treatment, at Day 5, it took another 5 days before the flora started to increase in number. After more than 10 days post-treatment, the

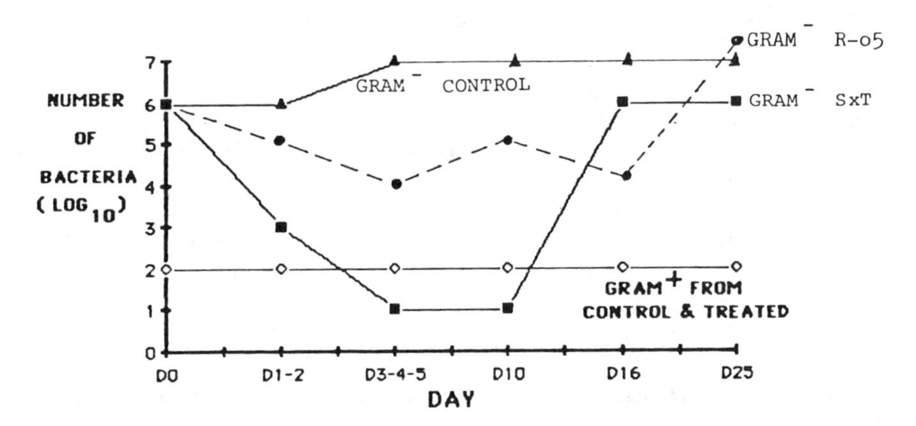

Figure 9.1.

Table 9.2 Bacterial Strains from Dr. Lallier's Laboratory

Bacterial Strain	Resistance Pattern[a]	Animal Source
A. hydrophila		
A-80-323	Am-Te-C-SSS-SxT-RO5	Canine
A-80-199	Am-Te-SSS-SxT-RO5	Human
A-80-153	Am-N-Te-SSS-RO5	Porcine
P-82-297 RNA[b]	Am	Fish
P-80-216 RNA	Am	Fish
A-79-124 RNA	Am	Porcine
A. sobria		
P-82-294 RNA	Am	Fish
A. salmonicida		
P-82-283	Te-SSS	Fish
P-82-285	Te-SSS	Fish
ASm-76-30 RNA	—	Fish
P-85-2 RNA	—	Fish
E. coli K12		
A RNA	—	
B RNA		

[a]Am, ampicillin; Te, tetracycline; C, chloramphenicol; SSS, triple sulfa; SxT, Sulfamethoxazole-Trimethoprim; RO5, sulfadimethoxine-ormetoprim; N, neomycin.
[b]RNA: bacterial strain resistant to 100 μg/mL of nalidixic acid.

Table 9.3 Resistance Patterns of the Gram-Negative Bacterial Strains Isolated from the Control Groups

		% of Strains Resistant to					
Day	Number of Strains Tested	Ampicillin (10 μg)[a]	Neomycin (30 μg)[a]	Tetracycline (30 μg)	Chloramphenicol (30 μg)	SSS (300 μg)	SxT (25 μg)
0	33	36[b]	0	33	36	0	0
1	28	0	0	0	0	0	0
2	30	30	0	0	0	0	0
3	38	3	3	0	0	0	0
4	0	0	0	0	0	0	0
5	1	0	0	0	0	0	0
10	3	0	0	0	0	0	0
15	5	60	0	20	0	20	0
20	2	0	0	0	0	0	0
25	1	0	0	0	0	0	0

[a]Antibiotic disc (concentration).
[b]% of strain resistant to the antibiotic tested. The sensitivity of each strain is determined by a modified Bauer et al. (1966) method described in the text.

Table 9.4 Resistance Patterns of the Gram-Negative Bacterial Strains Isolated from the SxT Group

Day	Number of Strains Tested	% of Strains Resistant to					
		Ampicillin (10 μg)[a]	Neomycin (30 μg)	Tetracycline (30 μg)	Chloramphenicol (30 μg)	SSS (300 μg)	SxT (25 μg)
0	26	46	12	0	0	0	0
1	15	20	0	0	0	0	0
2	16	19	0	0	0	0	0
3	0	0	0	0	0	0	0
4	0	0	0	0	0	0	0
5	6	100	0	0	0	0	0
10	39	72	0	10	18	0	18
16	56	70	0	23	0	16	0
25	33	52	0	42	0	42	0

[a] Antibiotic disc (concentration).

number of flora was back to normal. The drug then had no more effect on the development of the intestinal flora.

The group treated with 50 mg of RO5 per kilogram of fish per day gave a mean bacterial count of 10^4 to 10^6 during the treatment (Fig. 9.1). The flora slowly increased after treatment and were back to their initial number at Day 25. The bacterial count remained quite high during treatment but took approximately as long to reach the initial number as the flora of the fish treated with SxT.

3.2. Antibiotic Sensitivity Tests

The majority of the gram-negative strains isolated from the control groups were sensitive to all the antibiotics tested or resistant to ampicillin (Table 9.3). Nevertheless, resistances to other antibiotics were also observed. Only bacteria resistant to ampicillin were isolated during the SxT treatment (Table 9.4). The resistance to neomycin at Day 0 was not observed in the following days. After treatment, from Day 10 to Day 25, resistance to tetracycline and SSS was mainly observed, whereas resistance to chloramphenicol and SxT was observed only at Day 10. The RO5 treatment (Table 9.5) resulted in resistance to tetracycline and chloramphenicol mainly, and to neomycin and SxT. Ampicillin resistance was the only resistance observed from Day 15 to Day 25 in the RO5 treatment.

These results are complied in Tables 9.6 and 9.7. During treatment with SxT there were no strains resistant to three antibiotics, whereas with RO5 we observed up to 27% of the strains resistant to three or more antibiotics at Day 3 to 5. After treatment, the SxT group had 32% of the strains resistant to three or

Table 9.5 Resistance Patterns of the Gram-Negative Bacterial Strains Isolated from the RO5 Group

Day	Number of Strains Tested	% of Strains Resistant to					
		Ampicillin (10 μg)[a]	Neomycin (30 μg)	Tetracycline (30 μg)	Chloramphenicol (30 μg)	SSS (300 μg)	SxT (25 μg)
0	20	30	0	5	30	0	0
1	17	29	0	0	5	0	0
2	24	38	25	8	25	0	0
3	20	55	25	15	45	0	10
4	17	47	0	12	41	0	6
5	13	23	0	0	23	0	8
10	23	48	0	13	13	0	0
15	25	8	0	0	0	0	0
20	14	0	0	0	0	0	0
25	18	17	0	0	0	0	0

[a] Antibiotic disc (concentration).

more antibiotics by Day 25, whereas the RO5 group had mainly sensitive strains after Day 10.

The different resistance patterns were analyzed during and after the treatments (Tables 9.8 and 9.9).

3.3. Transfer of Antibiotic Resistance

Experiments were performed to verify the possibility of transfer of antibiotic resistance between bacteria of the same and different species. The strains used

Table 9.6 Resistance Patterns for the Bacterial Strains Isolated from the Speckled Trout During the Five-Day Treatment

Day	Number of Strains Tested	% of Strains		
		Sensitive to All ATB	Resistant to 1 or 2 ATB	Resistant to 3 or More ATB
		Gram-Negative Strains from SxT Group		
0	69	79	21	0
1–2	32	80	20	0
3–4–5	6	0	100	0
		Gram-Negative Strains from RO5 Group		
0	161	70	25	5
1–2	41	64	27	9
3–4–5	50	54	19	27

Table 9.7 Resistance Patterns for the Bacterial Strains Isolated from the Speckled Trout After Treatment

		% of Strains		
Day	Number of Strains Tested	Sensitive to All antibiotics	Resistant to 1 or 2 antibiotics	Resistant to 3 or more antibiotics
		Gram-Negative Strains from SxT Group		
10	39	28	56	15
16	59	30	54	15
25	38	50	18	32
		Gram-Negative Strains from RO5 Group		
10	23	52	35	13
16	25	92	8	0
20	14	100	0	0
25	18	83	17	0
		Gram-Positive Strains		
0–25	96	0–20	80–100	0

were obtained from Dr. Real Lallier's laboratory (Table 9.2) or isolated from the trout (Table 9.10). Two strains (*Aeromonas hydrophila* P-85-3 and *Pseudomonas* sp. P-85-4) were not used, since it was impossible to make them resistant to 100 μg/mL of nalidixic acid.

Tetracycline resistance was transferred from *A. hydrophila* to *A. hydrophila* and to *A. sobria* at 30°C and to *E. coli* at 37°C (Table 9.11). Resistance to SSS was also transferred from *A. hydrophila* to *A. salmonicida* at 18°C and between two strains of *A. salmonicida* at 18°C. At other temperatures, those transfers of resistance were not observed (Table 9.11).

Moreover, different donor–recipient mixtures were cultured after 3, 6, 24, and 48 h of incubation. There was no difference in transfer results between the incubation periods. An incubation of 18–24 h was selected for the transfer experiment at 22, 30, and 37°C. For those at 18°C, a minimum of 48 h of incubation was needed.

Antibiotic sensitivity testing was then performed on the conjugants. After the transfer experiment, all *Aeromonas salmonicida* were resistant not only to triple sulfa but also to tetracycline. However, the resistance transfer of tetracycline with the *A. hydrophila*, *A. sobria*, and *E. coli* strains was not coupled to any other resistance. The resistance transferred to *E. coli* was lost in one case.

Table 9.8 Resistance Patterns Observed in the Bacterial Strains Isolated from the Speckled Trout during the Five-Day Treatment

Day	Resistance Patterns
	Gram-Negative Strains from SxT Group
0	Am–N
1–2	Am
3–4–5	Am
	Gram-Negative Strains from RO5 Group
0	Am–C, Am–Te–C, Am–N
1–2	Am, Am–C, Am–N, Am–Te–C, Am–N–C, Am–C–SxT
3–4–5	Am, Am–C, Am–Te–C, Am–N–C, Am–C–SxT, Am–Te–SxT

4. DISCUSSION

The effect of antimicrobial compounds on the intestinal flora of speckled trout was evaluated by means of the variation in the number of bacteria per gram of intestinal content. The development of antibiotic resistance and the possibility of resistance transfer between bacteria of the same and different species were then analyzed.

Two experiments were performed on speckled trout. The control groups

Table 9.9 Resistance Patterns Observed in the Bacterial Strains Isolated from the Speckled Trout after Treatment

Day	Resistance Patterns
	Gram-Negative Strains from SxT Group
10	Am, Am–Te, Am–C–SxT
16	Am, Am–Te, Am–Te–SSS
25	Am, Am–Te–SSS
	Gram-Negative Strains from RO5 Group
10	Am, Am–Te–C
16	Am
20	(Sensitive to all antibiotics)
25	Am
	Gram-Positive Strains
0–25	Am, Te, Am–Te

Table 9.10 Bacterial Strains Isolated from the Experimental Trout

Bacterial Strain	Resistance Pattern[a]
A. hydrophila	
P-84-13 RNA	Am
P-84-26	Am-Te-SSS
P-85-1	Am-C-SxT-RO5
P-85-2	Am-Te-SSS
P-85-3	
Pseudomonas sp.	
P-85-4	

[a] Am, ampicillin; Te, tetracycline; C, chloramphenicol; SSS, Triple Sulfa; SxT, sulfamethoxazole-trimethoprim; RO5, sulfadimethoxine-ormetoprim; N, neomycin.

showed a dominant gram-negative intestinal flora in all fish. The gram-positive flora were lower in number and variable, depending on the fish. The variation in the gram-positive flora could suggest that these bacteria are more transient than the gram-negative bacteria. Similar observations are reported by Trust (1975). It would seem that the gram-positive bacteria, mainly the *Bacillus* species isolated from the digestive tract, come from the feed. The bacteria go through the digestive tract, come from the feed. The bacteria go through the digestive tract as spores but without colonizing it. Moreover, bile salts could have an inhibitory effect on gram-positive bacteria, causing the gram-negative dominance (Lésel, 1981).

Table 9.11 Transfer of Antibiotic Resistance

	Recipient			
Donor	*A. hydrophila* (4 strains) 30°C	*A. sobria* (1 strain) 30°C	*A. salmonicida* (4 strains) 18°C	*E. coli* (2 strains) 37°C
A. hydrophila (6 strains)[b]	Te[a] (6/11)	Te (2/4)	SSS (1/5)	Te (4/7)
A. salmonicida (1 strain)[c]	— (0/1)	— (0/1)	SSS (2/2)	— (0/1)

[a] Abbreviation indicates resistance transferred; number in parens is No. of transfers/No of conjugations.
[b] *A. hydrophila* with resistance Am-Te-C-SSS-SxT-RO5; Am-Te-SSS-SxT-RO5; Am-N-Te-SSS-RO5; Am-Te-SSS; Am-C-SxT-RO5; Am-Te-SSS.
[c] *A. salmonicida* with resistance Te-SSS.

Treatment with SxT and RO5 created a drop in the gram-negative population. The effect was greater with SxT, where 8 out of 15 trout showed no gram-negative flora during treatment. However all the trout from the RO5 group had important flora. The gram-positive flora remained quite stable during both treatment periods. After each treatment it took approximately two weeks before the flora count was back to normal.

The SxT drug did not provoke development of resistant strains during treatment. However, following treatment, once the flora was back to its initial quantity, different strains resistant to ampicillin, tetracycline, chloramphenicol, Triple Sulfa, and sulfamethoxazole–trimethoprim were isolated. These results could be explained by the fluctuation in the intestinal flora of the trout. The flora were almost undetectable during the 5-day SxT treatment. Resistance was therefore uncommon. The new bacteria with their resistance patterns could then develop freely in the trout intestine and be easily isolated.

The RO5 treatment gave different results with many resistance patterns observed during treatment, whereas most resistance was not observed after treatment. The disappearance of resistant bacteria after treatment could not be explained.

Transfers of triple sulfa resistance between two *Aeromonas salmonicida* strains were observed. This had already been cited by Aoki et al. (1983). The transfer of triple sulfa resistance from *A. hydrophila* to *A. salmonicida* was successful in one out of five trials. It is the first time to our knowledge that this transfer was successful. All these transfers occurred only at the optimal growth temperature of the recipient strains, 18°C. The transfer of SSS resistance was always coupled with a transfer of tetracycline resistance. Tetracycline resistance was also transferred from *A. hydrophila* to *A. hydrophila*, *A. sobria*, and *E. coli* in about one out of two experiments. The resistance transferred to *E. coli* was lost in one case.

Even though the transfer of SxT resistance was not observed, the SxT drug caused an important development of strains resistant to triple sulfa. Because the transfer of triple sulfa resistance from *A. hydrophila* to *A. salmonicida* was demonstrated, it is to be feared that prolonged utilization of SxT could be responsible for multiresistant pathogen *Aeromonas* strains.

In conclusion, use of antimicrobials cause an increase of resistant bacteria during or after treatment. The resistance can be transferred between different species of *Aeromonas* and *E. coli*. This fact could be responsible for the development of resistant bacteria in fish ponds and cause therapeutic difficulties in subsequent epidemics. Moreover, the transfer of resistance to bacteria that could develop in other animals, or even in humans, could become a public health hazard.

5. SUMMARY

This study attempted to evaluate the effects on the bacterial population's resistance patterns in speckled trout following treatment with SxT (sulfamethoxazole–trimethoprim) and RO5 (sulfadimethoxine–ormetoprim). A first *in vitro* part was performed, with 25-g speckled trout divided into a control, a SxT, and a RO5 group. The treated groups received 50 mg/kg of fish per day of SxT or RO5. Three trout from each group were sacrificed every day during the 5-day treatment and every five days post-treatment. The number of bacteria per gram of intestinal content was evaluated and the dominant colonies were identified to the genus or, when possible, to the species.

A second *in vitro* part consisted of evaluating the dominant colonies from each group each day for their antibiotic resistance patterns. The sensitivity of each strain was determined by a modified Bauer et al. (1966) method. The Mueller-Hinton culture media were supplemented with lysed horse blood and incubated at 20° and 30°C. Different sensitive or resistant bacterial strains were selected to verify the possibility of resistance transfer between bacteria. The results showed the control group had a bacterial flora composed almost exclusively of gram-negative rods, sensitive to the antibiotics tested except for ampicillin. Pellets medicated with SxT produced a drastic drop in the gram-negative population, while the effect was a lot less important with the RO5-coated pellets. At the end of the treatments, after the bacterial flora were back to their initial number, the flora from the SxT group showed different multiresistance patterns, whereas the RO5 group showed flora sensitive to all the antibiotics tested except for ampicillin.

The transfer experiments succeeded in transferring tetracycline resistance between different strains of *Aeromonas* and to *E. coli*, and transferring of the triple sulfa resistance between *A. salmonicida* strains and from *A. hydrophila* to *A. salmonicida*. The resistance transfers by uncontrolled antibiotherapy in fish ponds and hatcheries could probably become a source of resistance, causing therapeutic difficulties; it could even become a public health hazard.

ACKNOWLEDGMENTS

We thank Francine Bernard and Johanne Fontaine for their technical assistance. This research was supported by a grant from the Aquarium and Marine Centre of Shippagan and the Province of New Brunswick.

APPENDIX A

List of Antibiotics and Their Suppliers

Ampicillin	Ayerst Laboratories, Montréal, Québec
Neomycin	Rogar/STB (Pfizer), Kirkland, Québec
Tetracycline	Rogar/STB (Pfizer), Kirkland, Québec
Chloramphenicol	Rogar/STB (Pfizer), Kirkland, Québec
Sulfamethoxazole	Burroughs Wellcome Inc., Ville LaSalle, Québec

Trimethoprim	Burroughs Wellcome Inc., Ville LaSalle, Québec
Sulfadimethoxine	Hoffman-LaRoche Inc., Nutley, NJ
Ormetoprim	Hoffman-LaRoche Inc., Nutley, NJ
Nalidixic acid	Calbiochem, San Diego, CA
Oxytetracycline	Syndel Laboratories, Vancouver, B.C.
Triple sulfa	
Sulfamethazine	Syndel Laboratories, Vancouver, B.C.
Sulfathiazole	Anachemia, Montréal, Québec
Sulfanilamide	Anachemia, Montréal, Québec

APPENDIX B

Solvent for Trimethoprim and Ormetoprim

1 part 8.5% lactic acid
1 part *N*,*N*-dimethylformamide (DMF)

Only a few drops of the above mixture is necessary to dissolve the trimethoprim and ormetoprim powders. *Source:* Aoki et al. (1983); Bullock et al. (1974).

APPENDIX C

Sodium Phosphate Buffer

Solution A

NaH_2PO_4 0.2 M ($NaH_2PO_4H_2O$, 2.76 g/100 mL H_2O)

Solution B

NaH_2PO_4 0.2 M ($Na_2HPO_47H_2O$, 5.365 g/100 mL H_2O)

For solution at pH 8.0, 0.1 M

5.3 mL of A + 94.7 mL of B and complete to 200 mL with water

For solution at pH 6.0, 0.1 M

72.7 mL of A + 27.3 mL of B and complete to 200 mL with water

APPENDIX D

Components of Triple Sulfa

5.85 g sulfanilamide
5.85 sulfathiazole
3.9 g sulfamethazine

Source: Jean-Claude Panisset (Ed.), *Compendium Pharmaco-thérapeutique vétérinaire, 2d ed. Saint Hyacinthe, Qué: C.D.M.V. Inc., 1985–86.*

REFERENCES

American Public Health Association (1976). *Standard Methods for the Examination of Water and Wastewater*, 14th ed. Washington DC.

Aoki, T., T. Kitao, N. Iemura, Y. Mitoma, and T. Nomura (1983). The susceptibility of *Aeromonas salmonicida* strains isolated in culture and wild salmonids to various chemotherapeutics. *Bull. Jpn. Soc. Sci. Fish.* **49**: 17–22.

Barry, A.L. (1976). *The Antimicrobic Susceptibility Test: Principles and Practice*. Lea & Febiger, Philadelphia.

Barry, A.L., and C. Thornsberry (1980). Susceptibility testing: Diffusion test procedures. In E.H. Lennett (Ed.), *Manual of Clinical Microbiology*, 3d ed. Am. Soc. Microbiology, Washington, DC, pp. 463–474.

Bauer, A.W., W.M.M. Kirby, J.C. Sherris, and M. Turck (1966). Antibiotic susceptibility testing by a standardized single disk method. *Am. J. Clin. Pathol.* **45**: 493–496.

Bergey's Manual of Systematic Bacteriology (1984). Vol. 1, N.R. Krieg and J.G. Holt (Eds.). Williams & Wilkins, Baltimore/London.

Bourque, R., R. Lallier, and S. Larivière (1980). Influence of oral antibiotics on resistance and enterotoxigenicity of *Escherichia coli*. *Can. J. Comp. Med.* **44**: 101–108.

Bullock, G.L., H.M. Stuckey, D. Collis, R.L. Herman, and G. Maestrone (1974). *In vitro* and *in vivo* efficacy of a potentiated sulfonamide in control of furunculosis in salmonids. *J. Fish. Res. Board Can.* **31**: 75–82.

Bushby, S.R.M. (1969). Combined antibacterial action *in vitro* of trimethoprim and sulfonamides. *Postgrad. Med. J.* **45**: 10–18.

Compendium Pharmaco-thérapeutique Vétérinaire, 2d ed. (1985-86). Jean-Claude Panisset (Ed.). CDMV Inc., St-Hyacinthe, Quebec.

Dubourguier, H.C., M. Contrepois, and Ph. Gouet (1980). Antibiorésistance des *E. coli* isolés de veau diarrhéiques. *Bull. G.T.V. 80-4bis-B* **192**: 61–69.

Ferone, R., S.R.M. Bushby, J.J. Burchall, W.D. Moore, and D. Smith (1975). Identification of Harper-Cawston factor as thymidine phosphorylase and removal from media of substances interfering with susceptibility testing to sulfonamide and diaminopyrimidines. *Antimic. Agents Chemother.*, 91–98 (Jan. 1975).

Koch, A.E., and J.J. Burchall (1971). Reversal of the antimicrobial activity of trimethoprim by thymidine in commercially prepared media. *Appl. Microbiol.* **22**: 812–817.

Lennett, E.H. (Ed.) (1980). *Manual of Clinical Microbiology*, 3d ed. Am. Soc. Microbiology, Washington, DC.

Lésel, R. (1981). Microflore bactérienne du tractus digestif des poissons. In *La Nutrition des Poissons*. CNRS, Paris, pp. 89–100.

Smith, H.W. (1970). The transfert of antibiotic resistance between strains of Enterobacteria in chicken, calves, and pigs. *J. Med. Microbiol.* **3**: 165–180.

Trust, T.J. (1975). Facultative anaerobic bacteria in the digestive tract of chum salmon (*Oncorhynchus keta*) maintained in freshwater under defined culture conditions. *Appl. Microbiol.* **29**: 663–668.

10

TOXICITY TESTING OF SEDIMENTS: Problems, Trends, and Solutions

Martin R. Samoiloff

Bioquest International, Inc., Winnipeg, Manitoba, Canada

1. Introduction
2. Sample preparation
3. Biological tests
4. The application of biological tests to sediments
4.1. Comparison to chemical analysis
4.2. Applicable tests
References

1. INTRODUCTION

As a general rule, sediments act as natural adsorbents, binding a wide range of organic and inorganic compounds. In most aquatic ecosystems the sediments would be expected to contain both the highest concentration and the greatest diversity of contaminating compounds. However, each type of sediment has its own binding affinities, and will show selective binding of contaminating chemicals. Each sediment type will also show different desorption properties, releasing the trapped contaminants at varying rates under varying conditions. Having made this disclaimer on the diversity of sediments, I will discuss in this chapter the methods available and needed in the future for determining the toxic potential of sediments as a unified class of material.

In natural systems there are three primary routes by which toxic materials bound to sediments may enter biological systems. The first route is by contact between a target organism and the sediment, and absorption of contaminant through the integument. The second route is by ingestion of sediment by the target organism, desorption of the contaminant in the digestive system, and absorption of the contaminant through the digestive system into the general circulation of the host organism. The third route is by desorption of the contaminant from the sediment to the water column, and uptake to the target organism via the water. In each case the contaminant is active only if released from the sediment and passed to the target organism through some biological barrier; *in situ* toxicity may be considered to be a direct consequence of the desorption capacity of the sediment as well as a consequence of the actual toxicity of the contaminants. While detection of toxic chemicals bound to sediment suggests there is a toxic potential of the sediment, chemical analysis alone is insufficient to provide a realistic appraisal of the toxicity of those sediments. Rather, there must be demonstration of some adverse biological effect resulting from exposure to the sediments. Toxic materials must be present and biologically available to have any ecotoxicological significance.

At present the toxic potential of sediments is most often determined by analytical chemistry and detection of compounds on a previously established list of priority chemicals, but this method is not satisfactory from an ecotoxicological perspective. Real world sediment samples typically contain several hundred to several thousand different chemicals, each contributing alone and by synergistic and antagonistic interactions with other materials to the net toxic impact of the sediment. Priority chemical lists contain only those contaminants whose individual toxicity or environmental persistence has been well established by previous toxicological, epidemiological, or ecotoxicological studies. Toxic chemical lists exclude the vast majority of environmental contaminants not on the basis of demonstrated nontoxicity, but on the basis of little or no knowledge of the potential impact of these materials (National Research Council, 1984). Biomonitoring methods, using test organisms to evaluate the toxic potential of sediments, provide a far more comprehensive and realistic view of the toxic potential of environmental contaminants.

In ecotoxicology, determination of toxicity is often confounded by the question of target organism. For many, the sole objective of toxicological studies in the environment centers on the determination of possible adverse human health effects. Others focus on the protection of species that are of economic importance to the human population, including humans. I will take the broadest view and focus on determination of toxic effect to any living component of the impacted ecosystem. The focus on target organism determines the relative importance of the range of toxic effects detected. For those concerned with human health, the detection of carcinogens is a high priority, since few environmental contaminants are considered to pose an immediate threat to human life. This emphasis on carcinogens may be inappropriate, and may reflect the fact that the biological methods for detecting mutagens/carcinogens are readily available. Carcinogenesis is of little significance in natural nonhuman populations, because the frequency of affliction is low, and mortality from such causes would be masked by mortality due to competition and predation. Biological tests to establish the ecotoxicity of sediments must detect effects that would have a significant impact on natural populations.

Some ordering of priorities of biological effects must be established. Those toxic effects that reduce the probability of production of the next generation of organisms must be assigned the highest priority. These effects include lethality, sterilization, and physiological damage that prevents growth or significantly decreases the ability of the exposed population to adapt to normal environmental stress. Any set of conditions, resulting from single or multiple contaminants, that significantly decreases the overall fitness of an exposed population will place the next generation at risk.

Biological monitoring can be performed on field organisms *in situ*, although such monitoring normally yields results specific to each particular aquatic ecosystem. Population counts of organisms associated with specific contaminated sediments are of little value because these population levels are subject to variation caused by a wide variety of ecological factors other than contaminants, and baseline data are often not available. Another field approach is the determination of the population ratio between a highly sensitive species and a resistant species. This approach requires baseline data to monitor an increase in the relative size of the resistance species as a function of increasing contamination of sediments. Even when ubiquitous species are monitored, there may be major local population differences in overall sensitivity and resistance. A third *in situ* biomonitoring approach involves monitoring morphological abnormalities in field-exposed populations. This method can provide convincing evidence of adverse effects of contaminants, but again requires baseline data and can be highly skewed by differences in the sensitivities of local populations. A more universal approach is to collect contaminated sediment samples and biologically test them using a laboratory indicator organism.

2. SAMPLE PREPARATION

There are a number of methods available for the collection of sediments, and these will not be reviewed here. In the collection of sediments, methods must be chosen that meet the objectives of the study being carried out. Core samples are appropriate where chronic exposure to a particular contaminant has occurred or where a historical representation of the sediment toxicity is required. Where episodic exposure has taken place, the uppermost layers of the bottom sediment are of more importance; in dynamic systems, the suspended sediments may be of more significance.

Because the natural sediments are in an absorption–desorption equilibrium at the time of collection, care must be exercised to ensure that no major changes in this equilibrium occur prior to preparation. Heating, freezing, and drying are the most common carrying conditions that will alter the equilibrium. Also, storage vessels must be inert, neither contributing materials nor altering the contaminant–sediment equilibrium.

There are three basic methods by which the sediment samples may be prepared for bioassay: the intact sediment can be tested, material physically separated from the sediment can be tested, or the contaminants can be chemically separated from the sediments and then tested. Ideally the intact sediment should be bioassayed; however, most of the routine bioassays cannot be performed on a sediment bed. The two problems that restrict the use of bioassays on pure sediments are the difficulties in recovering the test organisms from the intact sediments, and the action of the natural biological populations of the sediments. Potential methods for bioassay of intact sediments will be discussed in a later section. Desorption of contaminants by physical methods isolates the subset of contaminants that are the most bioavailable. One method of desorbing loosely bound contaminants from sediments is elutriation, involving centrifugation of the sediment to obtain an aqueous supernatant fluid that can be tested for toxicity by bioassay or chemical analysis. The method is rapid and cost-effective, and can be readily standardized. Because elutriation yields an aqueous material for testing, a wide range of standard aquatic biological tests may be performed on elutriates. The major disadvantage of this method is that it primarily isolates only those water-soluble contaminants that are loosely bound to the sediment. Chemical extraction is the only method that gives an evaluation of the entire range of toxic contaminants in sediments.

One very comprehensive method for extracting and preparing the chemicals bound to contaminants for bioassay was developed by Birkholz (Birkholz, 1982; Samoiloff et al., 1983). The extraction protocol produces eight fractions for bioassay from each sediment sample. Each fraction contains a distinct suite of contaminants. The procedure involves an initial Soxhlet extraction with 1 : 1 acetone–hexane. This extract is partitioned into organic-free water, which is extracted with dichloromethane (DCl). This yields two extracts: the neutral compounds in the (DCM) extract, and the aqueous phase. The DCM extract is exchanged into hexane by evaporation of the DCM, and applied to a Florisil column. Four fractions of the original DCM extract are obtained by elution from the Florisil column with hexane, 1 : 1 hexane–dichloromethane, dichloromethane, and methanol, respectively. Three fractions are obtained from the aqueous extract. Two of the aqueous fractions are obtained by adjusting the pH to 11 and 2, respectively, and extracting with DCM. The remaining water-soluble material is the third aqueous fraction. The final fraction is a methanolic extract of the original extracted sediment. The fractions are exchanged into dimethylsulfoxide (DMSO) for biological testing. Recently, this method has been modified to yield only four fractions.

3. BIOLOGICAL TESTS

There are numerous biological assays for toxicity that can be applied to sediments or materials isolated from sediments. Selection of an appropriate biological indicator involves consideration of the ecological relevance of the test system, the sensitivity of the test, cost and time factors, availability of personnel capable of performing a specific test, and the nature of the information required. All of the biological assays provide quantitative measure of an adverse effect on some biological end point. These end points may range from short-term survival through to gene level effects manifested only after one or more generations. The end point should be such that effect on that end point indicates an effect on one or more biological processes common to all living systems. The actual test organism used is of less importance than the obtaining of ecotoxicologically relevant information about the toxic potential of the tested sample; what is important is the quantitative determination of a biologically significant effect on a biologically significant end point.

One widely applied end point is lethality, usually obtained by measuring the proportion of a test population that dies after a specific period of time (usually 96 h) of exposure to one of a series of concentrations of the tested material. By plotting the proportion dead at the end of the time against the concentration, an LC_{50} value is calculated. The LC_{50} is the calculated concentration at which 50% of the population will die during the exposure period. The lower the value of the LC_{50}, the more toxic the material tested. The LC_{50} can be used in two distinct fashions. Commonly, the LC_{50} is divided by some "application factor" ranging from 10 to 1000 to establish a "safe level" of a toxicant. This approach suggests that a dilution well below the concentration of material that causes 50% death will be without biological effect. There are several logical flaws in this approach, primarily that the lethal test used to establish the LC_{50} is usually performed over a time span that represents a small proportion of the mean life span of the test organism and ignores all sublethal effects. The other, more applicable use of the LC_{50} assay is for the ranking of a series of potentially toxic samples, that is, the determination of which of a series of samples poses the greatest risk. Death, however, while a biologically significant end point, represents the most extreme of such toxic end points and provides a very crude indication of the actual toxicity.

A number of biological tests are used to establish an EC_{50}, the concentration at which some biological function (growth, metabolism, reproduction) is reduced by 50%. The EC_{50} is better than the LC_{50} as an indication of the potential for decreased fitness of an exposed natural population. However, as demonstrated by Miller et al. (1985), the EC_{50} values for several standard test organisms may vary by several orders of magnitude. Both the EC_{50} and the LC_{50} provide information on the relative toxicity of the tested materials, useful for ranking, but not appropriate for making a decision on the toxicity/nontoxicity of the tested sample. This determination can be made by the establishment of toxic threshold criteria for each of these tests. I do not mean to suggest that there is a threshold for toxicity, but rather that for each test, an EC_{50} value must be defined a priori to represent an unacceptable level of toxic contamination. An EC_{50} below this value would be an indication of unacceptable levels of toxic contaminants.

A test developed by Samoiloff and co-workers (1980, 1984), using the nematode *Panagrellus redivivus*, tests three distinct biological end points: survival, growth, and maturation, reflecting lethal, sublethal, and gene level effects of the tested sample. The test is performed on the 96 h post-embryonic growth period of this short-lived organism, so that the exposure period covers a significant portion of the life cycle. The combined effects on these three end points can be expressed quantitatively as a single value termed "fitness." Rather than using a dose–response protocol, the test is usually performed at one standard concentration: 10% for water extracts or elutriates, 3% for methanol extracts, and 1% for other solvent extracts exchanged into dimethylsulfoxide. This provides a standard by which a series of similar samples may be compared. The test provides a rapid, cost-effective screening method for evaluating the toxic potential of complex mixtures.

4. THE APPLICATION OF BIOLOGICAL TESTS TO SEDIMENTS

4.1. Comparison to Chemical Analysis

Table 10.1 presents a summary of the results of the *Panagrellus redivivus* bioassay on water extracts and elutriates of sediments and soils previously identified as containing significant levels of one or more of the 129 priority chemicals. No significant toxicity was found in 58% of these samples, and lethal effects were found in only 16% of the samples. This shows a very poor relation between detection of toxicity using the nematode test and the presence of toxic chemicals from the priority list. This can be due to a failing of the bioassay or a failure of the presence of priority pollutants as a predictor of toxic effects. The comparison of one biological test with chemical analysis suggests that one or the other approach is flawed, but it cannot resolve which approach is at fault.

Table 10.2 compares three independent tests performed on extracts of Saskatchewan River sediments. The tests were (1) chemical analysis by gas

Table 10.1 ***Panagrellus redivivus* Test for Toxicity of 10% Water Extracts of 207 Sediments and Soils Containing Priority Chemicals**[a]

Effect	Number	Percentage
Lethal	35	16.9
Inhibitory	40	19.3
Gene level	11	5.3
No effect	121	58.4

[a] Although all the samples should show toxic effects, only 41.6% show any significant biological effect.

Table 10.2 Three Independent Tests for Toxicity of Chemical Fractions of Saskatchewan River Sediments[a]

GC/ECD	Bacteria	Nematode	Number	Percentage	Priority
Toxic	Toxic	Toxic	5	3	I
Toxic	Toxic	ND	0	0	II
Toxic	ND	Toxic	8	6	II
Toxic	ND	ND	10	7	III
ND	Toxic	Toxic	12	8	II
ND	Toxic	ND	14	10	III
ND	ND	Toxic	28	20	III
ND	ND	ND	66	46	0

[a]Gas chromatography with electron capture detectors (GC/ECD) was used to detect one or more of the 129 priority chemicals. The bacterial test was a modified *Salmonella typhimurium* test used to detect cytotoxic or mutagenic effects. The nematode test used *Panagrellus redivivus* to detect lethal, inhibitory of gene level effects. Toxic effect under each column refers to detection of a priority pollutant or a significant biological effect; not detected (ND) refers to a failure to detect priority pollutants, or to an absence of detected biological effects.

chromatography with electron capture detectors to detect priority pollutants in which toxicity was scored if one or more priority chemicals were present; (2) a test of *Salmonella typhimurium*, following the methods of Maron and Ames (1983), but testing for both mutagenesis and cell survival; and (3) the *Panagrellus redivivus* bioassay. Twenty-three of 143 samples (16%) were found to contain chemicals from the priority chemicals list. Of these, 5 were found to be toxic by both the nematode and bacterial bioassay, 8 were toxic only to the nematode, and 10 of these samples containing priority chemicals had no detectable effect on either bioassay. Twelve of the 120 samples that did not contain compounds on the priority list were toxic to both bacteria and nematode, while 14 were toxic only in the bacterial test, and 28 were toxic only in the nematode bioassay. In total, 54 of the 120 samples containing no priority chemicals (45%) produced a toxic effect in one or both biological tests. The absence of priority chemicals is not a good predictor of toxicity. Nor is the presence of priority chemicals a good predictor of toxicity, since 43% of the samples containing priority chemicals failed to evoke a toxic response in either bioassay.

Such studies as presented in Table 10.2, where several different tests are applied to each of a series of samples, provide optimal methods for establishing priorities. A determination of toxicity, either by detection of priority chemicals or by a significant biological effect can be considered to represent a "hit." Samples that produce a hit in each of the tests are assigned the highest priority, samples that produce a hit in two of the three tests are assigned the next highest priority, and those samples with only one hit are assigned a lower priority (Table 10.3). Those samples producing no hits are considered nontoxic.

Table 10.3 Three Classes of Priority of Toxicity Based on the Concordance between Three Independent Tests of Chemical Fractions of Sediments from the Saskatchewan River[a]

Priority	Total	Toxic by Chemical Analysis	Toxic by Biological Test
I	5	5	5
II	20	8	12
III	52	10	42

[a]The three tests are chemical analysis (gas chromatography), a modified *Salmonella typhimurium* bioassay, and the *Panagrellus redivivus* bioassay. Priority I samples show toxicity in all three tests, priority II samples show toxicity in 2 of 3 tests, and priority III samples are toxic in only one of the three tests.

There is presently no single biological test capable of detecting all possible toxic responses. As a result, the best possible protocol would employ several biological indicators for toxicity. For many of our studies we have employed the nematode bioassay, with its multiple biological end points, and the Microtox test (Bulich, 1983), which, by measuring the decrease in luminescence in the bacteria *Photobacterium fisherii* caused by a tested material, provides evidence of effects of the tested material on energy metabolism in the test population of bacteria. For purposes of comparison, we consider that an EC_{50} value less than 60 for an environmental sample indicates significant toxicity of that sample. That is, if a 6:4 dilution of the sample produces a 50% or greater reduction in luminescence after 15 min, that sample is considered toxic. In tests of 36 water extracts of sediment samples, using the nematode test and the Microtox test, 14 extracts were toxic to both tests, 14 were toxic to Microtox only, 1 was toxic to the nematode only, and 7 were not toxic. Those samples producing toxic effects in both tests were considered to be the highest priority class of contaminated samples.

This multiple bioassay approach can be carried out using a series of bioassays. In evaluating the toxicity of oil well sump fluids, three tests were used: a lethal test using 3-month-old rainbow trout, Microtox, and the nematode test. For the rainbow trout, the criterion for toxicity was death of 1/3 or more of the test population after 96-h exposure to the collected sample. The results of this series of tests is shown in Table 10.4. In 65 of the 117 samples, all three bioassays detected toxic effects, establishing a basis for considering these to be the most toxic samples. An additional 28 samples were toxic in two of the three tests, and 20 were toxic to only one of the three tests.

Because of the differing sensitivity of the three test organisms, not all samples will produce a toxic response in each test; however, comparing three such tests permits the establishment of a priority listing of samples at a cost well below that required for chemical analyses.

Table 10.4 Summary of a Fish LC_{50} Test, the Microtox Test, and Nematode Tests on 117 Oil Well Sump Samples

Fish	Microtox	Nematode	Number	%	Priority
Toxic	Toxic	Toxic	65	56	I
Toxic	Toxic	ND	8	7	II
Toxic	ND	Toxic	4	3	II
Toxic	ND	ND	2	2	III
ND	Toxic	Toxic	16	14	II
ND	Toxic	ND	2	2	III
ND	ND	Toxic	16	14	III
ND	ND	ND	2	2	0

4.2. Applicable Tests

Although a battery of bioassays is the preferred approach, often one test should be selected as a primary indicator. The selection of such a primary indicator organism will depend on the type of environmental monitoring and the nature of the sample. If monitoring is to be performed to establish risk to an economically important species, the primary indicator organism should be that species, or a closely related indicator species. Secondary bioassays using more standard tests should be used to verify the results with the primary indicator.

For laboratory bioassays on intact sediments, there are few bioassays that can be applied because the recovery of the test organisms from the sediment is often difficult. Flotation methods can be used to recover the bioassay organisms, but such methods are relatively inefficient and may be influenced by organisms present in the sample. Logically, the best approach for intact sediments is a seed germination test, where the test plant can be readily recognized, and the effects of the direct interaction with the sediment and the test organism can be readily quantified.

For extracts of sediments where the tested sample contains a solvent carrier system with its own set of biological effects, dose–response studies cannot be performed. Rather the biological test must evaluate a standard dilution of the extracts compared to a similar dilution of carrier. A test with multiple quantitative biological end points is of greatest applicability in such cases. The *P. redivivus* bioassay serves well for such analyses, while tests that typically establish an EC_{50} can be modified to give a relative toxicity value for a single concentration. For example, a Microtox test, can be used to establish the *time* required to reach a 50% reduction in luminescence at a single concentration, relative to a carrier reference standard. Such an ET_{50} is a valid comparative indicator of toxicity.

For studies on elutriates most of the standard aquatic bioassays can be used. Rather than using a single bioassay, however, a standard battery of tests, as described by Miller et al. (1985), is preferable. The suite of biological tests should cover a range of biological end points, from lethality to gene level effects, over a span of trophic levels. In my opinion, it is unfortunate that most batteries of tests presently under consideration focus primarily on LC_{50} data, because this measures only the most extreme toxic effects over a biologically short period. To apply only acute tests to a chronically contaminated system is unrealistic.

Many of the standard bioassays can be applied to fractions of sediments. However, the standard dose–response protocol cannot be followed due to solvent toxicity. This means either that standard dilutions must be used, such as those run in the *Panagrellus redivivus* test, or that a timed series of exposures must be used to calculate the time required to reach some predefined end point.

Whatever method of preparation, biological testing provides the most ecologically relevant method of analysis.

REFERENCES

Birkholz, D.A. (1982). The detection of contaminants in sediments: A coupled chemical fractionation/bioassay method. M.Sc. thesis, University of Manitoba.

Bulich, A.A. (1983). A practical and reliable method for monitoring the toxicity of aqueous samples. *Proc. Biochem.* **17**: 45–47.

Maron, D.M., and B.N. Ames (1983). Revised methods for the *Salmonella* mutagenicity test. *Mutat. Res.* **113**: 173–215.

Miller, W.E., S.A. Peterson, J.C. Greene, and C.A. Callahan (1985). Comparative toxicology of laboratory organisms for assessing hazardous waster sites. *J. Env. Qual.* **14**: 569–574.

National Research Council Board on Toxicology and Environmental Health Hazards. (1984). *Toxicity Testing: Strategies to Determine Needs and Priorities*. National Academy Press, Washington, DC.

Samoiloff, M.R., S. Schulz, K. Denich, Y. Jordan, and E. Arnott (1980). A rapid simple long-term toxicity assay for aquatic contaminants using the nematode *Panagrellus redivivus*. *Can. J. Fish. Aquat. Sci.* **37**: 1167–1174.

Samoiloff, M.R., J. Bell, D.A. Birkholz, G.R.B. Webster, E.A. Arnott, R. Pulak, and A. Madrid (1983). A combined bioassay-chemical fractionation scheme for the determination and ranking of toxic chemicals. *Env. Sci. Technol.* **17**: 329–334.

Samoiloff, M.R., and T. Bogaert (1984). The use of nematodes in marine ecotoxicology. In G. Persoone, E. Jaspers, and C. Claus (Eds.), *Ecotoxicological Testing for the Marine Environment*, Vol. 1. State University of Ghent, and Institute for Marine Scientific Research, Bredene, Belgium, pp. 407–426.

11

ROLE OF DISSOLVED OXYGEN IN THE DESORPTION OF MERCURY FROM FRESHWATER SEDIMENT

J.S. Wang and P.M. Huang

Department of Soil Science, University of Saskatchewan, Saskatoon, Saskatchewan, Canada

U.T. Hammer

Department of Biology, University of Saskatchewan, Saskatoon, Saskatchewan, Canada

W.K. Liaw

Saskatchewan Fisheries Laboratory, Department of Parks and Renewable Resources, Saskatoon, Saskatchewan, Canada

1. Introduction
2. Materials and methods
   2.1. Sample
   2.2. Treatment
   2.3. Analytical methods
3. Results and discussion
4. Summary and conclusions
References

1. INTRODUCTION

Dissolved oxygen in the bottom sediments of stratified eutrophic lakes can be largely depeleted by the microbial degradation of the decaying algae. The influence of the dissolved oxygen and redox potential (Eh) on the release of Hg (mercury) from the freshwater sediments has been studied by Feick et al. (1972), Kudo et al. (1975), Khalid et al. (1977), and Gambrell et al. (1980). The nature of the sediments (Feick et al., 1972) and the levels of Hg and pH values of the sediments (Khalid et al., 1977; Gambrell et al., 1980) seem to have an effect on the release of Hg from sediments, as influenced by the dissolved oxygen or Eh levels. However, the influence of dissolved oxygen on the rates and mechanisms of the Hg release from sediments is still obscure.

The objective of this study was to investigate the effect of the dissolved oxygen levels on the kinetics of the desorption of Hg from the sediment of a freshwater lake (Katepwa Lake) in the Canadian Prairies under different temperatures and the association of Hg release with the dissolution of Fe and Mn oxides in the systems.

2. MATERIALS AND METHODS

2.1. Sample

The sediment samples were collected with an Ekman dredge from selected sites of Katepwa Lake, Saskatchewan, Canada. The sediments from the individual sites were combined, thoroughly mixed, and stored at 4°C. The composite sample was used in the experiment. This lake sediment contains 122 and 20 mmol poorly crystalline oxides of Fe and Mn, and 46.6 g organic carbon per kilogram sediment.

2.2. Treatment

A wet sediment sample equivalent to 8 g of oven-dried weight was suspended in 40 mL distilled deionized water in a 500-mL Erlenmeyer flask. Sixty milliliters of 1.33×10^{-3} M $Hg(NO_3)_2$ solution was then added to the sediment suspension. The final sediment–solution ratio of the suspension was adjusted to 1 g to 20 mL by adding distilled deionized water. The suspension was equilibrated for 24 h at 25°C in a constant-temperature shaker. The sediment suspension at the end of the equilibration period was divided into eight equal parts and centrifuged at 1600 g for 20 min. The supernatant from each sediment suspension was removed by decantation and preserved by adding 1 mL of 18 M sulfuric acid for Hg determination.

The sediment obtained after centrifugation was dispersed in 75 mL distilled deionized water in a 125-mL Erlenmeyer flask. The sediment suspension was then flushed with N_2 plus 1% H_2 gas mixture at 60 mL/min for 30 min. The dissolved oxygen level of the suspension before and after the gas flushing treatment was measured by an Extech dissolved oxygen meter. The dissolved oxygen level decreased from 4.0 to 0.0 μg/mL. Traps of $KMnO_4$–H_2SO_4 (Miller et al., 1975) were connected to the gas outlet tube of the flask to determine the possible loss of gaseous Hg in the flushing process.

The N_2–H_2 gas-treated and the control samples were agitated at 25, 15, and 4°C for 0.25, 0.5, 1, 4, 8, 24, 48, and 96 h, respectively. At the end of each reaction period, the gaseous Hg in the Erlenmeyer flasks was flushed into a Hg trap with N_2 gas. The dissolved oxygen level, redox potential (Eh), and pH of the sediment suspensions were determined immediately. The sediment suspensions were then filtered through a 0.45-μm Millipore membrane filter.

2.3. Analytical Methods

The filtrates were digested with aqua regia and then with $KMnO_4$, and analysis for total Hg was carried out by the cold vapor technique using a UV Hg analyzer at the wavelength of 253.7 nm (U.S. Environmental Protection Agency, 1974). Dissolved Fe and Mn in the supernatant were determined by atmoic absorption spectrophotometry. The poorly crystalline oxides of Fe and Mn in the sediment samples were determined by extracting with acidified hydroxylamine hydrochloride (Chao, 1972; Ross et al., 1985). The dry combustion method (Tiessen et al., 1981) was used to determine the organic carbon content in the sediment.

3. RESULTS AND DISCUSSION

The rate curves in Fig. 11.1 show that the desorption of Hg from the Katepwa Lake sediment at 4.0 and 0.0 μg/mL dissolved oxygen levels obeyed the multiple first-order kinetics. An initial fast desorption process occurred in the first hour of the reaction period, followed by a slow desorption process. The slopes of the Hg desorption curves at 25, 15, and 4°C for the N_2–H_2-treated samples in the slow desorption process are substantially greater than those for the untreated samples. The rate constants (Table 11.1) show that, compared with the control samples, the rate of Hg release from the N_2–H_2-treated samples at the end of the first hour was increased by 1.2, 1.4, and 1.9 times at 25, 15, and 4°C, respectively, while the rate of the Hg release from the N_2–H_2-treated sample in the slow desorption at the end of 96 h was increased by 15, 17, and 24 times over the control samples at the three temperatures studied. This indicates that more Hg was released from the oxygen-depleted sediment and the influence of the dissolved oxygen levels on the rate of the Hg release became greater as the reaction proceeded.

The effects of temperature on the Hg desorption rates for the untreated samples may not be significant if the experimental errors are taken into account (Table 11.1). After the N_2–H_2 treatment, the rate constants of the Hg

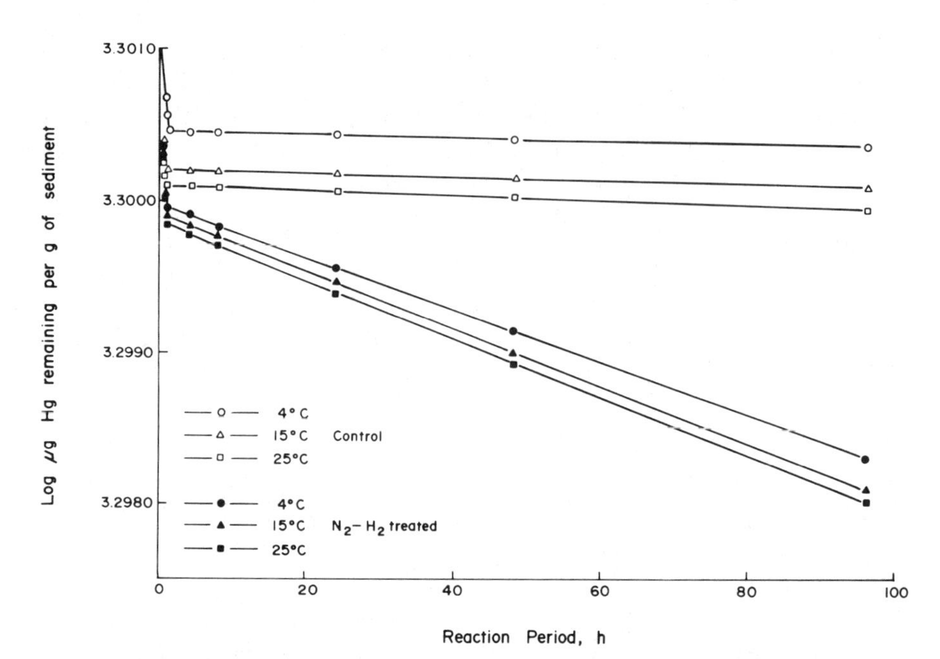

Figure 11.1. The rate curves of the desorption of Hg from the Katepwa Lake sediment as influenced by dissolved oxygen levels and temperatures. The initial dissolved oxygen levels of the control and N_2–H_2-treated samples were 4.0 and 0.0 µg/mL, respectively. For calculating the rate constants of the Hg release in the fast desorption process, the plottings were based on an expanded scale.

release from the sediment did not show any significant increase with increasing temperatures.

The data in Table 11.2 show that substantial amounts of Fe and Mn were dissolved in the solution of the N_2–H_2-treated samples upon the depletion of oxygen. In contrast, the dissolution of Fe and Mn from the untreated samples

Table 11.1 The Rate Constants of the Hg Release from the Katepwa Lake Sediment

Treatment	Reaction Period (h)	Rate Constant × 10^5 h^{-1} 4°C	15°C	25°C
Control	0–1	56 ± 15	80 ± 14	92 ± 21
	1–96	0.07 ± 0.01	0.11 ± 0.04	0.13 ± 0.02
N_2 + 1% H_2 gas mixture	0–1	106 ± 21	111 ± 18	113 ± 20
	1–96	1.7 ± 0.3	1.9 ± 0.4	1.9 ± 0.4

Table 11.2 The Release of Fe, Mn, and Hg from the Katepwa Lake Sediment as Influenced by the Dissolved Oxygen Levels at 25°C

Treatment	Reaction Period (h)	pH	Eh (mv)	O_2 (μg/mL)	μmol Released/kg Sediment		
					Fe	Mn	Hg
Control	1	7.50	441	4.0	nd[a]	nd	21 ± 2
	24	7.48	445	3.9	nd	nd	22 ± 3
	96	7.39	450	3.7	nd	nd	24 ± 3
N_2 + 1% H_2 gas mixture	1	7.75	11	0.0	10 ± 5	100 ± 20	27 ± 3
	24	7.72	5	0.0	140 ± 30	450 ± 80	36 ± 4
	96	7.65	−35	0.0	150 ± 30	570 ± 90	68 ± 7

[a] Not detectable.

was not detectable. The higher dissolution of Mn at lower Eh values and at pH of 7.39–7.75 (Table 11.2) is in agreement with the Eh–pH diagram for sedimentary manganese deposits (Krauskopf, 1957, p.63) and the pE–pH diagram for the synthesized manganese dioxide (Murry, 1974, p.369). The results of the dissolved Fe in Table 11.2 also show the same trend as in the Eh–pH diagram of several Fe oxides (Garrels and Christ, 1965). The increase in the Hg release in the oxygen-depleted samples can be partially attributed to the dissolution of Fe and Mn oxides. Since hydrous Fe and Mn oxides have high capacities to adsorb Hg (Thanabalasingam and Pickering, 1985; Wang et al., 1985), the adsorbed Hg may be released to the solution when the Fe and Mn oxides become unstable and partially dissolved upon the depletion of oxygen in the system. Because Hg was adsorbed on the surface of the hydrous Fe and Mn oxides, the dissolution of the oxides would result in the concomitant release of Hg. The data (Table 11.2) thus indicate that the increase in Hg released in the treated sample was partially associated with the dissolution of the Fe and Mn oxides from the sediments.

A light brown color was visually observed in the supernatant from the N_2–H_2-treated samples instead of the colorless supernatant from the untreated samples. The dissolution of organic materials could also take place in the system. Since the alternate aerobic and anaerobic conditions may increase the decomposition of the organic materials (Reddy and Patrick, 1975), the organically bonded Hg could also be released to the solution phase as the organic materials were partially dissolved upon the depletion of oxygen. The sulfides should not have been generated from the sediment samples under the pH (7.6) and Eh (−35 mv) condition (Garrels and Christ, 1965) in this study. Furthermore, if the sulfides were produced in the system, the Hg released would have been decreased in the N_2–H_2-treated samples because of the strong bonding between Hg and sulfides and the extremely low solubility of the HgS.

4. SUMMARY AND CONCLUSIONS

The impact of the oxygen depletion, which was induced by eutrophication, on the Hg dispersion from bottom sediment of stratified lakes has not been well understood. This study investigated the influence of the dissolved oxygen on the kinetics of the release of Hg from the sediment of a freshwater lake (Katepwa Lake) in Saskatchewan, Canada. The desorption process obeyed the multiple first-order reaction; a fast desorption occurred within one hour, and was followed by a slow desorption process. When the dissolved oxygen level decreased from 4.0 to 0.0 μg/mL (the Eh decreased from 450 to −35 mv), the rate of desorption was increased by 1.2 to 1.9 times for the fast desorption and by 15 to 24 times for the slow desorption process. Temperature had little effect on the Hg release under the level of the depleted dissolved oxygen and the temperature range studied. The data indicate that the dissolution of the Fe and Mn oxides is a significant mechanism involved in releasing Hg from the sediment to pore water when the dissolved oxygen is sufficiently depleted.

ACKNOWLEDGMENTS

This study was supported by the Natural Sciences and Engineering Research Council of Canada Strategic Grants G-1296 and G-1994.

REFERENCES

Chao, T.T. (1972). Selective dissolution of manganese oxides from soils and sediments with acidified hydroxylamine hydrochloride. *Soil Sci. Soc. Am. Proc.* **36**: 704–768.

Feick, G., Johanson, E.E., and Yeaple, D.S. (1972). Control of mercury contamination in freshwater sediments. U.S. EPA-R2-72-077, Washington, DC.

Gambrell, R.P., Khalid, R.A., and Patrick, Jr., W.H. (1980). Chemical availability of mercury, lead, and zinc in Mobile Bay sediment suspensions as affected by pH and oxidation reduction conditions. *Env. Sci. Technol.* **14**: 431–436.

Garrels, R.M., and Christ, C.L. (1965). *Solutions, Minerals, and Equilibria.* Harper & Row, New York, p. 450.

Khalid, R.A., Gambrell, R.P., and Patrick, W.H., Jr. (1977). Sorption and release of mercury by Mississippi river sediment as affected by pH and redox potential. In D. Drucker and R.E. Wildung (Eds.), *Biological Implications of Metals in the Environment*. Proc. 15th Annual Hanford Life Sci. Symp., Richland, Washington, pp. 297–314.

Krauskopf, K.B. (1957). Separation of manganese from iron in sedimentary processes. *Geochim. Cosmochim. Acta.* **12**: 61–84.

Kudo, A., Mortimer, D.C., and Hart, J.S. (1975). Factors influencing desorption of mercury from bed sediments. *Can. J. Earth Sci.* **12**: 1036–1040.

Miller, R.W., Schindler, J.E., and Alberts, J.J. (1975). Mobilization of mercury from freshwater sediments by humic acid. In F.G. Howell, J.B. Gentry, and M.H. Smith (Eds.), *Mineral Cycling in Southeastern Ecosystems*, Technical Information Center, U.S. Energy Research and Development Administration, Springfield, Virginia, pp. 445–451.

Murry, J.W. (1974). The surface chemistry of hydrous manganese dioxide. *J. Colloid Interface Sci.* **46**: 357–371.

Reddy, K.R., and Patrick, Jr., W.H. (1975). Effect of alternate aerobic and anaerobic conditions on redox potential, organic matter decomposition, and nitrogen loss in a flooded soil. *Soil Biol. Biochem.* **7**: 87–94.

Ross, G.J., Wang, C., and Schuppli, P.A. (1985). Hydroxylamine and ammonium oxalate solutions as extractants for Fe and Al from soils. *Soil Sci. Soc. Am. J.* **49**: 783–785.

Thanabalasingam, P., and Pickering, W.F. (1985). Sorption of mercury(II) by manganese(IV) oxide. *Environ. Pollut., Ser. B* **10**: 115–128.

Tiessen, H., Bettany, J.R., and Stewart, J.W.B. (1981). An improved method for the determination of carbon in soils and soil extracts by dry combustion. *Comm. Soil Sci. Pl. Anal.* **12**: 211–218.

U.S. Environmental Protection Agency. (1974). Methods for chemical analysis of water and waste. Office of Technol. Transfer, EPA-62576-74-003, Washington, DC.

Wang, J.S., Huang, P.M., Hammer, U.T., and Liaw, W.K. (1985). Influence of chloride on kinetics of the adsorption of mercury(II) by poorly crystalline Al, Fe, Mn, and Si oxides. *Water Poll. Res. J. Can.* **20**: 68–74.

12

INTEGRATED ECOTOXICOLOGICAL EVALUATION OF EFFLUENTS FROM DUMPSITES

R. Van Coillie, N. Bermingham, C. Blaise, and R. Vezeau

Environment Canada, Conservation and Protection Service, Quebec Region, Protection Directorate, Montreal, Quebec, Canada H3B 3H9

J.S.S. Lakshminarayana

Department of Biology, Université de Moncton, Moncton, New Brunswick, Canada E1A 3E9

1. Introduction
2. Spiral representation of various dimensions of ecotoxicity
3. Suggested experimental approach to the ecotoxicity spiral
3.1. Sampling
3.2. First-level analyses of environmental protection
3.2.1. Physicochemical characteristics
3.2.2. Microbiological characteristics
3.2.3. Lethal ecotoxicity
3.3. Second-level analyses of environmental protection
3.3.1. Short-term sublethal ecotoxicity
3.3.2. Middle-term sublethal ecotoxicity
3.4. Third-level analyses of environmental protection
3.4.1. Persistence and biodegradability of the ecotoxicity
3.4.2. Bioaccumulation of ecotoxicity
3.4.3. Genotoxicity
3.4.4. Estimation of chronic ecotoxicity
3.5. Integration
4. Operational procedure of the ecotoxicity spiral for effluent analyses
5. Conclusion
References

1. INTRODUCTION

Over the past thirty years several environmental disasters (involving mercury in Minimata, Japan; dioxins in Seveso, Italy; contaminants in Love Canal, United States; acid rain, and so on) have increased the public perception of the effects of toxic wastes on the environment. Increased public concern, as an outcome of new media and pressure groups, has gradually spurred more research on various ecotoxicological phenomena and new legislation aimed at environmental protection.

Environmental managers are aware that the information disseminated to the public will be of prime importance in reducing pollution. There are few proper ways of illustrating complex environmental issues and promoting the dissemination of balanced information. To better understand and appreciate the potential impact, ecotoxicity is conceptualized as progressing spirally.

2. SPIRAL REPRESENTATION OF VARIOUS DIMENSIONS OF ECOTOXICITY

Three levels of environmental protection, separated by two perpendicular axes, are depicted in the spiral (Fig. 12.1). The first axis separates lethal from sublethal effects, while the second separates directly acting phenomena (with acute and subacute toxic effects) from insidiously acting phenomena leading to chronic toxic effects. The two axes allow us to define the first, second, and third levels of environmental protection on the local, regional, and global scales and help us visualize and understand the cause and degradational impact of ecotoxicants (Table 12.1). Thus, at the first level, local impact will be considered, as well as the initial corrective methods to be applied. Within this perspective, causes of toxicity (contaminants, pathogens, etc.) and lethal effects are identified. At the second level, regional impact will be weighed against the assimilative potential of the receiving water. Insidious chronic effects make up the third level of environmental concern, and will further help us to take the necessary correctional steps, depending upon existing knowledge of aggression.

We considered all the aspects of the three levels of environmental protection, using three examples of the effects of toxic wastes (sulfuric acid, phenol, and PCB) released into the aquatic environment (Table 12.2), to study the various potential ecotoxic effects.

The spiral concept of ecotoxicity means that we must consider not only the concentration of the toxic substance, but also factors such as the scope of its toxicity and its persistence, biotransformation, and/or bioaccumulation. We tested this concept, using liquid wastes from dumps to determine its applicability with regards to ecotoxic testing and assessment at an acceptable cost.

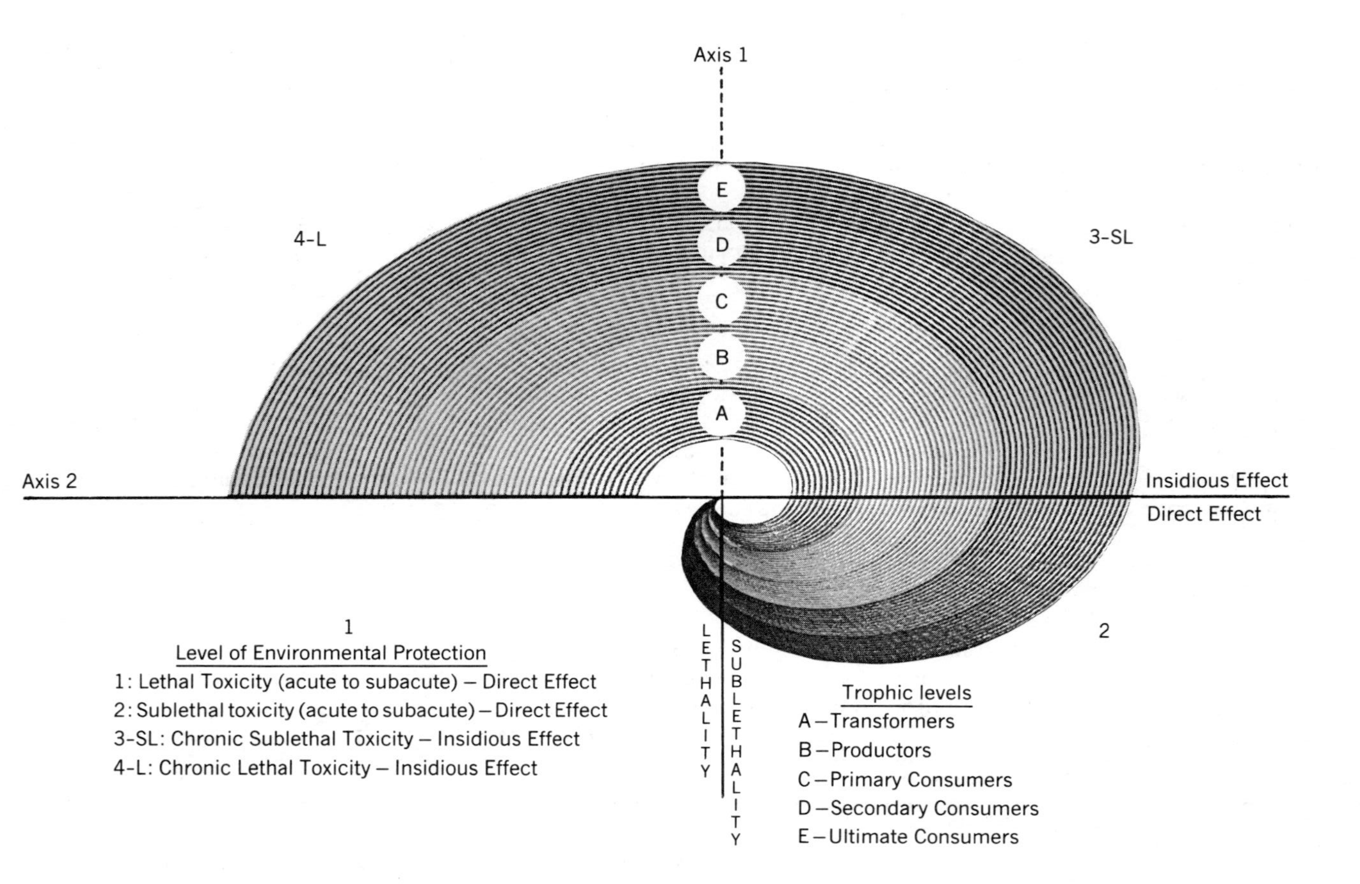

Figure 12.1. The ecotoxicity spiral.

Table 12.1 Three Levels of Environmental Protection for Aquatic Ecotoxicity

	Ecotoxic Impact		Environmental Degradation		
Level	Level of Toxicity	Scope	Cause	Consequences	Suggested Correction
First	Acute to subacute lethal toxicity	Local	Excessive presence of pollutants (contaminants, pathogenic agents, disturbed physicochemical conditions, etc.)	Nearly complete disappearance of biotic components of the environment, with the exception of ultratolerant species, (e.g., certain bacteria)	Primary treatment of waste
Second	Acute to subacute sublethal toxicity resulting from easily quantifiable, directly acting phenomena	Regional	Sublethal pollution causing an unbalance in the aquatic environment; these substances may be biodegradable, but not bioaccumulable: hence, safely assimilable.	Disturbance of biological diversity, and physiological stress of specific duration	Secondary biological treatment of wastewater, considering the assimilative capacity of the receiving environment.
Third	Chronic toxicity resulting from insidiously and dynamically acting phenomena, relatively difficult to quantify	Global	Insidious toxicity, often difficult to assess, not safely assimilated by the receiving environment, and acts through mecanisms such as bioaccumulation and/or genotoxicity	Reduction in long-term survival, growth rate, and reproduction of species; destabilization of structures and functions of aquatic ecosystems, with possible reoccurrence of lethal effects; introduction of genotoxic problems	Elimination of pollutants at source, following hazard assessment taking into account socioeconomic considerations

Table 12.2 Possible Toxic Effects of Three Different Environmental Discharges with or Without Treatment of Waste

Toxic substances released	Without treatment	Primary Treatment (e.g., neutralization of pH and removal of precipitate)	Secondary Treatment (e.g., biological treatment and removal of sludge)
H_2SO_4	Acute lethal toxicity at high concentration (1000 mg/1; Nriagu, 1978) Acute sublethal toxicity (10/1; Nriagu, 1978), Insidious toxicity, because of possibility of increase bioavailability of metals in the environment due to change in pH favoring bioaccumulation	Possibility of lingering weak sublethal toxicity, easily assimilated by the receiving environment Precipitate to be recycled	Useless for this type of pollutant
Phenol	Acute lethal toxicity at high concentration (10 mg/L; EPA, 1981) Acute sublethal toxicity at weak concentration (2000 to 10 μg/L; EPA, 1981) Insidious toxicity, because of possibility of formation of chlorophenols whose bioaccumulation will have organoleptic and possibly mutagenic consequences		Possibility of lingering weak sublethal toxicity, easily assimilated by the environment Biological sludge to be incinerated, buried or digested
PCB (Polychlorobiphenyls)	PCBs are toxic at very low levels; 10 to 100 μg/L can produce lethal effects, 1 to 10 μg/L affect reproduction, and 0.01 to 0.1 μg/L are noxious, owing to bioaccumulation (Roberts et al., 1979). Secondary treatment can be effective to concentrate the PCBs in biological sludge. However, the environmental problem remains, since PCBs are not easily biodegraded. Elimination at source is the only effective way to control these pollutants.		

Table 12.3 Physicochemical Quality of Effluents of Four Sites

Parameters or Contaminants	Site A	Site B	Site C	Site D	Water Favorable for Aquatic Life[a]
			General Parameters		
pH	7.2	7.1	7.3	6.9	6.5–8.0
Alkalinity (mg $CaCO_3$/L)	3053	515	645	1820	500
Radioactivity (mBq)	0.9	0.9	1.3	1.1	20
Calcium (mg/L)	64	77	106	154	100
Magnesium (mg/L)	55	33	50	88	100
Sulfides (μg/L)	16	42	10	680	20
Ammonia (mg/L)	386	34	8	94	0.02
Nitrites (mg/L)	0.23	0.25	1.58	0.39	1
Nitrates (mg/L)	14.5	1.8	7.3	6.9	5
Carbon (inorganic) (mg/L)	863	116	179	302	500
Carbon (organic) (mg/L)	282	184	39	709	30
Conductivity (μS/cm25°C)	8867	2133	4400	4100	1500
COD (mg O_2/L)	1086	404	94	1680	10
BOD (mg O_2/L)	155	283	6	1503	10
Hardness (mg/L)	385	327	470	747	120
Dissolved oxygen (mg/L)	nd	nd	nd	nd	4.0
Phosphorus (mg/L)	2.21	0.83	0.09	0.03	0.1
Suspended solids (mg/L)	83	51	8	61	25
Total solids (mg/L)	4143	1420	3363	3353	1000
			Inorganic Contaminants		
Mercury (μg/L)	0.3	0.3	0.3	0.3	0.1
Silver (μg/L)	<10	<10	<10	<10	10
Arsenic (μg/L)	<5	<5	<5	10	50
Barium (mg/L)	1.0	<0.3	<0.3	<0.3	1
Cyanides (μg/L)	8200	180	377	183	5
Cadmium (mg/L)	20	14	13	26	0.2
Chromium (mg/L)	0.09	0.04	0.04	0.05	0.04
Copper (μg/L)	<20	<20	<20	<20	5
Iron (mg/L)	7.0	15.1	0.4	13.5	0.3
Manganese (mg/L)	0.2	2.7	0.7	3.3	0.2
Nickel (μg/L)	100	<40	<40	<40	25
Lead (mg/L)	<0.1	<0.1	<0.1	<0.1	0.03
Selenium (μg/L)	<5	<5	<5	<5	5
Zinc (mg/L)	0.13	0.04	0.04	0.03	0.03
			Organic contaminants		
Oils and fats (mg/L)	47	2	18	3	0.2
Phenols (μg/L)	46	88	16	752	1

Table 12.3 Physicochemical Quality of Effluents of Four Sites (continued)

Parameters or Contaminants	Site A	Site B	Site C	Site D	Water Favorable for Aquatic Life[a]
Polychlorinated biphenyls (PCBs) (μg/L)					
Chlorinated pesticides	0.4	0.01	0.1	0.1	0.01
Lindane	0.02	—	—	—	0.01
Methoxychlor	0.1	0.1	0.1	0.1	0.03
Toxaphene	0.4	0.4	0.4	0.4	0.05
Endrin	0.04	0.04	0.04	0.04	0.02
Chlorinated phenoxy acids					
2.4D	2	2	2	2	1
2.4, 5T	1	1	1	1	—
Pesticides with triazine (μg/L)					
Atrazine	0.2	0.2	0.2	0.2	—
Cyanazine	0.4	0.4	0.4	0.4	—
Organophosphate pesticides (μg/L)					
Dimethoate	0.2	0.2	0.2	2	—
Azinphosmethyl	3	3	3	3	—
Phosmet	4	4	4	4	—
Carbamate pesticides (μg/L)					
Carbofurane	20	20	20	20	—
Aliphatic volatile hydrocarbons (μg/L)					
Dichloromethane	—	—	80	92	—
Trichloromethane (Chloroform)	—	—	18	—	—
Trichloethylene	—	—	—	10	—
Aromatic volatile hydrocarbons (μg/L)					
Benzene	—	—	—	13	—
Chlorobenzene	—	—	—	—	—
Dichlorobenzene	—	—	—	—	—
Methylbenzene (Toluene)	—	—	—	155	—
Dimethylbenzene (Xylene)	—	—	—	10	—
Ethylbenzene	—	—	—	—	—
Unidentifiable volatile hydrocarbons	150	—	—	—	—

[a]Environment Canada (1980a).

3. SUGGESTED EXPERIMENTAL APPROACH IN THE ECOTOXICITY SPIRAL

3.1. Sampling

Exhaustive experimental work was carried out with the samples of effluents from four different dumps; their ecotoxicologic features have already been published by Van Coillie et al. (1183a). Only essential data from this report were taken into consideration to meet the objectives of our present study. To maintain the confidentiality of the four sample sites, they were named A, B, C, and D. They were chosen to represent city–industrial, domestic, agricultural, and mixed sources, respectively. Triplicate samples were collected at each site after sufficient mixing, and *in situ* tests were conducted for temperature, pH, salinity, and dissolved oxygen. The sample aliquots were kept at 4°C in different containers with proper preservatives, and carried to the laboratory for further chemical analyses as recommended by the EPA (1979), Environnement Quebec (1979, 1980), Environment Canada (1979a, 1981), and APHA et al. (1980). Separate samples were taken in sterile glass bottles for bacteriological analyses (APHA et al., 1980); the samples were analyzed within 24 hours, while the other chemical and bioanalyses were undertaken within 48 hours.

3.2. First-Level Analyses of Environmental Protection

3.2.1. Physicochemical Characteristics

All chemical parameters were analyzed according to the Environment Canada (1979a), manual except mercury (Environnement Quebec, 1979), triazines and organophosphates (EPA, 1979), and volatile hydrocarbons (mass spectrometry, *Federal Register*, 1979). The average values for each parameter at each sampling station were compared with those of the favorable water quality criteria as given by Environment Canada (1980a) (Table 12.3). From this, the anomalous concentrations of some parameters (ammonia, cyanides, phenols, cadmium, oil and fats, iron, chemical oxygen demand, biochemical oxygen demand, volatile hydrocarbons) at all four sampling sites can be picked up (Table 12.4). When we combine the specific degrees of physicochemical excess for each of the four sites, the overall picture becomes:

	Site A	Site B	Site C	Site D
Sum (Flagrant degrees of excess) (>10×)	21,545×	2035×	656×	6097×
Sum (Relative degrees of excess) (<10×)	40×	22×	28×	71×
Sum (Physicochemical degrees of excess)	21,585×	2057×	684×	6118×

From these results the physicochemical degree of excess of the samples could be considered to be extremely elevated at site A, very high at sites B and D, and high at site C. However, these conclusions are limited to the time and space of sampling and should be considered with prudence.

Table 12.4 Flagrant Degrees of Excess of Physicochemical Parameters

Sites		Contaminants	Flagrant degrees of Excess in the Effluents (>10×) in Relation to Natural Waters[a]
A	Ammonia	(386 mg/L)	19,300×
	Cyanides	(8200 μg/L)	1,640×
	Oils and fats	(47 mg/L)	235×
	COD	(1086 mg O_2/L)	109×
	Cadmium	(20 μg/L)	100×
	Phenols	(46 μg/L)	46×
	Polychlorobiphenyls	(0.4 μg/L)	40×
	Phosphate	(2.2 mg/L)	22×
	Iron	(3.0 mg/L)	22×
	BOD	(155 mg O_2/L)	15×
	Volatile hydrocarbons	(150 μg/L)	15×[b]
B	Ammonia	(34 mg/L)	1,700×
	Phenols	(88 μg/L)	88×
	Cadmium	(14 μg/L)	70×
	Iron	(15 mg/L)	50×
	COD	(404 mg O_2/L)	40×
	Cyanides	(180 μg/L)	36×
	BOD	(283 mg O_2/L)	28×
	Manganese	(2700 μg/L)	13×
	Oils and fats	(2 mg/L)	10×
C	Ammonia	(8 mg/L)	400×
	Oils and fats	(18 mg/L)	90×
	Cyanides	(377 μg/L)	75×
	Cadmium	(13 μg/L)	65×
	Phenols	(16 μg/L)	16×
	Volatile hydrocarbons	(98 μg/L)	10×[b]
D	Ammonia	(94 mg/L)	4,700×
	Phenols	(752 μg/L)	752×
	COD	(1680 mg O_2/L)	168×
	BOD	(1503 mg O_2/L)	150×
	Cadmium	(26 μg/L)	130×
	Iron	(13.5 mg/L)	45×
	Cyanides	(183 μg/L)	37×
	Sulfides	(680 μg/L)	34×
	Volatile hydrocarbons	(260 μg/L)	26×[b]
	Organic carbon	(709 mg/L)	24×
	Manganese	(3300 μg/L)	16×
	Oils and fats	(3 mg/L)	15×

[a] Standard information: see Environment Canada (1980a).

[b] Different volatile hydrocarbons revealed mutations at or near 10 μg/L (Rokosh and Lovasz, 1979) and we took this value as our criteria for comparison.

3.2.2. *Microbiological Characteristics*

Microbiological quality of the wastewaters at the four sampling sites was determined according to methods reported by Washington et al. (1971) and APHA et al. (1980). We also tested the samples for the presence of *Pseudomonas aeruginosa*, an adventitious pathogen (Santos Ferreira, 1977), *Aeromonas hydrophila*, a well-known high pathogen (Shotts and Rimler, 1973), and *Proteus* species (responsible for gastrointestinal and urinary infections), which were subjected to sensitivity tests with eight clinically used antibiotics (Tenant et al., 1975). The microbiological characteristics of the samples at the four sites showed (Table 12.5):

1. Numerous heterotrophic bacteria, indicative of biodegradation potential (Blaise, 1973); 20°C to 35°C count ratios, suggesting a predominance of psychrophiles (Stokes and Redmond, 1966).
2. Varying degrees of bacterial contamination and the presence of specific pathogen showing antibiotic resistance.

Table 12.5 Bacterial Quality of Effluents of the Four Sites

Indicators or Species	Site A	Site B	Site C	Site D	Drinking Water and/or Natural Water Standards[a]
Total coliforms (*n*/100 mL)	223	13	397	<10	10
Fecal coliforms (*n*/100 mL)	<10	<10	83	<10	1–10
Fecal streptococci (*n*/100 mL)	170	13	47	10	1–10
Heterotrophic bacteria (*n*/100 mL at 35°C)	114×10^5	3.6×10^5	32×10^5	1×10^5	
Pseudomonas aeruginosa (*n*/100 mL)	1600	37	666	5	0
Aeromonas hydrophila (*n*/mL)	100	100	100	100	0
Colonies of *Proteus* (*n*)	32	—	18	—	0
Bacterial resistance to 3 antibiotics or more	100%	—	28%	—	—

[a]Environment Canada (1980a).

Table 12.6 Potential Lethal Ecotoxicity for the Effluents at the Four Sites

	Site A	Site B	Site C	Site D
LC_{50}[a]	2%	25%	nd	8%
TU_l[b]	50	4	nd	12

[a] LC_{50} 96 h: The lethal concentrations were calculated after the "Probit" calculation (Stephan, 1977).
[b] Toxic lethal units = 100%/LC_{50}.

3. The microbiological degrees of excess at sites A, B, C, and D were 1671, 40, 337, and 5, respectively.

3.2.3. Lethal Ecotoxicity

To verify whether the physicochemical and bacteriological contaminants in the samples at the four sites were capable of producing lethality, trout (*Salmo gairdneri*) bioassays were conducted (Environment Canada, 1980b). Lethality was decreasingly observed at sites A, D, and B, respectively. Site C waters, on the other hand, proved to be nonlethal in the fish test (Table 12.6).

3.3. Second-Level Analyses of Environmental Protection

After identifying chemical and microbiological contaminants and their potential as lethal ecotoxicants at the four sampling sites, it was important to consider their impact on receiving waters. The effluents at the four sampling sites can be diluted to produce no more than sublethal effects; we have given particular attention to this aspect in our studies.

A simple, effective program of sublethal ecotoxic bioassays utilizing different physiological parameters was conceived to test the diluted wastewaters. Trophic and exposure levels were considered to be two important variables in this respect.

1. *Exposure*. In aquatic toxicology, short-, medium-, and long-term toxicities must be distinguished. Sprague (1971), EPA (1973), APHA et al. (1980), Maciorowski et al. (1981), and others have shown that short- and medium-term bioassays have duration of 4 and 20 days, respectively. Our sublethal tests took into account these exposure periods.

2. *Ecological stage*. Often, toxicants will not affect different aquatic trophic levels with equal intensity (EPA, 1973); for instance, bacterial, phytoplankton, zooplankton, and consumers (fish). Ecotoxicity testing with different biological indicators is therefore recommended.

3.3.1. Short-Term Sublethal Ecotoxicity

The following biotests were used to demonstrate acute sublethal effects of water samples:

Trophic level	Biological indicator	Duration of the bioassay	Physiological parameter examined	References
Bacteria	*Photobacterium phosphoreum*	5 min	Bioluminescence 15°C	Beckman Instruments (1980)
Phytoplankton (algae)	*Chlamydomonas variabilis*	24 h	Mobility 24°C	Van Coillie et al. (1982)
			Fluorescence 24°C	Couture et al. (1982)
	Selenastrum capricornutum	24 h	Fluorescence 24°C	Couture et al. (1982)
Zooplankton	*Daphnia pulex*	96 h	Mobility 24°C	Environment Canada (1979b)
Fish	*Salmo gairdneri*	96 h	Swimming capacity 15°C	Webb and Brett (1973) Thellen and Van Coillie (1982)

The results of these bioassays (Table 12.7) showed that:

(a) The ecotoxicological sensitivity was unequal for the different types of short-term sublethal bioassays. As a result, when toxic units were totaled for each type of bioassay, decreasing biotest sensitivity was as follows:

Daphnia pulex (mobility, 96 h, 24°C), 37.8 TU_{sl}
Photobacterium phosphoreum (bioluminescence, 5 min, 15°C); 29.5 TU_{sl}
Chlamydomonas variabilis (mobility, 24 h, 24°C), 19.9 TU_{sl}
Chlamydomonas variabilis (fluorescence, 24 h, 24°C): 15.9 TU_{sl}
Selenastrum capricornutum (fluorescence, 24 h, 24°C): 7.5 TU_{sl}
Salmo gairdneri (swimming capacity, 96 h, 15°C); 0 TU_{sl}

Sublethal ecotoxicological sensitivity varied, therefore, with the ecological level studied, the species tested, and the physiological parameter examined. Such results demonstrate that it is essential to perform several types of sublethal bioassays.

Table 12.7 Potential Short-Term Sublethal Ecotoxicity Effluents at the Four Sites

Biological Indicators (Physiological Parameters)	Site A	Site B	Site C	Site D	Total
Photobacterium phosphoreum (Bioluminescence) 15°C					
IC_{50} 5 min	8.8%	76.9%	nd	5.9%	
TU_{sl}	11.3	1.3	nd	16.9	29.5
Chlamydomonas variabilis (Fluorescence) 24°C					
IC_{50} 24 h	14.9%	94.7%	nd	12.4%	
TU_{sl}	6.7	1.1	nd	8.1	15.9
Clamydomonas variabilis (Mobility) 24°C					
IC_{50} 24 h	8.5%	nd	nd	12.3%	
TU_{sl}	11.8	nd	nd	8.1	19.9
Selenastrum capricornutum (Fluorescence) 24°C					
IC_{50} 24 h	69.0%	59.1%	69.0%	33.8%	
TU_{sl}	1.4	1.7	1.4	3.0	7.5
Daphnia pulex (Mobility) 24°C					
IC_{50} 96 h	4.1%	80.0%	nd	8.2%	
TU_{sl}	24.4	1.2	nd	12.2	37.8
Salmo gairdneri (Swimming capacity) 15%					
IC_{50} 96 h	nd	nd	nd	nd	
TU_{sl}	nd	nd	nd	nd	0

Note: nd, not detected. IC_{50}, inhibitory concentrations; calculated after the "Probit" calculation (Stephan, 1977). TU_{sl}, sublethal toxic units = 100%/IC_{50}.

(b) Short-term sublethal ecotoxic potential was different in waste waters at the four sites studied; the values were:

	Site A	**Site B**	**Site C**	**Site D**
Sum of TU_{sl}:	53.6	5.3	1.4	48.3
Midpoint of TU_{sl} for estimating short-term sublethal ecotoxic potential	11	1	0	10

*TU_{sl}, sublethal toxic units.

3.3.2. Medium-Term Sublethal Ecotoxycity

Could there be a manifestation of delayed toxicities after many days of entry of the dump effluents into the receiving water environments? To answer this question, medium-term sublethal ecotoxicological tests (8 to 20 days of exposure) were undertaken.

Trophic level	Biological indicator	Duration of the bioassay	Physiological parameter examined	References for methods
Phytoplankton	*Selenastrum capricornutum*	8 days	Growth 24°C	Joubert (1980)
Phytoplankton	*Selenastrum capricornutum*	8 days	Fluorescence 24°C	Couture et al. (1982)
Zooplankton	*Daphnia pulex*	14 days	Mobility 24°C	Environment Canada (1979b)
			Reproduction 24°C	OECD (1980)
Fish	*Salmo gairdneri*	20 days	Swimming capacity 15°C	Webb and Brett (1973) Thellen and Van Coillie (1982)

Results of the above tests (Table 12.8) when compared to those given in Table 12.7 indicated the following:

(a) As before (Table 12.7), the medium-term sublethal ecotoxicological sensitivity varied with the ecological level of species and parameters studied, but the order of variation differed:

Selenastrum capricornutum (growth, 8 d, 24°C); 427.8 TU_{sl}
Selenastrum capricornutum (fluorescence, 9 d, 24°C); 211 TU_{sl}
Salmo gairdneri (swimming capacity, 20 d, 15°C); 208 TU_{sl}
Daphnia pulex (reproduction, 14 d, 24°C); 60 TU_{sl}
Daphnia pulex (mobility, 14 d, 24°C); 36 TU_{sl}

(b) The waste waters of sites A and D proved to be more toxic at the short-term and medium-term exposure than those of sites B and C, as shown below:

	Site A	Site B	Site C	Site D
Total of TU_{sl}	338.6	148.3	7.3	348.3
Midpoint of TU_{sl} for estimating the medium-term sublethal ecotoxicity (compare with 3.31b)	70	30	1	70

Table 12.8 Potential Middle-Term Sublethal Ecotoxicity for Effluents at the Four Sites

Biological Indicators (Physiological Parameters)	Site A	Site B	Site C	Site D	Total
Selenastrum capricornutum (Growth) 24°C					
IC_{50} 8 days	2.0%	0.8%	35.7%	0.4%	427.8
TU_{sl}	50	125	2.8	250	
Selenastrum capricornutum (Fluorescence) 24°C					
IC_{50} 8 days	2.5%	<30.0%	60.0%	0.6%	211
TU_{sl}	40.0	>3.3	1.7	166.0	
Daphnia pulex (Mobility) 24°C					
IC_{50} 14 days	3.2–5.6%	<10%	50–100%	nd	
TU_{sl}	17.9–31.2	>10	1.0–1.8	nd	>28.9–43.0 mean 36
Daphnia pulex (Reproduction) 24°C					
IC_{50} 14 days	3.2–5.6%	<10%	56–100%	3.2–5.6%	46.8–74.2
TU_{sl}	17.9–31.2	>10	1.0–1.8	17.9–31.2	mean 60
Salmo gairdneri (Swimming capacity) 15°C					
IC_{50} 20 days	0.5%	nd	nd	12.0%	
TU_{sl}	200	nd	nd	8	208

Note: nd, not detected. IC_{50}, inhibitory concentrations; calculated after the "Probit" calculation (Stephen, 1977). TU_{sl}, sublethal toxic units = 100%/IC_{50}.

(c) Sublethal ecotoxicity was more intense at medium-term compared to short-term particularly for the fish and algae:

	Short term	**Middle term**
TU_{sl} of 4 sites for *Selenastrum capricornutum* (fluorescence, 24°C)	7.5	211
TU_{sl} of 4 sites for *Daphnia pulex* (mobility) 24°C	37.8	36
TU_{sl} of 4 sites for *Salmo gairdneri* (swimming capacity, 15°C)	0	208
	45.3	455

3.4. Third-Level Analyses of Environmental Protection

Because of the persistence of toxicants, the sublethal ecotoxicity induced by the diluted effluents at the four sites has greater long-term consequences than those accepted for the second level of environmental protection. Second-level effects

Table 12.9 Biodegradation of the Ecotoxicity of Effluents of the Four Sites

Biological Indicators (Physiological Parameters)		Site A	Site B	Site C	Site D
Photobacterium phosphoreum (Bioluminescence) 15°C					
IC_{50} 5 min	Day 0 before	8.8%	76.9%	nd	5.0%
TU_{sl}	biodegradation	(11.3)	(1.3)	(nd)	(16.9)
IC_{50} 5 min	Day 31 after	20%	nd	nd	90.0%
TU_{sl}	biodegradation	5.0	nd	nd	1.1
Initial toxicity (Day 0) / Final toxicity (Day 31) =		2.3×	1.3×	nd	15.4×
Selenastrum capricornutum (Fluorescence) 24°C					
IC_{50} 8 days	Day 0 before	2.5%	<30%	60%	0.6%
TU_{sl}	biodegradation	(40.0)	>(3.3)	(1.7)	(166.0)
IC_{50} 8 days	Day 31 after	10%	100%	100%	42%
TU_{sl}	biodegradation	(10.0)	(1.0)	(1.0)	(2.4)
Initial toxicity (Day 0) / Final toxicity (Day 31) =			>3.3×	1.7×	69.2×
Salmo gairdneri (Survival) 15°C					
LC_{50} 4 days	Day 0 before	2.0%	25.0%	nd	8.0%
TU_{l}	biodegradation	(50)	(4.0)	(nd)	(12.0)
IC_{50} 4 days	Day 31 after	4.4%	nd	nd	14.5%
TU_{l}	biodegradation	(22.6)	(nd)	(nd)	(6.9)
Initial toxicity (Day 0) / Final toxicity (Day 31) =		2.2×	4.0×	nd	1.7×

Note: nd, not detected. IC_{50} and LC_{50}, inhibitory concentrations at 50% (a physiological parameter or lethal concentration for 50% of individuals during experimental exposure for each type of test. TU_{sl} and TU_{l}, sublethal and lethal toxic units = 100%/IC_{50} and 100%/LC_{50} respectively.

are likely to be "acute" or "subacute" compared to the long-term effects, which are often chronic (see Section 2; Fig. 12.1). Actually, the distinction between the two types of effects appears fairly general and is debatable in terms of the multiple significance accorded to them in ecotoxicology today. But chronic ecotoxicity should be investigated at a third level of environmental protection after testing for medium-term sublethal ecotoxicity at the second level.

Long-term testing is complicated and expensive. However, it is possible to conduct a minimum number of short bioassays to establish the presence of different phenomena, and then to estimate their importance.

3.4.1. Persistence and Biodegradability of the Ecotoxicants

Biodegradability tests were undertaken to verify whether the ecotoxicity of the wastewaters of the four sites changed with the activity of microbial agents. The microbes consist of psychrophilic heterotrophic bacteria, which are well known as agents of biodegradability (Blaise, 1973). To demonstrate whether the ecotoxicity of the samples could be modified by the metabolic action of similar microbes, the samples were enriched with psychrophilic bacteria originating from three wastewaters that were similar in chemical and microbiological characteristics to those of the four sites under investigation. Microbial seed present in the three domestic wastewaters was collected and gradually acclimatized to the samples. One liter of the microbial seed was added to 200 liters of the sample. The biodegradability tests were carried out for 31 days at 15°C with a photoperiod of 16 h light and 8 h dark, continuous aeration, and manual weekly agitation. Waters used for rainbow trout were chosen as controls. At the end of each experiment, the ecotoxicity of the samples were determined with the following organisms: *Photobacterium phosphoreum*, *Selenastrum capricornutum*, and *Salmo gairdneri*. Table 12.9 compares the resulting ecotoxicity with that initially shown (See Sections 3.2.3, 3.3.1, and 3.3.2) and indicates that the ecotoxicity of waters of the four sites decreased following biodegradation of toxicants. Furthermore, the decrease varied with the biological indicators and sites, as revealed below, in a table summarizing the difference between the toxicity before and after the biodegradation of the samples at the 4 sites (see Table 12.9 for details):

Biological indicators (physiological parameters; duration of test; temperature)	**Site A**	**Site B**	**Site C**	**Site D**	**Mean**
Photobacterium phosphoreum (Bioluminescence; 5 min; 15°C)	2.3×	1.3×	nd	15.4×	4.7×
Selenastrum capricornutum (Fluorescence; 8 days; 24°C)	4.0×	>3.3×	1.7	69.2×	19.6×
Salmo gairdneri (Survival; 4 days; 15°C)	2.2×	4.0×	nd	1.7×	2.1×
Midpoint	2.8×	>2.9×	nd	28.8×	8.9×

3.4.2. Bisaccumulation of Ecotoxicants

Could the toxic agents present in the effluents not only act directly on the species of different trophic levels of receptor aquatic environments, but also bioaccumulate and later transfer to different trophic levels? To test for this possibility we adopted the method given by Blaise et al. (1982).

Volumes of culture of the alga *Selenastrum capricornutum*, which were in exponential phase of growth at 24°C, were mixed (1 : 1) with volumes of the samples from the four sites. After 24 h we recovered the algae by centrifugation, then the cell walls of the algae were ruptured by ultrasound, and we extracted the internal cellular lysates. The ecotoxicity of the extracts was tested using *Photobacterium phosphoreum* (see Section 3.3.2). The experiments were limited to bioaccumulation of ecotoxicants at the primary trophic level. The results were expressed in the form of bioaccumulation toxic factor (i.e., the ratio of the toxic units found in the algae to the toxic units of the samples tested):

$$\text{BF} = \frac{(\text{TU of contaminated lysate} - \text{TU of control lysate})/\text{g of algae}}{\text{TU of the sample tested/mL}}$$

Table 12.10 shows the bioaccumulation of ecotoxicants in effluents of the four sites. The toxicity of the waters of sites was bioaccumulated in the algae in different degrees as follows:

- very pronounced in sites B and C.
- moderately pronounced at sites A and D in decreasing order.

The bioaccumulation of heavy metals in the algae is slow in the wastewaters (14 days required for a reduction of 50% of their initial concentration, according to Becker, 1983.) Thus, it is probable that the toxic agents bioaccumulated during 24 h in the effuents of dumpsites were principally composed of relatively stable organics. However, we cannot ignore the fact that the heavy metals in the effluents can bioaccumulate and adversely affect the algae in the receiving ecosystem as reported by Rai et al. (1981) and De la Noue et al. (1983).

3.4.3. Genotoxicity

Some waste effluents are believed to be genotoxic (see Rokosh and Lovasz, 1979). To verify whether the effluent waters of the four sites contain a genotoxic potential, mutagenicity tests using six mutant strains of *Salmonella typhimu-*

Table 12.10 Results Related to Bioaccumulation Test

	Site A	Site B	Site C	Site D
Accumulation of the ecotoxicity in TU_{sl}/g algae after 24 h exposure	7400	9,800	4,200	8000
Factors of bioaccumulation	2200	30,000	21,000	1900

Table 12.11 Estimation of Chronic Ecotoxicity of Effluents of the Four Sites

Chronic Aspect—Ecotoxicity after Biodegradation of 31 days for:	Chronic Toxic Units			
	Site A	Site B	Site C	Site D
Photobacterium phosphoreum[a]	5	—	—	1.1
Selenastrum capricornutum[a]	10	1	1	2.4
Salmo gairdneri[b]	71.2	—	—	21.7
Midpoint (1)	28.7	0.3	0.3	8.4
Bioaccumulation of ecotoxicity in mg/algae[c] (2)	7.4	9.8	4.2	8.0
Ecotoxicity of genotoxic potential[d] (3)	—	5.2	—	—
Midpoint (1), (2), (3)	12.0	5.1	1.5	5.5

[a] dee Table 12.9.
[b] Table 12.9 mentioned lethal toxic units for *Salmo gairdneri* after biodegradation. The transfer to chronic toxic units was estimated basing on the relation between the total TU_{sl} at middle term and TU_l for this species: 208/66 = 3.15 (see Sections 3.2.3 and 3.3.2).
[c] Bioaccumulated sublethal toxic units in the contaminated lysate of *Selenastrum capricornutum* after bioaccumulation, which was tested on *Photobacterium phosphoreum* (see Table 12.10).
[d] The mutagenic concentration of the effluents of site B was four times greater than LC_{50}, which is similar to results of Rapson et al. (1980). We based our calculation on this statement for a mutagenic potential at 50% of water from site B (see Section 3.4.3) from their LC_{50} (see Section 3.2.3). Next, for passing the TU_l level to TU_{sl} we considered the relation between the total of TU_{sl} middle-term means to mean of TU_l: 171/66 = 2.6 (see Sections 3.2.3 and 3.3.2).

rium for histidine synthesis were done according to the recommended techniques (Ames et al., 1975). The site B samples showed, on an average, 3 genetic mutations in the 6 strains. No mutagenic manifestation was observed in the other samples.

3.4.4. Estimation of Chronic Ecotoxicity

The three series of tests mentioned in Section 3.4 suggest that the ecotoxicity of the effluents at the four sites has a chronic dimension. Ecotoxicity persists after 31 days of biodegradations (see Table 12.9) and, moreover, the toxicants can bioaccumulate and provoke genotoxicity in biological organisms.

Table 12.11 gives an estimation of the chronic ecotoxicity of effluents based on toxic units of the three series of tests. From Table 12.11 the following two inferences can be made:

(a) The chronic ecotoxicity of effluents of the four sites after the three bioassays showed the total chronic toxic units (TU_c) in the following decreasing order:

Bioassay for biodegradability. Ecotoxicity persistent (total of midpoints for 3 indicators): 37.7 TU_c
Bioassay for bioaccumulation. Resultant ecotoxicity: 29.4 TU_c
Bioassay for mutagenicity. Estimated ecotoxicity: 5.2 TU_{sl}

(b) The chronic ecotoxicity varied for every site as follows:

	Site A	**Site B**	**Site C**	**Site D**
Midpoint of TU_c for the site	12.0	5.1	1.5	5.5

3.5. Integration

Finally we regrouped the results of the different analyses of effluents from the four sites in Table 12.12, and from this, four general statements could be made:

(a) The methodological steps in three levels adopted from the spiral of

Table 12.12 Ecotoxicological Characterization of Effluents of Four Sites

Ecotoxicologic Dimensions	Site A	Site B	Site C	Site D	Total
First-Level Environmental Protection					
Excess chemical & physicochemical	21,585×	2057×	684×	6118×	30,444×
Microbiological excess	1,671×	40×	337×	5×	2,053×
Lethal ecotoxicity (TU_l)	50	4	0	12	66
Second-Level Environmental Protection					
Midpoint of short-term sublethal toxicity (TU_{sl})	11	1	0	10	22
Midpoint of middle-term sublethal toxicity (TU_{sl})	70	30	1	70	171
Third-Level Environment Protection					
Midpoint of chronic ecotoxicity (TU_c)	12	5	1	6	24

ecotoxicity permitted nearly complete characterization of the ecotoxicological noxiousness of the effluents.

(b) Compared to the values obtained for 96-h lethal toxicity, the results obtained for the short-term sublethal toxicity (≥96 h) appeared to be too low. This is surprising because the toxic units normally increase from the lethal level (LC_{50}), to the sublethal (IC_{50}, inhibitory concentration of 50%) according to Sprague (1971). The duration of sublethal tests was probably too short for *Photobacterium phosphoreum* (5 min) or *Chlamydomonas variabilis* and *Selenastrum capricornutum* (24 h). Nevertheless, the short-term sublethal tests can be useful for prescreening. However, the medium-term sublethal bioassay can yield the toxicity values below the lethal levels.

(c) The similarity in the order of the toxicity for all the tests tends to validate the proposed methodology.

Level 1	Degrees of excess concentration for chemical parameters	$A > D > B > C$
	Degrees of excess toxicity for microbiological parameters	$A > C > B > D$
	Lethal ecotoxicity	$A > D > B > C$
Level 2	Flagrant noxiousness	$A > D > B > C$
	Short-term sublethal ecotoxicity	$A \geq D > B \geq C$
	Medium-term sublethal ecotoxicity	$A \geq D > B \geq C$
Level 3	Sublethal ecotoxicity	$A \geq D > B > C$
	Insidiuous noxiousness	$A > D > B > C$
Total	Global noxiousness	$A > D > B > C$

(d) It should be noted that certain types of ecotoxicity were particularly observed at sites B and C. Remember that the effluent of site B showed only mutagenic effects and the effluent of site C was classed in the second position for microbiological excess. We would like to point out that ecotoxicological evaluation should be made with a large spectrum of known parameters. Otherwise, certain excesses and/or toxicities are not properly revealed or are badly estimated.

Statements (a)–(d) above imply that the experimental approach for verifying the concept of the ecotoxicity spiral is pertinent. With this understanding we present an operational procedure for this approach that is relatively simple and economic.

Table 12.13 Operational Procedure for a Global Ecotoxicological Assessment

Operations Analyses Bioessays	Level of Environmental Protection	Pertinence and/or Levels of Ecotoxicity	Volume Required per Sample[a]	Methodological References	Cost[b]
		A. First Operational Step (see Fig. 12.2)			
A-1 Sampling No. 1, including *in situ* analyses (ph, *T*, etc.)	1, 2	Basic activity	N/A	Environment Canada (1979a, 1980a)	8.0 pd
A-2 Distribution and routing of subsamples, filtration, and freeze-drying of residues	N/A	In preparation for subsequent steps	N/A	Environment Canada (1979a, 1981)	1.0 pd
A-3 Microbiological analyses • Fecal coliforme • *Aeromonas hydrophila*	1, 2	Human pathogen Fish pathogen	0.5 L	APHA et al. (1980), Shotts & Rimler (1973)	1.5 pd
A-4 Biodegradability 5 d	2, 3	Chronic persistence	7.0 L	Van Coillie et al. (1983)	1.0 pd
A-5 Screening bioassays:					
• Bacteria bioluminescence 5 min *Photobacterium*	1,[c] 2	Lethal to acute sublethal		Beckman (1980)	1.5 pd
• Algae growth 4 d *Selenastrum capricornutum*	2	Acute to subacute sublethal	1.0 L f	Blaise et al. (1982)	1.0 pd
• Protozoa ingestion 1 d *Colpidium campylum*	2	Acute to subacute sublethal		Dive & Leclerc (1975)	1.5 pd
A-6 Accumulation of ecotoxicity in algae over 24 h	3	Chronic bioaccumulation		Blaise el al. (1982)	5 pd

A-7 Genotoxicity bioessays	3	Chronic mutagenicity		Ames et al. (1975)	5 pd
Mutagenicity test					
A-8 General physicochemical analyses • Ammonia • Total organic carbon • Hardness • Total and suspended solids	1, 2	Essential parameters to determine significance of toxic effects	2.5 1 and 2.0 L f	APHA et al. (1980)	3 pd
A-9 Initial assessment	1, 2, 3	Coordination and decision concerning further tests or report	N/A	N/A	1.5 pd
Sub-total A	1, 2, 3		13 L		30 pd
		B. Second Operational Step (see Fig. 12.2)			
B-1 Sampling included in A-1					
B-2 Identification of pathogenic agents and assessment of their resistance to antibiotics	1, 2, 3	Specify extent of microbial problem	N/A	Washington et al. (1971)	0–5 pd
B-3 Assessment of toxicity on samples on samples after 5 d of biodegradation	1, 2, 3	Check persistence of lethal and/or chronic ecotoxicity identified	1.0 L f Included in A-4	See A-5–A	1–16 pd

Table 12.13 Operational Procedure for a Global Ecotoxicological Assessment (continued)

Operations Analyses Bioessays	Level of Environmental Protection	Pertinence and/or Levels of Ecotoxicity	Volume Required per Sample[a]	Methodological References	Cost[b]
B-4 Chemical dosage of contaminants after standard preservation • Heavy metals • Mercury • Cyanide • Sulfide • Phenols • Nonvolatile organic substances • Others	1, 2, 3	Verify and identify chemical causes of ecotoxicity identified	5.75 L, 4.0 L, and 2.0 L f included in A-4	APHA et al. (1980)	1–17 pd and possibly $6K
B-5 Complementary assessment	1, 2, 3	Coordination and decision on further activity or report	N/A	N/A	1–2 pd
B-6 Exploratory evaluation of lethal toxicity on fish	1	Acute lethal	15 L		0–3 pd
Subtotal B lethal toxicity on fish	1, 2, 3		20.75 L		3–40 pd and possibly $6K
C. Third Operational Step					
A second sampling sequence can be planned and carried out as necessary to obtain the results of bioassays conducted at higher biological levels or to acquire pertinent information from nonperservable, costly chemical parameters such as volatile organic substances. During the second sampling sequence, the activities described in the first and second operational steps may be repeated if necessary (see Fig. 12.3)					
C-1 Sampling 2	1, 2	Basic activity	N/A	See A-1	1–13 pd
C-2 Activities described in first step	1, 2, 3	As necessary only	0–13 L	See A-2–A-8	0–22 pd

C-3 Activities described in second step	1, 2, 3	As necessary only	0–20.75 L	See B-2–B-4	0–40 pd
C-4 Ecotoxicological bioassays at higher biological levels					
• *Salmo gairdneri*					
Mortality 4 d	1	Acute lethal	400–9000 L	Environment Canada (1980b), Blaise (1984)	0–25 pd
Muscular ATP 4 d	2	Acute sublethal			0–15 pd
Muscular ATP	2, 3	Subacute sublethal leading to chronic			0–35 pd
Bioaccumulation and cytotoxicity 21 d	2, 3				
• *Daphnia pulex*					
Mobility	2	Acute sublethal	1 L	Environment Canada (1979b)	0–5 pd
Reproduction	3	Chronic			0–10 pd
C-5 Carcinogenicity bioassays					
Test with mammal or cytotoxicityité	3	Chronic	10 L		Possibly $10K
C-6 Determination with costly, nonpreservable chemicals					
• Volatile organic parameters	1, 2		40 mL	*Federal Register* (1979)	Possibly $3K
C-7 Report	1, 2, 3		N/A	N/A	1–5 pd
Subtotal C	1, 2, 3		40 mL–9000 L		5–170 and possibly $13K

[a]Volume per sample. "f" indicates that it must be filtered at 0.22*u* following which the filtrate and residue may be freeze-dried for later analysis.
[b]Cost of 10 samples in pd (person-days) and/or in financial support (K = 1000).
[c]The EC_{50} (effective concentration at 50%) determined with *Photobacterium phosphoreum* (parameter: bioluminescence in 5 min) for various products and effluents is often fairly close to their LC_{50} (lethal concentration at 50%) determined with *Salmo gairdneri* (parameter: cumulative mortality in 96 h). This has often been proven in the EPS laboratories in Longueuil (unpublished data). It follows that the bioassay with *Photobacterium phosphoreum* not only can reveal acute sublethal ecotoxicity, but is also useful in estimating lethal ecotoxicity.

4. OPERATIONAL PROCEDURE OF THE ECOTOXICITY SPIRAL FOR EFFLUENT ANALYSES

To establish this procedure, the following considerations should be taken into consideration:

1. The results for the three levels of environmental protection.
2. Limited delay between sampling, chemical and microbiological analyses, and bioassays.
3. Different types of bioassays, based on the following factors:

- Levels of ecotoxicity (lethal, sublethal, and chronic).
- Ecological level (bacteria, phytoplankton, zooplankton, fish, humans).
- Species and, if possible, chemical and physical parameters.

4. Cost of analyses and bioassays (to be carried out in triplicate for statistical reasons).

The reconciliation of all the variables could probably be realized within a methodology of three operational steps. Table 12.13 presents such an operational procedure. The decision to proceed from one operational step to another depends on the following factors:

- Observation of noxiousness at a particular step.
- Desired degree of evaluation.
- The cost in relation to the estimates of effort, shown below for 10 samples:

First sequence of sampling (35 L/sample):

Step A. 30 pd (person-days)

Step B. 3 to 40 pd, and possibly $6,000 depending upon the problems to be resolved.

Second sequence of sampling, only as needed (40 mL to 9000 L of sample):

Step C. 5 to 170 pd and possibly $13,000.

For three years, the Environmental Protection Service, Quebec Region, has been evaluating this procedure (see Figs. 12.2 and 12.3) as a practical and economical protocol. The aim of this approach is to predict the ecotoxicologic problems that may be engendered by different liquid wastes whose ecotoxicity is not known.

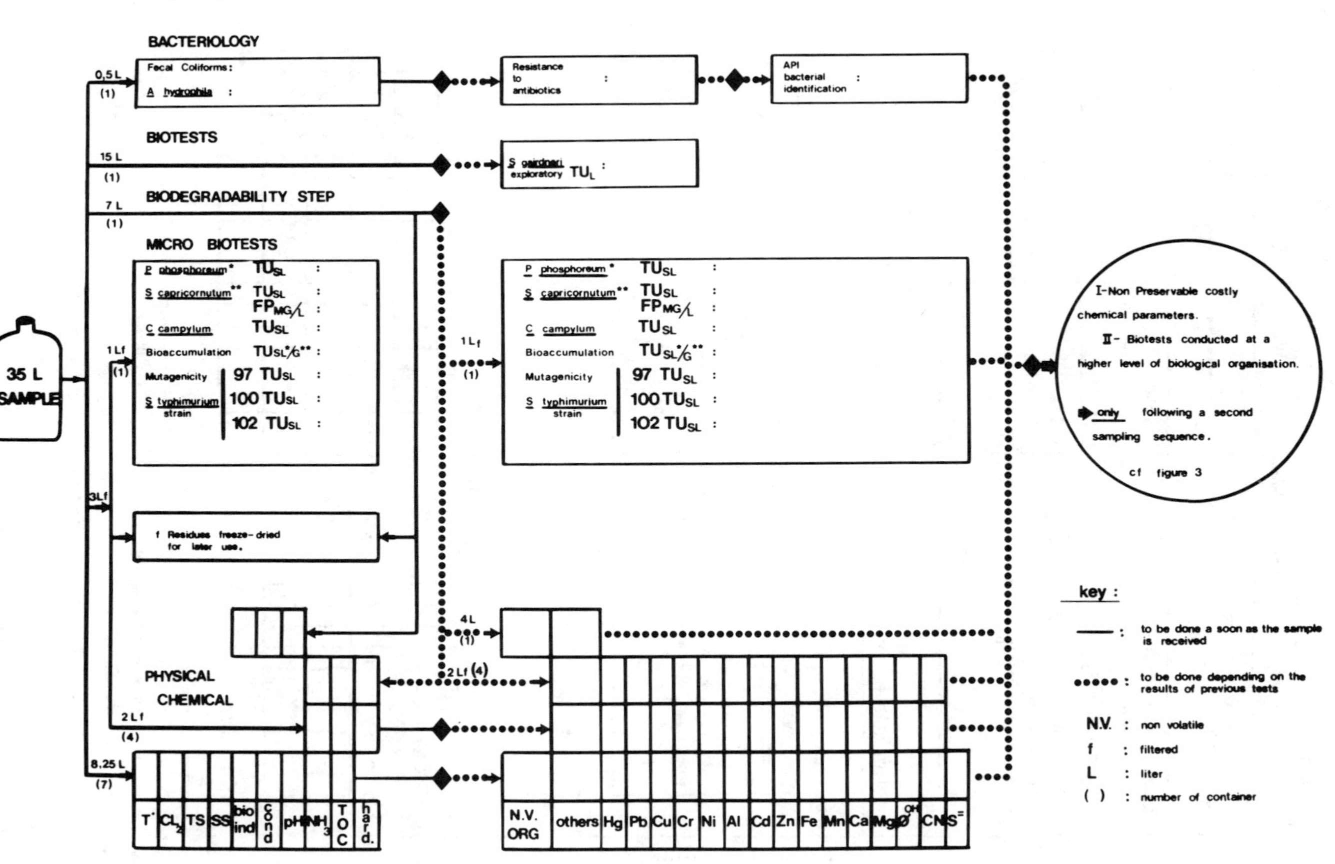

Figure 12.2. Simultaneous integrated ecotoxicological assessment of wastewater. Operational steps following the initial sampling sequence (35 L/sample).

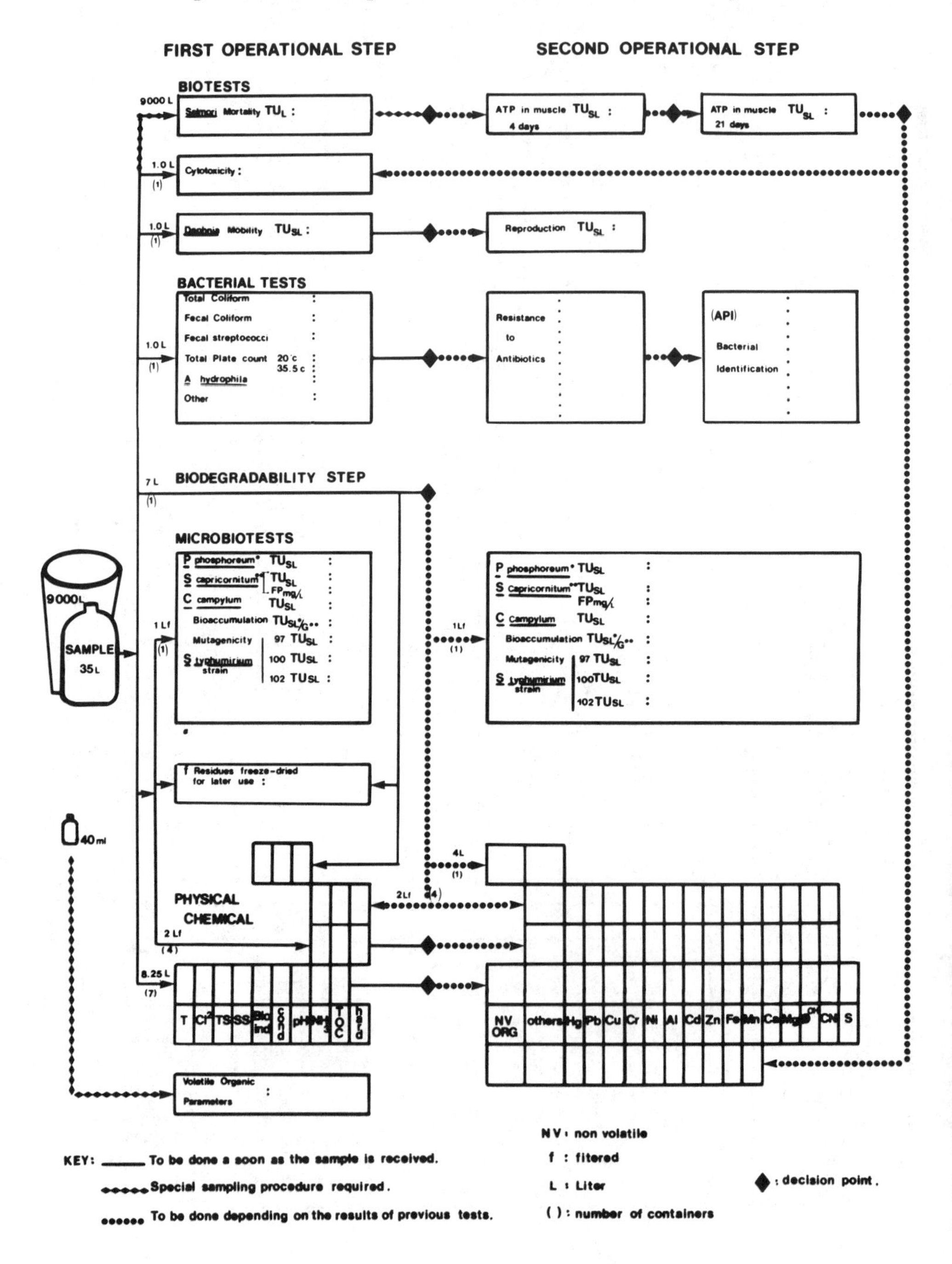

Figure 12.3. Simultaneous integrated ecotoxicological assessment of wastewater. Operational steps following a second sampling sequence (40 mL and/or 35–9000 L).

REFERENCES

Ames, B.N., McCann, J., and Yamasaki, E. (1975). Methods for detecting carcinogens and mutagens with the *Salmonella*/mammalian-microsome mutagenicity test. *Mutation Res.* **31**: 347–384.

APHA (American Public Health Association), AWWA (American Water Works Association), and WPCF (Water Pollution Control Federation) (1980). *Standard Methods for the Examination of Water and Wastewater*, 15th ed. APHA editor, Washington, DC.

Becker, E.W. (1983). Limitations of heavy metal removal from waste water by means of algae. *Water Res.* **17**: 459–466.

Beckman Instruments Inc. (1980). Operating Instructions of Microtox T.M. Toxicity Analyzer System, Model 2055. Beckman Instrument Inc., U.S.A., Manual 110679.

Blaise, C. (1973). Degradative microorganisms in the Ottawa River. M. Sc. Thesis, University of Ottawa.

Blaise, C. (1984). Potentiel des microtests avec bactéries et algues pour l'évaluation des écotoxicités. Thèse de doctorat, Université de Metz.

Blaise, C., Ska, B., Sabatini, G., Bermingham, N., and Van Coillie, R. (1982). Potentiel de bioaccumulation de substances toxiques d'eaux résiduaires industrielles à l'aide d'un bioessai utilisant des algues et des bactéries. Dans: Les tests de toxicité aiguë en milieu aquatiques. *INSERM* (Institut National de la Santé et de la Recherche Médicale), Paris, Vol. 106, 155–165.

Couture, P., Couillard, D., and Croteau, G. (1981). Un test biologique pour caractériser la toxicité des eaux usées. *Environ. Pollut. Ser. B* **2**: 217–222.

Couture, P., Van Coillie, R., Campbell, P.G.C., and Thellen, C. (1982). Le phytoplancton, un réactif biologique sensible pour détecter rapidement la présence de toxiques. Dans: Les tests de toxicité aiguë en milieu aquatique. *INSERM* (Institut National de la Santé et Recherche (Médicale), Paris, Vol. 106, pp. 255–272.

De La Noue, J., Thellen, C., and Van Coillie, R. (1983). Traitements tertiaires d'eaux usées municipales par production de biomasses d'algues. Agriculture Canada, Direction du recyclage des déchets, Ottawa.

Dive, D., and Leclerc, H. (1975). Standardized test method using protozoa for measuring water pollutant toxicity. *Prog. Water Technol.* **7**: 67–72.

Environment Canada. (1979a). *Analytical Methods Manual*. Environment Canada, Inland Waters Directorate, Ottawa.

Environment Canada. (1979b). Culture and toxicity testing methods using *Daphnia pulex* at the Aquatic Toxicology Laboratory, EPS, Northwest Region. Environment Canada, Environmental Protection Service, Northwest Region, Edmonton.

Environment Canada. (1980a). Références sur la qualité des eaux. Guide des paramètres de la qualité des eaux. Environment Canada, Direction Générale des Eaux Intérieures, Ottawa.

Environment Canada. (1980b). Méthode normalisée de contrôle de la toxicité aiguë des effluents. Environment Canada, Direction Générale de la Pollution des Eaux, Ottawa.

Environment Canada. (1981). *Analytical Methods Manual 1979: Update 1981*. Environment Canada, Inland Waters Directorate, Ottawa.

Environnement Quebec. (1979). Méthodes d'analyse du mercure dans l'eau, les sédiments, les boues, les sols, les milieux biologiques, l'air et les hydrocarbures. Environnement Québec, Bureau d'Etudes sur les Substances Toxiques, Quebec.

Environnement Quebec. (1980). Méthodes d'analyse des pesticides organochlorés et des biphényles polychlorés dans l'eau, les sédiments, les boues, les sols, les milieux biologiques, l'air et les hydrocarbures. Environnement Québec, Bureau d'Etudes sur les Substances Toxiques, Quebec.

EPA (Environmental Protection Agency) (1973). *Water Quality Criteria Book*, Vol. 3, *Effects of Chemicals on Aquatic Life*. Environmental Protection Agency, Washington, DC.

EPA (Environmental Protection Agency) (1979). Methods for benzidine, chlorinated organic compounds, pentachlorophenols and pesticides in water and wastewater. In *Test Methods for Evaluating Solid Waste. Physical/Chemical Methods*, Appendix 3. Environmental Protection Agency, Washington, DC.

EPA (Environmental Protection Agency) (1981). *Treatability Manual*. Vol. 1. Environmental Protection Agency, Washington, DC.

Federal Register (1979). Purgeables: Method 624. *Federal Register*, Washington, DC, **44**: 69532–69539.

Goss, L.B., and Wyzga, R. (1982). A conceptual framework for ecological risk analysis. 63rd Meeting Pacific Division Annu. Assoc. Advancement Sci., University California, Santa Barbara, June 20–25, 1982.

Joubert, G. (1980. A bioassay application for quantitative toxicity measurements, using the green algae *Selenastrum capricornutum*. *Water Res*. **14**: 1759–1763.

Maciorowski, A.F., Sims, J.L., Little, L.W., and Gerard, F.O. (1981). Bioassays, procedures and results. *J. Water Pollut. Control Fed*. **53**: 974–993.

Nriagu, J.O. (1978). *Sulfur in the Environment*, Part II; *Ecological Effects*. Wiley, New York.

OECD (Office of Economical Cooperation and Development) (1980). Guidelines for the Testing of Chemicals. Section 2: Effects on Biotic Systems. Office of Economical Cooperation and Development, Paris.

Rai, L.C., Gaur, J.P., and Kumar, H.D. (1981). Phycology and heavy metal pollution. *Biol. Rev*. **56**: 99–151.

Rapson, W.H., Wazar, M.A., and Butsky, V.V. (1980). Mutagenicity produced by aqueous chlorination of organic compounds. *Bul. Environ. Contam. Toxicol*. **24**: 590–596.

Roberts, J.R., Rodgers, D.W., Bailey, J.R., and Porice, M.A., (1979). Polychlorobiphényles: Critères biologiques pour évaluer leurs effets dans l'environnement. Conseil National des Recherches du Canada, Ottawa.

Rokosh, D.A., and Lovasz, T.N. (1979). Detection of Mutagenic Activity: Screening of Twenty-Three Compounds of Industrial Origin in the St. Clair River.

Santos Ferreira, M.O. (1977). Marqueurs épidémiologiques de *Pseudomonas aeruginosa*. Thèse de Doctorat d'Etat, Université Paris Sud.

Schoenert, R., Couture, P., Thellen, C., and Van Coillie, R. (1982). The sensitivity of six strains of unicellular algae *Selenastrum capricornutum* to six reference toxciants. *Can. Tech. Rep. Fish; Aquat. Sci*. **1151**: 200–202.

Shotts, E.B., and Rimler, R. (1973). Medium for the isolation of *Aeromonas hydrophila*. *Appl. Microbiol*. **26**: 550–553.

Sprague, C.E. (1971). Review paper: Measurement of pollutant toxicity to fish, II. Utilizing and applying bioassay results. *Water Res*. **5**: 245–266.

Stephan, C.E. (1977). Methods for calculating a LC_{50}. In Aquatic Toxicology and Hazard Evaluation. American Society for Testing and Materials, STP **634**: 65–84, Washington, DC.

Stokes, J.L., and Redmond, M.L. (1966). Quantitative ecology of psychrophilic microorganisms. *Appl. Microbiol*. **14**: 74–78.

Tennant, A., Toxopeus, R., Bastien, J.A.P., Beauchamp, M., and Wandevint, J. (1975). *Salmonella in Poultry Packing Plant Effluents*. Environment Canada, Environmental Protection Service, Ontario Region, Toronto.

Thellen, C., and Van Coillie, R. (1982). Effects toxiques d'une préparation opérationnelle de fénitrothion sur trois espèces de salmonidés. Ministère Energie et Ressources du Québec, Direction de l'Environnement.

Van Coillie, R., Couture, P., Schoenert, R., and Thellen, C. (1982). Mise au point d'une évaluation rapide de la toxicité originale des effluents et de leurs composants à l'aide d'algues. Environment Canada, Service de Protection de l'Environnement, Région du Québec, Montréal.

Van Coillie, R., Bermingham, N., Blaise, C., Couture, P., and Denizeau, F. (1983a). Caractérisation écotoxicologique des eaux provenant de 4 dépôtoires d'enfouissement. Environnement Canada, Service de Protection de l'Environnement, Région du Québec, Montréal.

Van Coillie, R., Couture, P., and Visser, S. (1983b). The use of algae in aquatic ecotoxicology. In *Aquatic Ecotoxicology*. Wiley, New York, pp. 488–502.

Vezeau, R. (1982). Protocoles d'échantillonnage, de préservation et de préparation des échantillons pour les analyses des polluants prioritaires. Environment Canada, Service de Protection de l'Environnement, Région du Québec, Montréal.

Washington, J.A., Yu, P.K.W., and Martin, W.J. (1971). Evaluation of accuracy of multitest micromethod system for the identification of enterobacteriaceal. *Appl. Microbiol.* **22**: 267–269.

Webb, P.W., and Brett, J.R. (1973). Effets of sublethal concentrations of sodium pentachlorophenate on growth rate, food conversion efficiency and swimming performance in underyearling sockeye salmon *Oncorhyynchus nerka*. *J. Fish. Res. Board Can*. **30**: 499–507.

13

THE USE OF A FUGACITY MODEL TO ASSESS THE RISK OF PESTICIDES TO THE AQUATIC ENVIRONMENT ON PRINCE EDWARD ISLAND

L.E. Burridge and K. Haya

Department of Fisheries and Oceans, Marine Chemistry Division, Biological Station, St. Andrews, New Brunswick, Canada

1. **Introduction**
2. **Material and Methods**
3. **Results and Discussion**
4. **Summary**

Appendices A and B

References

1. INTRODUCTION

In 1982 over 300 pesticide formulations of over 100 active ingredients were sold for agricultural use on Prince Edward Island (PEI) which has an area of 5700 km^2. Over 30 of these formulations were sold in quantities exceeding 1000 kg or 1000 L (Gill, 1984; Anon., 1982a). Many of these pesticides were introduced years ago and environmental impact data on them do not meet current registration requirements. Many registered pesticides are being reevaluated by Agriculture Canada. To aid this process we explored the applicability of a fugacity model to help assess the risk to the aquatic environment associated with pesticide use in PEI.

Fugacity is interpreted as the escaping tendency of a chemical from a phase (or compartment); its dimension is a unit of pressure (atm) (MacKay, 1979). MacKay (1979), MacKay and Paterson (1981, 1982), and MacKay et al. (1983, 1985) described the use of a model based on fugacity that allowed the prediction of the distribution of a chemical in six compartments (air, soil, water, sediment, suspended solids, and aquatic biota).

There are four levels of fugacity models. Each succeeding level is capable of modeling increasingly complicated environmental conditions. Each level is well described in the literature (e.g., MacKay and Paterson, 1981). The level I model predicts equilibrium distribution of a fixed amount of pesticide without transformation (MacKay, 1979).

The fugacity model is a tool with which researchers and regulators predict how certain compounds might act in the aquatic environment. In PEI the possibility exists that pesticides will find their way into marine as well as freshwater ecosystems.

Zitko and McLeese (1980) used a level I fugacity model to predict environmental distribution of pesticides and pesticide formulations in a forest environment. They calculated expected equilibrium concentrations in each compartment and incorporated available lethality data to derive hazard indices and to assess relative hazard to the aquatic environment.

We used the fugacity model of MacKay (1979) as modified by Zitko and McLeese (1980). The level I fugacity model does not account for degradation or advective loss of pesticide from a model ecosystem. It does, however, lend itself to the analysis of large sets of data such as ours. Our purpose was to identify pesticides that may be a hazard to aquatic biota in PEI, to identify gaps in the current literature, and to indicate where further research is needed.

2. MATERIALS AND METHODS

We defined two model ecosystems (freshwater and marine), each composed of six environmental compartments within a 1-km^2 area. In the freshwater model the size of each compartment is expressed as a volume (Table 13.1). The freshwater system is 95% soil and 5% water by area; air volume is 1 km^3. The soil compartment is 20 cm deep; depth of water is 1 m. Sediment which was 5 cm deep occupies the same area as water. The suspended solids and aquatic biota compartments relate to the water compartment by volume fractions. In our case suspended solids were arbitrarily set at concentrations of 2.5 g/m^3 and aquatic biota at 0.75 g/m^3 to reflect the high productivity of PEI ponds (Smith, 1963). The proportion of organic carbon in the soil including vegetation was estimated at 10% and lipid in the aquatic biota at 10%.

The model seawater ecosystem was assumed to be 50% soil and 50% water by area. The air ecosystem, soil and sediment depths, organic carbon content, and lipid concentration were the same as for the freshwater ecosystem. The average depth of the water was 3 m and the volume fraction of aquatic biota

Table 13.1D Size of Compartments in Equilibrium Distribution Model

Compartment	Fresh water (m^3)	Seawater (m^3)
Air	1.0E+09	1.0E+09
Water	5.0E+04	1.5E+06
Suspended solids	5.0E+04	1.5E+06
Sediment	2.5E+03	2.5E+04
Aquatic Biota	5.0E+04	1.5E+06
Soil	1.9E+05	1.0E+06

was 0.25 g/m^3. The concentration of suspended solids was high at 3.5 g/m^3, reflecting the effect of tidal action.

The model required the following information about each pesticide: solubility in water, Henry's constant, molecular weight, adsorption coefficient (K_{oc}), octanol/water partition coefficient (K_{ow}), and an estimate of the application rate. Such information was found in Anon. (1982a) and Worthing and Walker (1983). Much of the necessary information was not readily found in the literature. For these compounds parameters were estimated as follows:

$$\text{Henry's constant} = \frac{\text{vapor pressure (atm)}}{\text{solubility in water (mol/m}^3)}$$

$$\log K_{oc} = 3.64 - 0.55 \log s$$

$$s = \text{solubility in water (mg/L) (Kenaga, 1980)}$$

$$\log K_{ow} = 5.00 - 0.67 \log s$$

$$s = \text{solubility in water } (\mu\text{mol/L}) \text{ (Chiou } et\ al. \text{ 1977)}$$

All calculations were performed using an HP-3000 computer. The program was written by Dr. V. Zitko.

Zitko and McLeese (1980) incorporated lethality data (LC_{50}) to determine lethality index, hazard index, and relative hazard index. Briefly, the lethality index is the expected equilibrium concentration (EEC) of a pesticide in the water compartment divided by its LC_{50}. Information about lethality of pesticides to aquatic animals was taken from Pimentel (1971) and Worthing and Walker (1983). The hazard index is the lethality index multiplied by the bioconcentration factor (BCF) for that pesticide where

$$\text{BCF} = \frac{\text{EEC in aquatic biota}}{\text{EEC in water}}$$

The relative hazard index is the hazard index of a pesticide divided by the hazard index of a benchmark pesticide. Zitko and McLeese (1980) chose fenitrothion as their reference compound. Fenitrothion is an organic phos-

Table 13.2D Relative Hazard Indices of Pesticides Used on PEI (Freshwater Model)[a]

Chlorpyrifos	7.1E+01
Azinphosmethyl	5.9E+01
Methidathion	1.0E+01
Phorate	5.9E+00
Chlordane	4.5E+00
Endosulfan	2.8E+00
Cypermethrin	2.0E+00
Trifluralin	1.2E+00
Malathion	1.0E+00
Captan	8.8E–01

[a]Malathion hazard index = 6.8E+04.

phrous insecticide used in the control of agricultural and forest pests. It has been the subject of considerable research and therefore was the natural choice as the reference pesticide in their work. Malathion, also an organophosphate insecticide, has broad use in agriculture and was chosen as our reference pesticide.

The maximum suggested application rate was used to calculate EECs. Lethality data were collected for fish, where possible. However, in the seawater model, data for marine invertebrates were used since there are few available data for lethality of pesticides to marine fish.

3. RESULTS AND DISCUSSION

Of the 100 pesticides, sufficient chemical and biological data were available to assess only 60 in the freshwater ecosystem and 8 in the seawater ecosystem. Appendix A includes the raw data collected for each pesticide assessed. Appen-

Table 13.3D Relative Hazard Indices of Pesticides Used on PEI (Seawater model)[a]

Malathion	1.0E+00
Chlordane	3.8E–01
Endosulfan	2.8E–01
Fenvalerate	1.6E–01
Carbaryl	1.2E–03
Lindane	4.8E–04
Permethrin	9.5E–05
EPTC	1.4E–07

[a]Malathion hazard index = 3.9E+06.

dix B is a list of the compounds for which complete data sets were not found.

The ten compounds having the highest relative hazard indexes in freshwater are listed in Table 13.2. Table 13.3 lists the relative hazard indices of the eight compounds assessed in seawater.

Chlorpyrifos has the highest relative hazard index (Table 13.2). It is an organophosphate insecticide (OP) that is toxic to fish and aquatic invertebrates (Worthing and Walker, 1983). It is not sold in large quantities on PEI (Anon., 1982b; Gill, 1984). Chlorpyrifos has been the subject of considerable research (see, for example, Marshall and Roberts, 1978). Similarly, azinphosmethyl (another OP) has a large relative hazard index. Like chlorpyrifos, its use, handling, and storage should be closely monitored.

Two of the compounds listed in Table 13.2 were sold in quantities exceeding 1000 kg or 1000 L in 1982: methidathion and phorate (Gill, 1984). Methidathion was introduced in 1966. It is an organophospate insecticide–acaricide with long residual effect (Thomson, 1985). Phorate, introduced in 1954, is an organophosphate insecticide–acaricide, often applied as granules, capable of protecting plants for 4–12 weeks (Thomson, 1985). No aquatic lethality data relating these pesticides to marine species are available (see Table 13.3).

The remaining pesticides in Table 13.2 represent three more classes of pesticides. The organochlorines include endosulfan, chlordane, and captan. These are toxic to fish and invertebrates and may persist in aquatic environments (Thomson, 1985). Trifluralin is a dinitroanaline herbicide. It is applied by incorporation in the soil. It has long-term effectiveness and is very toxic to fish (Thomson, 1983).

Cypermethrin is a pyrethroid insecticide. The pyrethroids have become popular due to their low mammalian toxicity; however, they are toxic to fish and more toxic to aquatic arthropods (e.g., Zitko et al., 1979; Stephenson, 1982). It is noteworthy that fenvalerate, another pyrethroid, ranks eleventh of the 60 pesticides (see Appendix A). Fenvalerate is one of the pesticides sold in large quantities (Gill, 1984). This class of compounds is currently under temporary registration in Canada and warrants further study.

In the seawater model, there are no pesticides having relative hazard indices greater than that of malathion (Table 13.3). Chlordane and endosulfan have lower hazard indices. The acute toxicity of malathion to invertebrates and its high water solubility account for this change in relative ranking. The most important observation from our review of the seawater data (Table 13.3) is the lack of assessments performed (8/100). The potential for these compounds to affect the marine environment in PEI is high. In many cases, fields where pesticides are used border on the ocean. The need for research into effects on marine and estuarine species is urgent.

The physical constants of many of these pesticides are not available or easily found. Vapor pressure data are the most difficult to find and this is the main reason only 60 of 100 pesticides could be assessed in the freshwater model (see Appendix B).

Some pesticides used in PEI present hazards not only to the aquatic

APPENDIX A

Compound	Molecular Weight	Vapor Pressure (atm)	Water Solubility (mol/m^3)	Application Rate (mol/ha)	LC_{50} (mol/L)	Species	Relative Hazard Rank
2,4-D(amine)	221.0	5.3E−4*	2.8E+0	1.6E+0	2.3E−2*		60
Acephate	183.2	2.3E−9	3.5E+3	6.1E+0	5.5E−3	R. trout	59
Alachlor	269.8	2.9E−8	8.9E−1	1.7E+1	6.6E−6	R. trout	16
Aldicarb	190.3	1.3E−7	3.2E+1	2.9E+1	4.6E−5	R. trout	37
Atrazine	215.7	4.0E−10	1.4E−1	7.8E+0	4.0E−5	R. trout	34
Azinphosmethyl	317.1	1.0E−11	1.0E−1	2.1E+1	1.3E−8	R. trout	2
Bensulfide	397.5	1.3E−9	6.3E−2	1.3E+1	1.8E−6	R. trout	13
Brodifacoum	523.4	1.3E−9	1.9E−2	2.0E−2	9.7E−5	R. trout	32
Captan	300.6	1.3E−8	1.1E−2	5.6E+0	2.3E−7	R. trout	10
Carbaryl	201.0	6.7E−6	6.0E−1	5.9E+0	9.9E−6	R. trout	35
Carbaryl(SW[a])	—	—	—	—	1.5E−7	Shrimp	5
Carbofuran	221.3	2.7E−8	3.2E+0	6.0E−1	1.3E−6	R. trout	29
Chlordane	409.8	1.3E−8	2.4E−4	2.7E+1	2.2E−7	R. trout	5
Chlordane(SW)	—	—	—	—	5.0E−9	Shrimp	2
Chlorfenvinphos	359.6	5.3E−9	4.2E−1	1.1E+0	1.5E−6	Guppie	21
Chlorothalonil	265.9	1.3E−5	2.3E−3	9.5E+0	9.4E−7	R. trout	28
Chloroxuron	290.7	2.4E−12	1.4E−2	1.9E+1	9.6E−5	Bluegill	31
Chlorpyrifos	350.6	2.5E−8	5.7E−3	1.6E+1	9.0E−9	R. trout	1
Cyanazine	240.7	2.0E−12	7.1E−1	1.3E+1	3.7E−5	R. trout	30
Cypermethrin	416.3	1.9E−12	3.6E−5	3.4E−1	7.0E−9	R. trout	7
Deltamethrin	505.2	2.0E−11	4.0E−6	2.0E−2	2.0E−9	R. trout	12
Demeton-O	258.3	3.8E−7	2.3E−1	3.3E+0	2.3E−6	R. trout	22
Demeton-S	230.3	3.5E−7	8.7E−3	3.7E+0	4.3E−5	R. trout	39
Diazinon	304.3	9.7E−10	1.3E−1	5.5E+0	8.5E−6	R. trout	27
Dicamba	221.0	4.5E−8	2.9E+1	2.5E+1	1.3E−4	R. trout	42
Dichlobenil[b]	172.0	7.3E−10	1.0E−1	1.2E+2	1.0E−4	Guppies	23
Dichlorvos	221.0	1.6E−5	4.5E+1	4.5E+0	3.9E−6	Bluegill	36
Diclofop	327.2	3.4E−10	9.2E−3	3.4E+0	1.1E−6	R. trout	15
Dimethoate	229.2	1.1E−8	1.1E+2	3.1E+0	2.7E−5	R. trout	50
Diquatdibromide	344.0	1.3E−10	2.0E+3	2.4E+0	7.3E−4	Bluegill	58
Disulfoton	274.4	2.4E−7	9.1E−2	1.2E+1	6.9E−6	R. trout	18
Endosulfan	406.9	1.2E−5	8.1E−4	1.1E+1	3.0E−9	R. trout	6
Endosulfan(SW)	—	—	—	—	4.9E−10	Shrimp	3
EPTC	189.3	4.5E−5	2.0E+0	3.6E+1	1.0E−4	R. trout	45
EPTC[c](SW)	—	—	—	—	1.1E−4	Crab	8
Fenvalerate	419.9	3.7E−10	2.4E−3	2.3E−2	9.0E−9	R. trout	11
Fenvalerate(SW)	—	—	—	—	9.5E−11	Shrimp	4
Fluazifop butyl	383.3	5.5E−10	5.2E−3	5.2E+0	1.4E−6	Bluegill	14
Hexazinone	252.3	2.7E−10	1.3E+2	4.8E+0	1.1E−3	F. minnow	54
Lindane	290.8	5.6E−8	2.4E−2	1.2E−1	7.6E−8	R. trout	19
Lindane(SW)	—	—	—	—	1.5E−8	Shrimp	6

APPENDIX A (continued)

Compound	Molecular Weight	Vapor Pressure (atm)	Water Solubility (mol/m³)	Application Rate (mol/ha)	LC_{50} (mol/L)	Species	Relative Hazard Rank
Linuron	249.1	2.0E−8	3.3E−1	1.4E+1	6.4E−5	R. trout	33
Malathion	330.3	5.3E−8	4.5E−1	1.0E+1	3.0E−7	Bluegill	9
Malathion(SW)	—	—	—	—	3.9E−10	Shrimp	1
MCPA(amine)	200.6	2.0E−9	4.1E+0	1.1E+1	5.0E−5	R. trout	38
Metalaxyl	279.3	2.9E−9	2.5E+1	1.1E+0	3.6E−4	R. trout	56
Methamidophos	141.1	4.0E−7	1.4E+4	7.9E+0	3.6E−4	R. trout	57
Methidathion	302.3	1.9E−9	8.3E−1	2.6E+0	7.0E−9	Bluegill	3
Methomyl	162.2	6.7E−8	3.6E+2	1.2E+1	6.2E−7	Goldfish	24
Methylisocyanate	73.1	2.7E−2	1.0E+2	3.4E+3	5.1E−5	R. trout	43
Metolachlor	283.8	1.7E−8	1.9E+0	8.8E+0	7.0E−6	R. trout	25
Metribuzin	214.3	1.3E−8	5.6E+0	5.2E+0	3.5E−4	R. trout	52
Monolinuron	214.6	6.4E−5	3.4E+0	7.0E+0	2.3E−4	Goldfish	49
Oxamyl	219.3	3.1E−7	1.3E+3	5.1E+0	1.9E−5	R. trout	53
Pebulate	203.3	4.7E−5	3.0E−1	3.0E+1	3.6E−5	R. trout	44
Permethrin	391.3	2.6E−6	5.1E−4	1.8E−1	2.3E−8	R. trout	26
Permethrin(SW)	—	—	—	—	2.0E−9	Lobster	7
Phorate	260.4	8.5E−7	1.9E−1	1.3E+1	5.0E−8	R. trout	4
Phosmet	317.3	1.3E−6	6.9E−2	3.2E+0	9.5E−7	R. trout	17
Pirimicarb	238.3	4.0E−8	1.1E+1	1.2E+0	1.2E−4	R. trout	55
Prometryne	241.4	1.3E−9	1.4E+2	6.2E+0	1.0E−5	R. trout	40
Simazine	201.7	8.1E−12	2.5E−2	2.2E+1	4.9E−4	R. trout	41
Tebuthiuron	228.3	2.7E−9	1.1E+1	3.9E+1	6.1E−4	R. trout	48
Terbacil	216.7	6.3E−10	3.3E+0	1.9E+1	4.0E−4	P. Sunfish	46
Triadimenfon	293.8	1.0E−9	8.8E−1	1.9E+0	1.7E−4	Goldfish	51
Triallate	304.0	1.6E−7	1.3E−2	5.5E+0	3.9E−6	R. trout	20
Trichlorfon	257.4	1.0E−8	5.8E+2	4.4E+0	7.0E−6	R. trout	47
Trifluralin	335.5	1.4E−7	3.0E−3	5.0E+0	1.2E−7	R. trout	8

[a] SW, toxicity data for marine organisms.
[b] 48-h LC_{50}.
[c] 24-h LC_{50}.

environment but also to people working with the pesticides. While it is not the intent of this paper to discuss human health effects, one compound deserves particular attention. Dinoseb is a dinitrophenol herbicide used heavily on PEI. It was introduced in 1940 (Thomson, 1983). Dinoseb is extremely toxic to aquatic organisms (Zitko et al., 1976). Its relative hazard was not assessed, because Henry's constant could not be estimated (see Appendix B). Recently the use of dinoseb was suspended by the United States Environmental Protec-

APPENDIX B

Compound	Molecular Weight	Vapor Pressure (atm)	Water Solubility (mol/m³)	Application Rate (mol/ha)	96 h LC_{50} (mol/L)	Species
2,4-DB	249.1		1.8E−1	1.2E+1	1.6E−5	R. trout
Amitrole	84.1		3.3E+3	2.7E+2	1.2E−3	R. trout
Benomyl	290.3	Neg.	6.9E−3	1.2E+0	5.9E−7	R. trout
Bromoxynil[a]	403.0		3.2E−1	1.0E+0	3.7E−7	R. trout
Calciferol	396.7		1.3E−1			
Captafol	349.1	Neg.	4.0E−3	6.4E+0	1.4E−6	R. trout
Carboxin	235.5	1.3E−3	7.2E−1		5.1E−6	Bluegill
Chloramben	206.0	9.2E−6	3.4E+0	2.2E+1		
Chlorophacinone	374.8	Neg.	Spar.			
Chloropicrin	164.4	3.2E−2	1.6E+1	7.3E+3		
Chlorprophram	213.7		4.2E−1	4.2E+1	2.3E−5	R. trout
Chlorthal	332.0	6.7E−7	1.5E−3	4.2E+1		
Crotoxyphos	314.3	1.8E−8	3.2E+0		2.3E−7	Bluegill
Dalapon	143.0		6.3E+3	2.6E+2	7.0E−4	R. trout
Dichloropropene + Dichloropropane		4.6E−2				Harlequin
Dichlorprop[a]	235.1	Neg.	1.5E+0	1.2E+1	2.1E−5	Bluegill
Dicofol	370.5		Insol.	1.2E+1	1.4E−6	R. trout
Difenzoquat	249.3		3.1E+3	3.4E+0	2.8E−3	R. trout
Dinocap	364.3		Spar.	9.2E+0	4.1E−8	R. trout
Dinoseb	240.2		4.2E−1	1.7E+1	2.0E−4	C. trout
Diphenamid	239.3		1.1E+0	2.5E+1	2.0E−4	R. trout
Dodine	287.4		2.2E−1	1.6E+1	1.8E−6	R. trout
Ethephon	144.5		6.9E+3	6.1E+0	2.4E−3	R. trout
Fensulfothion	308.3		5.0E+0	1.4E+2	2.9E−5	R. trout
Ferbam	416.5	Neg.	3.1E−1	4.0E+1	3.1E−5	Catfish
Glyphosate	169.1		7.1E+1	2.5E+1	5.1E−4	R. trout
Ioxynil	370.9		1.3E−1			
Iprodione	330.2	1.3E−9	3.9E−2	3.2E+0		
Maleic hydrazine	112.1		5.3E+1	3.0E+1	1.3E−2	R. trout
Mancozeb			Insol.			
Maneb[a]	265.3		Insol.		6.8E−6	Carp
MCPB	250.7		1.8E−1	1.4E+1	2.5E−4	F. minnow
Mecoprop	214.6		2.9E+0	1.3E+1		
Methoxychlor	345.7		2.8E−4	1.6E+0	1.9E−7	R. trout
Metiram[a]		1.0E−10	Insol.			Harlequin
Mevinphos	224.1	1.7E−7	Misc.	2.2E+0	5.4E−8	R. trout
Naphthalene acetic acid	186.2		2.2E+0	3.4E+1		
Paraquat	257.2	Neg.	Sol.	3.3E+0	9.7E−7	B. trout
Phosalone	367.8	Neg.	2.7E−2	2.2E+0	4.1E−7	R. trout

APPENDIX B (continued)

Compound	Molecular Weight	Vapor Pressure (atm)	Water Solubility (mol/m³)	Application Rate (mol/ha)	96 h LC_{50} (mol/L)	Species
Piperonyl butoxide	338.4	Neg.	Insol.		1.0E−7	R. trout
Pyrethrins	328.4			1.6E+0		
Rotenone	394.4		3.8E−2		7.9E−8	R. trout
Thiabendazole	201.2		2.5E−1	4.7E+0		
Thiophanate methyl[a]	342.4		Spar.	4.6E+0	3.2E−5	Carp
Thiram	240.4	Neg.	1.2E−1	2.8E+1	5.4E−6	R. trout

Note: Neg., negligible; Spar., Sparingly soluble; Insol., insoluble; Sol., soluble; Misc., Miscible.
[a] 48-h LC_{50}.

tion Agency. The reason for this was an apparent serious risk of birth defects in children of women exposed while pregnant to dinoseb during or shortly after application (Anon., 1986).

More lethality data are required to use the MacKay model. Many of the compounds registered before 1970 have not been tested for effects on freshwater invertebrates. Most of the 100 pesticides are insecticides and it is not unreasonable to expect that their effect on aquatic invertebrates would be significant. Much of the information presently lacking will become available as pesticides are reevaluated. However, compounds such as phorate and methidathion (because of their use and high relative hazard indices) should be examined on a priority basis. Dinoseb, for reasons already stated, demands immediate attention.

The level I fugacity model does not account for degradation or nonequilibrium conditions. This no doubt simplifies the PEI situation: Because degradation processes are not accounted for, the predicted equilibrium concentrations are unlikely to match those that might be observed. A first step in studying compounds that we have identified as potentially hazardous would be to reexamine them using a higher level fugacity model.

The pesticides listed in Table 13.1 and in the appendixes represent a broad range of chemical classes and use patterns. The use of the fugacity model has eased the burden of ranking these compounds. We feel that these rankings are reasonable. Compounds that rank high are either very toxic, very persistent, or both. The relative rankings will change according to input parameters and level of modeling. For example, it is improbable that the organophosphates would rank as high if degradation were considered, although we contend that they are a substantial hazard nonetheless. While there are obvious short-comings to this procedure, we believe that it is an objective screening tool with which to select pesticides for priority testing and for further biological assessment.

4. SUMMARY

More than 100 pesticides in 300 formulations are sold in Prince Edward Island each year. A level I fugacity model was used to assess the relative hazard of these pesticides to freshwater and marine environments in PEI. Only 60 pesticides could be assessed in the freshwater model. Chlorpyrifos had the highest relative hazard index. Two of the top 10 compounds ranked by relative hazard index, phorate and methidathion, are sold in large quantities yearly. There is very little information available on the effects of pesticides in marine systems; therefore, only 8 compounds were assessed. This lack of data must be addressed with particular attention to compounds such as phorate and methidathion, which rank high in the freshwater model. The model provides a means to conveniently compare large data sets.

ACKNOWLEDGMENTS

The authors thank V. Zitko and D.J. Wildish for reviewing the manuscript. Dr. Zitko also provided the computer program. Officials from the Province of Prince Edward Island, Department of Cultural and Community Affairs, provided sales statistics for pesticides in PEI.

REFERENCES

Anon. (1982a). A compendium of information on agricultural pesticides used in the Atlantic Region. Environmental Protection Service, Atlantic Region.

Anon. (1982b). An inventory of pesticide usage on Prince Edward Island and Nova Scotia. Seatech Investigation Services Ltd.

Anon. (1986). Another pesticide ban by environmental protection agency. *Chem. Eng. News* **64** (42): 10.

Chiou, C.T., Freed, V.H., Schmedding, D.W., and Kohnert, R.L. (1977). Partition coefficient and bioaccumulation of selected organic chemicals. *Environ. Sci. Tech.* **11**: 475–478.

Gill, M. (1984). Major-use pesticides on Prince Edward Island. Internal report of the Government of P.E.I., Department of Community and Cultural Affairs, p. 4.

Kenaga, E.E. (1980). Predicted bioconcentration factors and soil sorption coefficients of pesticides and other chemicals. *Ecotoxicol. Environ. Safety* **4**: 26–38.

MacKay, D. (1979). Finding fugacity feasible. *Environ. Sci. Tech.* **13**: 1218–1223.

MacKay, D., and Paterson, S. (1981). Calculating fugacity. *Environ. Sci. Tech.* **15**: 1006–1014.

MacKay, D., and Paterson, S. (1982). Fugacity revisted. *Environ. Sci. Tech.* **16**: 645A.

MacKay, D., Joy, M., and Paterson, S. (1983). A quantitative water, air, sediment interaction (QWASI) fugacity model for describing the fate of chemicals in lakes. *Chemosphere* **12**: 981–997.

MacKay, D., Paterson, S., Cheung, B., and Neely, W.B. (1985). Evaluating the environmental behavior of chemicals with a level III fugacity model. *Chemosphere* **14**: 335–374.

Marshall, W.K., and Roberts, J.R. (1978). Ecotoxicology of chlorpyrifos. Pub. No. 16079 of the environmental secretariat of the *Nat. Res. Coun. Can.*, 314 pp.

Pimentel, D. (1971). Ecological effects of pesticides on non-target species. United States Exc. Off. of Pres.: Office of Science and Technology, Washington, DC.

Stephenson, R.R. (1982). Aquatic toxicology of cypermethrin, I. Acute toxicity to some freshwater fish and invertebrates in laboratory tests. *Aquatic Toxicol.* **2**: 175–185.

Smith, M.W. (1963). The Atlantic Provinces of Canada. *In* D.G. Fry (Ed.), *Limnology in North America.* University of Winconsin Press, Madison WI, pp. 527–534.

Thomson, W.T. (1983). *Agricultural Chemicals Book, II. Herbicides*, 1983–1984 revision. Thomson, Fresno, CA.

Thomson, W.T. (1985). *Agricultural Chemicals Book, I. Insecticides*, 1985–1986 revision. Thomson, Fresno, CA.

Worthing, C.R. and Walker, S.B. (Ed.) (1983). *The Pesticide Manual, A World Compendium*, 7th ed. The British Crop Protection Concil, Thornton, Heath, U.K.

Zitko, V., and McLeese, D.W. (1980). Evaluation of hazards of pesticides used in forest spraying to the aquatic environment. *Can. Tech. Rep., Fish. Aquat. Sci.*, No. 985.

Zitko, V., McLeese, D.W., Carson, W.G., and Welch, H.E. (1976). Toxicity of alkyldinitrophenols to some aquatic organisms. *Bull. Environ. Contam. Toxicol.* **16**: 508–515.

Zitko, V., McLeese, D.W., Metcalfe, C.D., and Carson, W.G. (1979). Toxicity of Permethrin, Decamethrin, and related pyrethroids to salmon and lobster. *Bull. Environ. Contam. Toxicol.* **21**: 338–343.

14

ASSESSMENT OF THE INORGANIC BIOACCUMULATION POTENTIAL OF AQUEOUS SAMPLES WITH TWO ALGAL BIOASSAYS

S. Bisson

Sciences de l'Environnement, Université du Québec à Montréal, Montréal, Québec, Canada

C. Blaise and N. Bermingham

Laboratoire Capitaine Bernier, Direction de la Protection, Conservation et Protection, Environnement Canada, Longueuil, Québec, Canada

1. **Introduction**
2. **Materials and methods**
   2.1. Culture and growth conditions
   2.2. Bioaccumulation protocol
   2.3. Chemical determination of algal metal uptake
   2.4. Photooxidation procedure
   2.5. Bio-analytical determination of algal metal uptake
3. **Results and discussion**
   3.1. Metal uptake by *C. variabilis*
   3.2. Photooxidation toxicity
   3.3. Bio-analytical expression of metal uptake
4. **Summary and test prospects**
   Acknowledgments
   References

1. INTRODUCTION

Environmental protection strategies have greatly evolved over the past few decades as have the biological tests used to detect and monitor impact on aquatic ecosystems subjected to myriads of man-made chemicals and to diverse industrial discharges (Korte et al., 1985; Leclerc and Dive, 1982; Persoone et al., 1984). The reductionist approach which focused on isolated compartments of aggression (for example, fish lethality) is now giving way to a more holistic manner of viewing ecotoxicity. In this light, hazard assessment schemes (HAS) are marking the 1980s as effective means of examining different environmental milieux (Blaise et al., 1985; Chapman and Long, 1983; Miller et al., 1985; Mount et al., 1984). While such schemes will vary in outlook and scope of assessment, all seek to address, in the most relevant and cost-effective way, those perturbations that relate to the fate and effects of chemicals present in the samples under investigation. In this sense, bioaccumulation is one important factor that must be appraised in any integrated HAS employing a combination of biological (effects detection) and chemical (cause identification) testing.

The sensitive place that algae hold in the food web (Christensen and Scherfig, 1979; Neuhold and Ruggerio, 1976), their versatility in ecotoxicological applications (Couture et al., 1985; van Coillie et al., 1983), and their remarkable uptake characteristics (Friant and Sherman, 1981; Hassett et al., 1981; Rai et al., 1981), make them ideal organisms to work with in an attempt to develop practical and economical bioaccumulation screening tests for eventual incorporation in a presently used comprehensive test battery and approach for the ecotoxicological assessment of wastes and wastewaters (Blaise et al., 1985). Our overall objectives in this research initiative are to develop experimental protocols reflecting a potential of samples for toxic inorganic, organic, and genotoxicant uptake by algae. While traditional bioaccumulation procedures with algae basically involve exposure to an aqueous sample followed by chemical quantification of uptake (Canteford et al., 1978; Hassett et al., 1981), our method is to assess algal cell contents after exposure with specific low-volume-requiring biotests whose toxic expression is indicative of a particular type of uptake. In this venture, we have successfully elaborated a procedure to screen for the accumulation of toxic organic molecules in algae (Blaise et al., 1981; Ska et al., 1984) and are presently in the process of evaluating another procedure to detect genotoxic uptake in algae. In this paper, we present the results of work undertaken to establish an experimental protocol to specify the uptake of toxic inorganics by algae.

Developing and applying biotests where noted effects can be attributed to a particular chemical class of aggression can be advantageous for hazard assessment of any previously unstudied liquid sample. In fact, evidence that biotests can prove useful in this way is already apparent (Blaise et al., 1981; Couture et al., 1981; McFeters et al., 1983; Ska et al., 1984; Thomas et al., 1986). Capitalizing on biotesting information in this manner could help to reduce the time and cost associated with blind chemical scanning for a wide array of (in)organic contaminants by providing guidance for subsequent chemical characterization efforts. Noting effects, therefore, through biological testing followed by enlightened chemical testing is certainly a more profitable strategy, we feel, from both a scientific and cost-efficient point of view.

During initial investigations undertaken to develop a simple technique to detect the bioaccumulation potential of wastewaters with algae (Blaise et al., 1981), the green alga *Selenastrum capricornutum* was used as an indicator of accumulation. After a 24-h exposure, cells were collected, lysed by sonication, and a reconstituted lysate was assayed with the Microtox bacterial bioluminescence procedure (Beckman, 1982). Resulting Microtox toxicity was interpreted as the expression of some (in)organic toxicant accumulation in the algae. Relating accumulated toxicity to the toxicity determined for each aqueous sample studied then allowed us to calculate toxic accumulation factors. This (Blaise et al., 1981) and later work (Ska et al., 1984) confirmed that lysate toxicity expressed by the Microtox system was organic in nature and that the microbial assay was unable to display toxic responses to lysates known to contain metals. It was theorized that any mineral uptake by the algae was organically bound by cell constituents and that lysate toxicity could not express itself via this luminescent bioassay. This masking effect, reported for work performed with $HgCl_2$ (Ska et al., 1984), led us to investigate different ways of unbinding accumulated inorganic molecules in the hope that a suitable bioassay organism could then indicate a positive toxic response after being challenged with a treated accumulation mixture. With this objective, we have turned to irradiation, a well-documented procedure employed for several applications when organic matter oxidation in water samples is required (Batley and Farrar, 1978; Blazka and Prochazkova, 1983; Campbell et al., 1983; Malaiyandi et al., 1980). In our recent work in this area, photooxidation experiments involving UV irradiation of algal cells exposed to inorganic compounds appears to be a promising technique to liberate organically bound metals by destruction (i.e., mineralization) of cell constituents. Irradiated cell solutions are then profitably assayed with a sensitive microtest that confirms the uptake of inorganic toxicants by algae. The results of this work are described herein.

2. MATERIALS AND METHODS

2.1. Culture and Growth Conditions

Chlamydomonas variabilis Dangeard (obtained from the Institut national de la recherche chimique appliquée, Vert-le-Petit, France), a green mobile unicellular alga easily cultured under laboratory conditions and representative of a genus well known for its metal affinity (Bates et al., 1983; Hassett et al., 1981; Sakaguchi et al., 1979), was chosen as the test organism for bioaccumulation.

Incubation conditions of the algae in their AAM (algal assay medium) 30% synthetic medium were similar to those described by Miller et al. (1978). Because of the relatively high biomasses required (see Section 2.2), late log/early stationary phase cells served as inocula for all toxicity and bioaccumulation experiments with *C. variabilis*. The cells were grown in 4-L Erlenmeyer flasks, each containing 1 L of 2× AAM 30% medium, and obtained after 10 days from a 50,000 cell $\cdot$ mL^{-1} seed. Cell counts determined with an electronic particle counter (Coulter Counter Model TA II, 70μ cell opening) were correlated with algal dry weight (USEPA, 1971), so that metal accumulation factors (expressed per gram dry weight) could be later calculated. In general, 1 mg dry weight of algae corresponded to 10^7 cells for cultures used as inocula.

2.2. Bioaccumulation Protocol

While long-term plans are to use this bioaccumulation test to appraise unknown liquid waste samples, it was necessary for us to demonstrate the suitability of the test for the purposes intended by first working with known metal solutions. With individual working solutions of 1000 $\mu g \cdot L^{-1}$ for Cd^{2+} ($CdCl_2$), Zn^{2+} ($ZnCl_2$), and Cu^{2+} ($CuSO_4 \cdot 5H_2O$), a prebioaccumulation test was conducted to determine the maximum concentration of each metal that would not inhibit growth (i.e., no drop in cell count in relation to control) after a 24-h exposure period. From this information, bioaccumulation concentrations were fixed at 250 $\mu g \cdot L^{-1}$ and lower for each metal tested. Both pre- and definitive bioaccumulation tests were undertaken under the experimental conditions described below:

Working volume: 40 mL in 125-mL flask (prebioaccumulation test) and 1 L in 4-L flask (definitive test)
Algal inoculum: 70 $mg \cdot L^{-1}$
Nutrient spike: 2× AAM (Miller et al., 1978)
Buffer to control pH at 6.7 ± 0.1: 6.8 $g \cdot L^{-1}$ KH_2PO_4 and 1.2 $g \cdot L^{-1}$ NaOH (Hassett et al., 1981)
Exposure period: 24 ± 1 h
Lighting: 95u $E \cdot m^{-2} \cdot s^{-1}$ (continuous)
Temperature: 24 ± 2°C
Shaking speed: 90 rpm
Metal addition: 0, 50, 100, 250 $\mu g \cdot L^{-1}$

Cells prepared for inocula as well as those collected after bioaccumulation testing are centrifuged (500*g*, 10 min) and washed in sterile buffered (15 $mg \cdot L^{-1}$ $NaHCO_3$) Millipore Super Q water three consecutive times. Postbioaccumulation cells are finally concentrated in a 10-mL volume and subdivided into two portions. An aliquot corresponding to 40 mg algal dry weight is pipetted into a borosilicate glass test tube and reconstituted to a 10-mL volume with $NaHCO_3$ buffer addition. This portion is reserved for the photooxidation assay (Section 2.4). The remaining aliquot (corresponding approximately to 30 mg algal dry weight) is similarly reconstituted to 10 mL and preserved with an addition of 0.2 mL concentrated HNO_3 (Ultrex Baker) for chemical analysis of metal uptake (Section 2.3). Both aliquots are held refrigerated (4°C) overnight.

2.3. Chemical Determination of Algal oetal Uptake

Metal content of the algal cell concentrate is determined by atomic absorption spectrophotometry (AAS Perkin-Elmer Model 3030B) according to a recommended procedure for assaying biological materials (BEST, 1981). Algal metal uptake is finally reported in $\mu g \cdot g^{-1}$ algal dry weight.

2.4. Photooxidation Procedure

Irradiation is performed by immersing an ultraviolet probe (Pen Ray Model 3SC-9, Ace Glass Inc., NJ, 2.5 W with 80% of energy emitted at 254 nm) into each 10-mL algal cell concentrate solution. Photooxidation occurs for 3 h in the presence of 3% H_2O_2, after which time the concentrate solution is completely colorless. The pH of the irradiated solution is then adjusted to 7.0 ± 0.1 with 1 or 10% HNO_3. A TOC check (Dohrmann DC-80 total organic carbon analyzer) is undertaken to assess the efficiency of photooxidation. The remaining volume is bioanalyzed to determine its relative toxicity (Section 2.5).

2.5. Bioanalytical Determination of Algal Metal Uptake

The toxicity of irradiated algal cell concentrate solutions was estimated with two low-volume-requiring microbiological tests. Fifteen-minute EC_{50}'s were determined with the Microtox bacterial luminescence assay (Beckman, 1982), while 4-day EC_{50}'s were reported with an algal microplate assay employing *Selenastrum capricornutum* as bioindicator (Blaise et al., 1986). Toxicity values obtained from irradiated solutions were transformed to sublethal toxic units ($TU_{sl} = 100\%/EC_{50}$ v/v%) and correlated with nominal metal bioaccumulation exposure concentrations.

3. RESULTS AND DISCUSSION

3.1. Metal Uptake by *C. variabilis*

Cadmium, copper, and zinc uptake was chemically determined in *C. variabilis* following exposure conditions to each of the three nominal metal concentrations investigated (Fig. 14.1). Metal exposure concentrations were precisely determined by AAS immediately prior to the onset of experimentation and

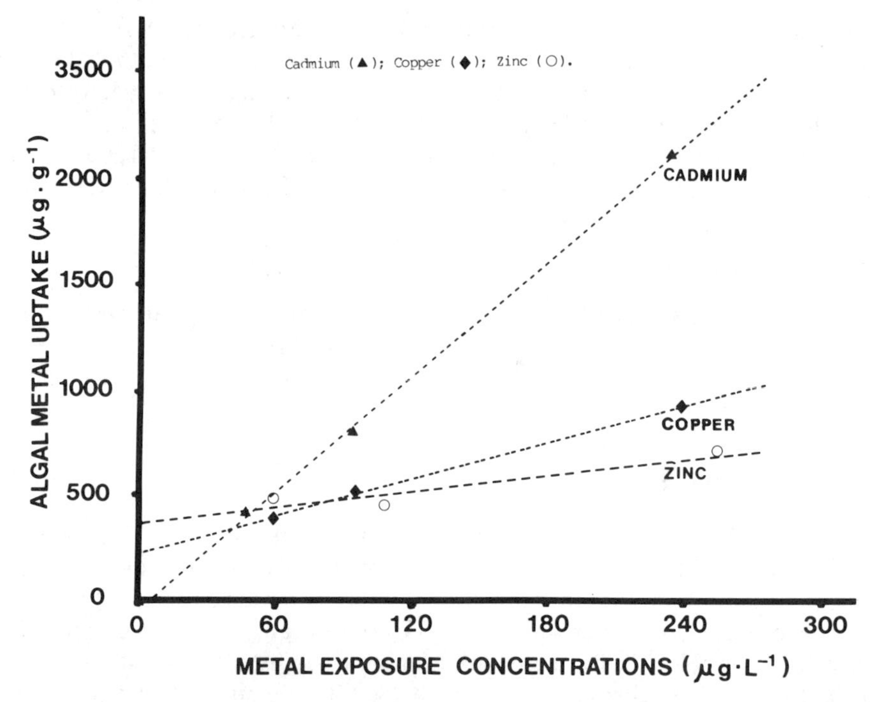

Figure 14.1. Metal uptake by *C. variabilis*.

reflect an average of three values obtained from three separate bioaccumulation tests. While metal uptake efficiency is classically shown to be a function of metal type and exposure concentration (Rai et al., 1981), the results in Fig. 14.1 clearly show the assimilative capacities of *C. variabilis* and validate the experimental accumulation test protocol employed. Metal bioaccumulation factors for *C. variabilis* are indicated in Table 14.1. Those reported for Cd were independent of exposure concentrations tested. Cu and Zn, on the other hand, exhibited bioaccumulation factors that are inversely related to exposure concentration. Barring one coefficient of variation showing marked variability (i.e., 42% for the Cu bioaccumulation factor at an exposure concentration of 250 $\mu g \cdot L^{-1}$), all results reflect acceptable methodological reproducibility. Although it is beyond the scope of this paper to discuss bioaccumulation mechanisms, it is interesting to observe that the uptake characteristics shown here by *C. variabilis* are similar to those reported for other microalgae (e.g., Canteford et al. (1978), for Cu and Zn; Truhaut et al. (1980), for Cd).

3.2. Photooxidation Toxicity

In considering a totally bioanalytical way of detecting inorganic accumulation in algae, it was first necessary to assess the toxicity of unexposed (control) algal cell concentrates after irradiation. Significant toxicity was, in fact, generated by

Table 14.1 Metal Bioaccumulation Factors[a] for *C. variabilis*

Nominal Metal Exposure Concentrations ($\mu g. \cdot L^{-1}$)	Metal[b]		
	Cd	Cu	Zn
50	8400 ± 850(10)	7400 ± 660(9)	9500 ± 1600(17)
100	8000 ± 500(6)	5100 ± 430(8)	4600 ± 740(16)
250	8400 ± 570(7)	2400 ± 1000(42)	2800 ± 340(12)

[a] Bioaccumulation factor: ratio of metal concentration in algae ($\mu g \cdot g^{-1}$ dry weight) to that in tested metal solution ($\mu g \cdot mL^{-1}$).
[b] Means, standard deviation, and coefficients of variation for three separate tests.

the *S. capricornutum* microtest when challenged with such irradiated control solutions (Table 14.2). Attempts to eliminate this toxicity by varying irradiation time, algal biomass, and/or oxidizing agent concentration proved unsuccessful. Further investigations with these preservation and irradiation techniques allowed us to minimize background toxicity and to account for it reproducibly (Table 14.2). Procedure 3 was thus retained for all subsequent experiments. While Procedure 1 did not yield higher background toxicity than Procedure 3, its reproducibility was poorer (CV = 26%), probably because of sample frothing due to O_2 bubbling, which caused some volume loss. For reasons unexplainable by us, freezing the algal cell concentrate prior to irradiation (Procedure 2) increased background toxicity significantly.

Although control toxicity is taken into consideration by subtracting it from total toxicity expressed by exposed irradiated solutions, we have not yet sought to explain its origin. It is probably caused by one of three (in)dependently acting factors. First, photooxidation liberates the inorganic content of the algal biomass itself. Second, residual concentrations of the oxidizing agent may contribute toxicity. Third, toxic by-products may result from irradiation. pH is not a factor, since irradiated solutions invariably gave values of 8.0 ± 0.5 and all were adjusted to 7.0 ± 0.1 prior to bioanalysis.

3.3. Bioanalytical Expression of Metal Uptake

Bioanalytical indication of metal uptake in *C. variabilis* is presented in histogram form in Fig. 14.2 following 4-day EC_{50} determinations of irradiated solutions with the *S. capricornutum* microplate assay. Here again, toxicity in control irradiated solutions has been deducted from that in metal-exposed irradiated solutions and bars reflect an average (± standard deviation) metal uptake from three separate experiments. It is clear that toxicity has been significantly expressed in all exposed irradiated solutions, regardless of metal type and test concentration. Hence, the *S. capricornutum* microtest appears to show adequate resolution and sensitivity in detecting inorganic toxicity accumulated by *C. variabilis* after applying the irradiation protocol employed. In fact, results obtained with this microbioassay closely parallel the chemically

Table 14.2 Toxicity of Control Algal Cell Concentrate Solution Measured with the *S. capricornutum* Micro Test, Following Different Preservation and Irradiation Techniques

Procedure	Concentrate Preservation Mode (Prior to Irradiation)	Oxidizing Agent during UV Irradiation	Sublethal Toxic Units		
			Mean	CV[c]	N[d]
1	Refrigerated (4°C)	O_2 (pure)[a]	10.2[b]	26	5
2	Frozen (−20°C)	H_2O_2(3%)	15.4	19	9
3	Refrigerated (4°C)	H_2O_2(3%)	10.7[b]	10	10

[a]Frothing of solution undergoing irradiation results using O_2 as an oxydizing agent.
[b]Not significantly different from each other ($P > 95\%$, Student t test).
[c]Coefficient of variation = (standard deviation × 100)/Mean.
[d]Number of samples tested.

determined metal uptake data presented in Fig. 14.1. Cell content of cadmium and copper increase markedly with exposure concentration (Fig. 14.1), and cell toxicity augments similarly (Fig. 14.2). With zinc, metal uptake appears relatively independent of exposure concentration (Fig. 14.1), and the toxicity response displays an analogous trend (Fig. 14.2).

Microtox assays on irradiated test solutions proved totally insensitive to detecting inorganic uptake by *C. variabilis* at all metal exposure concentra-

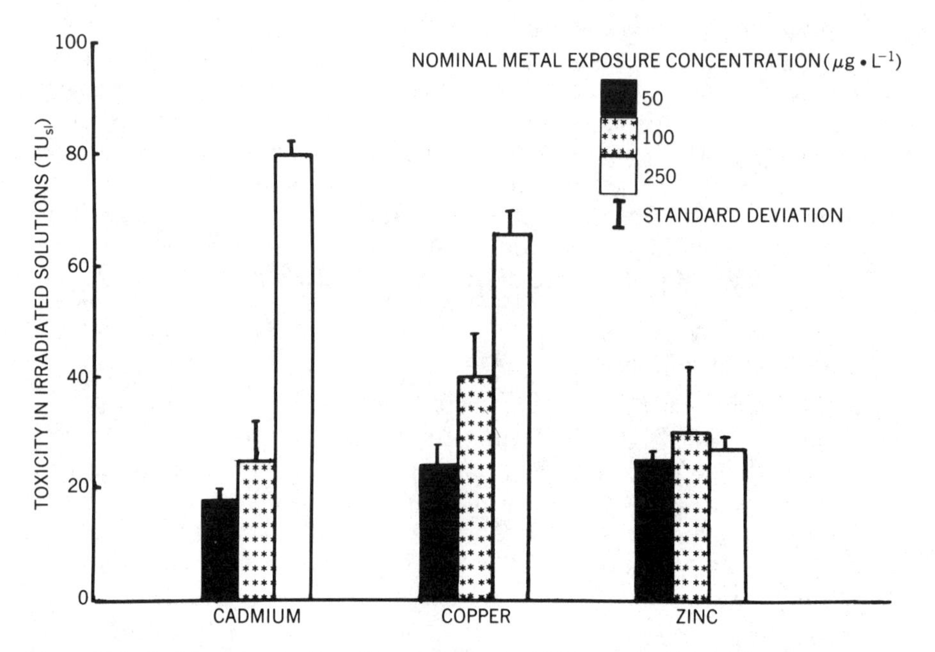

Figure 14.2. Toxicity (expressed in sublethal toxic units, TU_{sl}) generated by the *S. capricornutum* microtest in metal-exposed algal cell concentrate solutions after irradiation.

tions. Residual organic matter after irradiation may be at fault because of its complexing action on metals. While 1500 ± 500 $mg \cdot L^{-1}$ of TOC content are present in the algal biomass before irradiation, 49 ± 21 $mg \cdot L^{-1}$ were found to remain after irradiation. Although organic matter destruction efficiency is high (97% on average), it has not been possible to eliminate it totally. Other workers have reported similar findings in this respect (Batley and Farrar, 1978; Higgins, 1982; McDonald, 1980). The Microtox test may therefore be unreactive to metals bound by remaining organic matter. While the same argument can apply to the *S. capricornutum* microtest, the exposure period of the latter to irradiated solutions is much longer (4 days versus 15 min) and metal availability and reactivity is probably not impeded as a result.

4. SUMMARY AND TEST PROSPECTS

A novel approach to detecting inorganic accumulation in algae is described. While traditional means of indicating metal uptake in algae include exposure followed by chemical analysis of cell content, the procedure developed here is entirely biological. It draws upon the metal uptake properties of *C. variabilis*, the partial unbinding of accumulated inorganics by photooxidation of cell constituents, and the sensitivity of a low-volume algal microbioassay with *S. capricornutum* whose toxic expression to irradiated *C. variabilis* solutions reflects mineral sorption.

The *C. variabilis/S. capricornutum* inorganic bioaccumulation/toxicity indicator test may prove useful within the framework of an integrated hazard assessment scheme to screen previously unstudied industrial effluents and other unknown liquid wastes. To explore this possibility, future research efforts will be directed as follows:

- Investigations with other metals
- Metal mixture experimentation to assess interaction effects on uptake and subsequent toxicity responses
- Application to chemically well-characterized effluent/liquid waste samples to appraise test performance.

ACKNOWLEDGMENTS

The authors are grateful for the technical assistance provided by the Environmental Protection Laboratory staff and extend their thanks to regional management of Environment Canada, Quebec Region, for supporting this research initiative. We are also indebted to Diane Brûlé for typing the manuscript.

REFERENCES

Bates, S.S., Létourneau, M., Tessier, A., and Campbell, P.G.C. (1983). Variation in zinc adsorption and transport during growth of *Chlamydomonas variabilis* (Chlorophyceae) in batch culture with daily addition of zinc. *Can. J. Fish. Aquat. Sci.* **40**: 895–904.

Batley, G., and Farrar, Y. (1978). Irradiation technique for release of bound heavy metals in natural waters and blood. *Anal. Chim. Acta* **99**: 283–292.

Beckman Instruments Inc. (1982). *Microtox™ System Operating Manual, Beckman Instructions 015-555879*. Beckman Instruments Inc., Microbics Operations, Carlsbad, CA 92008.

BEST (Bureau d'étude sur les substances toxiques). (1981). Les méthodes d'analyse du cadmium, chrome, cobalt, cuivre, fer, manganese, nickel, plomb et zinc dans l'eau, les sédiments, les milieux biologiques et l'air. Comité de normalisation des méthodes d'analyse, Environnement Québec, BEST 29.

Blaise, C., Ska, B., Sabatini, G., Bermingham, N., and Legault, R. (1981). Potentiel de bioaccumulation de substances toxiques d'eaux résiduaires industrielles à l'aide d'un essai utilisant des algues et des bactéries. *Inst. Nat. Santé Rech. Méd.* **106**: 155–165.

Blaise, C., Bermingham, N., and van Coillie, R. (1985). The integrated ecotoxicological approach to assessment of ecotoxicity. *Water Qual. Bull.* **10**: 3–10, 60–61.

Blaise, C., Legault, R., Bermingham, N., van Coillie, R., and Vasseur, P. (1986). A simple microplate algal assay technique for aquatic toxicity assessment. *Toxicity Assessment* **1**: 161–181.

Blazka, P., and Prochazkova, L. (1983). Mineralization of organic matter in water by U.V. irradiation. *Water Res.* **17**: 355–364.

Campbell, P., Bisson, M., Bougie, R., Tessier, A., and Villeneuve, J.P. (1983). Speciation of aluminium in acidic freshwaters. *Anal. Chem.* **55**: 2246–2252.

Canteford, G.S., Buchanan, A.S., and Ducker, S.C. (1978). Accumulation of heavy metals by the marine diatom *Ditylum brightwellii* (West) Grunow. *Aust. J. Mar. Freshwater Res.* **29**: 613–622.

Chapman, P., and Long, R. (1983). The use of bioassays as part of a comprehensive approach to marine pollution assessment. *Marine Pollut. Bull.* **14**: 81–84.

Christensen, E.R., and Scherfig, J. (1979). Effects of manganese, copper and lead on *Selenastrum capricornutum, Chlorella stigmatophora. Water Res.* **13**: 79–82.

Couture, P., Couillard, D., and Croteau, G. (1981). Un test biologique pour caractériser la toxicité des eaux usées. *Environ. Pollut.* **2**: 217–222.

Couture, P., Visser, S.A., van Coillie, R., and Blaise, C. (1985). Algal bioassays: Their significance in monitoring water quality with respect to nutrients and toxicants. *Schweiz. Z. Hydrol.* **47**: 127–158.

Friant, S.L., and Sherman, J.W. (1981). The use of algae as biological accumulators for monitoring aquatic pollutants. In E.D. Kennedy (Ed.), Proceedings of second interagency workshop on *in situ* water quality sensing: biological sensors, held at Pensecola Beach, Florida, U.S.A., on 28–30 April, 1980. NOAA/National Marine publication, pp. 185–206.

Hassett, J.M., Jennett, J.C., and Smith, J.E. (1981). Microplate technique for determining accumulation of metals by algae. *Appl. Environ. Microbiol.* **41**: 1097–1106.

Higgins, H.W. (1982). Determination of the low levels of ^{14}C radioactivity in cell exudates by means of U.V. photo-oxidation and a rapid $^{14}CO_2$—Collection technique. *Anal. Biochem.* **127**: 121–133.

Korte, F., Klein, W., and Sheehan, P. (1985). The role and nature of environmental testing methods. In P. Sheehan, F. Korte, W. Klein and P. Bourdeau (Eds.), *Appraisal of Tests to Predict the Environmental Behaviour of Chemicals*. SCOPE, Wiley, New York, pp. 1–11.

Leclerc, H., and Dive, D., (Eds.) (1982). Les tests de toxicité aiguë en milieu aquatique. *Inst. Nat. Santé Rech. Méd.* **106**: 600 pages.

Malaiyandi, M., Sadar, M., Lee, P., and O'Grady, R. (1980). Removal of organics in water using hydrogen peroxide in presence of ultraviolet light. *Water Res.* **14**: 1131–1135.

McDonald, K.L. (1980). Photocombustion for black liquor analysis. *Tappi* **63**: 79–80.

McFeters, G., Bond, P., Olson, S., and Tchan, Y. (1983). A comparison of microbial bioassays for the detection of aquatic toxicants. *Water Res.* **17**: 1757–1762.

Miller, W., Greene, J., and Shiroyama, T. (1978). The *Selenastrum capricornutum* Printz algal bottle test: Experimental design, application, and data interpretation protocol. U.S. Environmental Protection Agency Report No. EPA-600/9-78-018, Corvallis, OR.

Miller, W.E., Peterson, S.A., Greene, J.C., and Callahan, C.A. (1985). Comparative toxicology of laboratory organisms for assessing hazardous waste sites. *J. Environ. Qual.* **14**: 569–574.

Mount, D., Thomas, N., Norberg, T., Barbour, M., Roush, T., and Brandes, W. (1984). Effluent and ambient toxicity testing and instream community response on the Ottawa River, Lima, Ohio. EPA-600/3-84-080.

Neuhold, J., and Ruggerio, L. (1976). Ecosystem processes and organic contaminants. National Science Foundation, Washington, DC, NSF-RA-760008.

Persoone, G., Jaspers, E., and Claus, C. (Eds.) (1984). Ecotoxicological Testing for the Marine Environment, Vol. 1. State Univ. Ghent and Inst. Mar. Scient. Res., Bredene, Belgium.

Rai, L.C., Gaur, J.P., and Kumar, H.D. (1981). Phycology and heavy-metal pollution. *Biol. Rev.* **56**: 99–151.

Sakuguchi, T., Tsu–i,T., Nakajima, A., and Horikoshi, T. (1979). Accumulation of cadmium by green microalgae. *Eur. J. Appl. Microbiol. Biotechnol.* **8**: 207–215.

Ska, B., Delisle, C., and Blaise, C. (1984). Evaluation de la bioaccumulation à l'aide d'un test biologique combiné algue-bactérie. *Environ. Technol. Lett.* **5**: 69–74.

Thomas, J., Skalski, J., Cline, J., McShane, M., Simpson, J., Miller, W., Peterson, S., Callahan, C., and Greene, J. (1986). Characterization of chemical waste site contamination and determination of its extent using bioassays. *Environ. Toxicol. Chem.* **5**: 487–501.

Truhaut, R., Férard, J.F., and Jouany, J.M. (1980). Cadmium IC_{50} determinations on *Chlorella vulgaris* involving different parameters. *Ecotoxicol. Environ. Safety* **4**: 215–223.

USEPA (United States Environmental Protection Agency) (1971). Algal assay procedure bottle test. National Eutrophication Research Program, Corvallis, OR.

Van Coillie, R., Couture, P., and Visser, S.A. (1983). Use of algae in aquatic ecotoxicology. In J.O. Nriagu (Ed.), *Advances in Environmental Science and Technology*, Vol. 13, *Aquatic Toxicology*. Wiley, New York, pp. 487–502.

15

BIODEGRADATION OF PETROLEUM IN THE MARINE ENVIRONMENT AND ITS ENHANCEMENT

Kenneth Lee and Eric M. Levy

Department of Fisheries and Oceans, Bedford Institute of Oceanography, Dartmouth, Nova Scotia, Canada

1. **Introduction**
2. **Factors limiting the rate of biodegradation**
   2.1. Hydrocarbon-degrading microorganisms
   2.2. Physical/chemical properties of oil
   2.3. Temperature
   2.4. Nutrients
   2.5. Oxygen
   2.6. Salinity/pressure
3. **Enhancement of microbial degradation**
   3.1. Chemical dispersion
   3.2. Seeding
   3.3. Oxygenation
   3.4. Nutrient enrichment
4. **Summary**
   References

1. INTRODUCTION

When crude oil is spilled in an aquatic environment, most of the volatile lower molecular weight compounds are usually lost within a few days, leaving chemical and biological processes of decomposition to operate upon the remaining fraction (Jordan and Payne, 1980). Although biodegradation plays a major role in the natural "weathering" of petroleum (National Academy of Sciences, 1985; Atlas, 1981), it is generally a slow and incomplete process. For example, Miget (1973) reported that while regions contaminated by oil spills are usually colonized by large populations of oil-degrading bacteria within one or two weeks, two months or more may be necessary before appreciable biodegradation is observed. Since present knowledge of the interactions and/or the sequence of events that occur during microbial degradation of hydrocarbons in the environment is far from complete, this review summarizes the major factors that influence or limit the interaction between microorganisms and hydrocarbons in the marine environment and presents a synopsis of the various approaches that may be developed to enhance natural biodegradation of hydrocarbons.

2. FACTORS LIMITING THE RATE OF BIODEGRADATION

The persistence of petroleum contaminants within the marine environment is dependent on a combination of factors that include the nature of the biota, the physical/chemical properties of seawater and of the oil, and the ambient environmental conditions. Because of the multitude of possible interactions among these factors, it is difficult to predict accurately the rates at which a specific crude oil or refined product will be degraded in the environment. Nevertheless, since the potential effect of each of these factors has been examined, the general direction in which they will collectively alter the oil can be predicted qualitatively.

2.1. Hydrocarbon-Degrading Microorganisms

Petroleum-degrading microorganisms appear to be ubiquitous in the marine environment (Atlas, 1978; Roubal and Atlas, 1978; Robertson et al., 1973; Tagger et al., 1976; Walker and Colwell, 1976b; Bunch and Harland, 1976; Mulkins-Phillips and Stewart, 1974c). Since hydrocarbons are naturally occurring organic compounds as well as anthropogenic contaminants, indigenous microorganisms have adapted themselves to metabolize these compounds and the role of microorganisms in mitigating the impact of petroleum spills on the environment has been the subject of numerous reviews (Atlas, 1986, 1981; Bossert and Bartha, 1984; Colwell and Walker, 1977; Floodgate, 1984; Zobell, 1964; National Academy of Sciences, 1975; Karrick, 1977; Van der Linden, 1978; Jordan and Payne, 1980). While biodegradation usually does not mineralize all of the constituents in a complex hydrocarbon mixture such as crude oil, the residues resulting from microbial attack are relatively harmless because of their low biological activity (Hughes and Stafford, 1983).

No single microbial species appears to be able to completely degrade any given oil. Since microorganisms are selective, many different microbial species in a mixed culture may be required for significant overall degradation. Mulkins-Phillips and Stewart (1974c) reported finding hydrocarbon-degrading bacteria of the genera *Nocardia*, *Pseudomonas*, *Flavobacterium*, *Vibrio*, and *Achromobacter* in coastal waters and sediments from the northwest Atlantic. Based on the frequency of isolation, Bartha and Atlas (1977) stated that the predominant genera of aquatic hydrocarbon-degrading microorganisms were *Pseudomonas*, *Achromobacter*, *Arthrobacter*, *Micrococcus*, *Nocardia*, *Vibrio*, *Acinetobacter*, *Brevibacterium*, *Corynebacterium*, *Flavobacterium*, *Candida*, *Rhodotorula*, and *Sporobolomyces*. Although yeasts and filamentous fungi may be very active in degrading hydrocarbons (Cerniglia and Perry, 1973; Walker et al., 1975c), their scarcity in the open ocean suggests that in global terms their activity is probably insignificant (Floodgate, 1984). On the other hand, in inshore waters, on beaches, in salt marshes, and in sand dunes, they may be locally important as degraders of oil.

Walker et al. (1975a, 1975b) reported that a strain of the achlorophyllous alga *Prototheca zopfi*, isolated from sediments collected from Baltimore Harbour, was capable of degrading 10% of a motor oil and 40% of a crude oil—degradative capabilities comparable to those of bacteria. Cerniglia et al. (1980), found that species of cyanobacteria and algae including *Oscillatoria* spp., *Microcoleus* sp., *Anabaena* spp., *Agmenellum* sp., *Coccochloris* sp., *Nostoc* sp., *Aphanocapsa* sp., *Chlorella* spp., *Dunaliella* sp., *Chlamydomonas* sp., *Ulva* sp., *Cylindretheca* sp., *Amphora* sp., *Porphyridium* sp., and *Petalonia* sp., were capable of metabolizing naphthalene. Since only 19 cultures were assessed in this study, it is evident that the ability to oxidize aromatic hydrocarbons may be widely distributed among cyanobacteria and algae.

Although the numbers of hydrocarbon-degrading microorganisms are generally very low in areas of the open ocean away from sources of petroleum, population levels increase dramatically along shipping lanes (Mironov, 1970), and in petroleum-contaminated coastal waters (Buckley et al., 1976; Oppenheimer et al., 1977). The indigenous microbial flora respond to elevated oil concentrations by increasing their oil-degrading capacity by selectively promoting the growth of oil-degrading strains and by inducing enzymatic mechanisms within the existing strains. For example, laboratory studies have demonstrated that the numbers of hydrocarbon-degrading microorganisms within environmental samples generally increase when exposed to petroleum (Zobell, 1973; Soli, 1973; Colwell and Walker, 1977), and this response has been verified in sediments and waters of pristine environments in both experimental (Atlas et al., 1978) and accidental oil spills (Kator and Herwig, 1977; Colwell et al., 1978; Ward et al., 1980; Atlas et al., 1980a, 1980b; Atlas and Bronner, 1981). In contrast, Bunch et al. (1983b) have reported that the number of total viable heterotrophs or the relative abundance of oil-degrading bacteria in the microbial community of the water column may not always increase in response to short-term (acute) exposure to high levels of oil. Atlas and Bartha (1973b) suggested that the size of microbial populations may decrease in the presence of oil, if no induction of oil-degrading enzymes has occurred from prior exposure to hydrocarbons.

It is evident, however, that in general the distribution of hydrocarbon-utilizing microorganisms quantitatively reflects the degree or extent of exposure of the ecosystem to hydrocarbons. At sites where oil is already present, either from acute or long-term exposures, there will be greater numbers of oil-degrading bacteria, presumably because of continuous enrichment (Zobell and Prokop, 1966; Roubal and Atlas, 1978; Walker and Colwell, 1974; Gunkel et al., 1980; Gunkel and Gassman, 1980). In uncontaminated ecosystems, oleoclastic or hydrocarbon-degrading microorganisms generally constitute less than 0.1% of the microbial community, whereas they can constitute up to 100% of the viable microorganisms in hydrocarbon-contaminated ecosystems (Azoulay et al., 1983; Atlas, 1981; Mulkins-Phillips and Stewart; 1974c; LePetit et al., 1977).

It is now widely accepted that biodegradation of different classes of petroleum compounds occurs simultaneously, but at different rates (Walker et al., 1976b; Walker and Colwell, 1976b). In the marine environment, normal alkanes are degraded most rapidly, followed by isoalkanes, cycloparaffins, aromatics, and larger polycyclic aromatics (Bartha and Atlas, 1977; Wong et al., 1984). However, in most cases, the rates of hydrocarbon biodegradation in the marine environment have been estimated by the questionable procedure of measuring the amount of oil present at the beginning and the end of an experiment, and assuming that the rates are linear and similar for all components.

The biodegradation rate of hydrocarbons in the marine environment is highly dependent on the composition of the biota, the chemical nature of the petroleum mixture, and environmental conditions. From a review of existing literature, Zobell (1969) estimated that the rates of degradation of crude oils, lubricating oils, cutting oils, and oily wastes in well-oxygenated coastal waters at temperatures between 24 and 30°C ranged from 0.02 to 2.0 g m^{-2} d^{-1}. Atlas and Bronner (1981), estimated that microbial degradation could remove up to 0.05 g of hydrocarbons m^{-2} d^{-1} from intertidal sediments contaminated by the Amoco Cadiz oil spill. Arhelger et al. (1977) estimated rates of petroleum biodegradation in arctic and subarctic waters on the basis of the oxidation of *n*-[^{14}C]dodecane: in Port Valdez, 0.7 g L^{-1} d^{-1}; in Chukchi Sea, 0.5 g L^{-1} d^{-1}; and in the Arctic Ocean, 0.001 g L^{-1} d^{-1}. The observed rates of hydrocarbon degradation showed a definite link to climatic conditions and, in general, low rates of microbial degradation and long persistence times for the hydrocarbon contaminants have been observed in studies of experimental oil releases and spills conducted in the arctic and subarctic.

Walker and Colwell (1976a) reported that the rates of petroleum uptake and mineralization were higher in samples collected from an oil-contaminated harbor than in similar samples from a relatively uncontaminated region. Turnover times of 15 and 60 min were reported for the contaminated and uncontaminated areas respectively, using *n*-[^{14}C]hexadecane. It must be emphasized, however, that the "heterotrophic potential" measured by the uptake and/or mineralization of ^{14}C-labeled substrates is only a relative index of bacterial activity with specific substrates and is not an absolute measure of heterotrophic production and total oil degradation.

2.2. Physical/Chemical Properties of Oil

Differences in the composition and concentration of the hydrocarbons in petroleum influence their susceptibility to biodegradation. For example, Mulkins-Phillips and Stewart (1974a) reported that, under similar experimental conditions, a *Nocardia* sp. degraded 94% of the *n*-alkanes and 35% of the unresolved fraction of an Arabian crude oil, but only 77 and 13%, respectively, of a Venezuelan crude. Walker et al. (1976a) found that the extent of biodegradation of two crude oils (South Louisiana and Kuwait) and two refined fuel oils (Nos. 2 and 6) by indigenous bacteria in Baltimore Harbour sediments differed greatly with 22% of the South Louisiana crude oil, 49% of the Kuwait crude oil, 45% of the No. 2 fuel oil, and 89% of the No. 6 fuel oil (Bunker C) remaining after 7 weeks. Analyses of samples by column chromatography and mass spectrometry indicated that some components of the petroleum were selectively degraded, whereas others were more persistent. Furthermore, in support of the results obtained by Mulkins-Phillips and Stewart (1974a), they reported significant differences in the degradation of identical compounds in the different oils. As a general trend, the rates of biodegradation processes tend to decrease as recalcitrant substances form a greater proportion of the remaining residue, that is, as the substrate gains a higher proportion of condensed polyaromatic compounds, condensed cycloparaffins, asphaltenes, and resin compounds.

The rate of biodegradation of oil in the marine environment is also influenced by alterations in its physical state; for example, the formation of "mousse" (water-in-oil emulsions) and tar balls inhibits the biodegradation and abiotic weathering of oil by restricting the availability of oxygen and mineral nutrients, and by reducing photooxidative and evaporative losses (National Academy of Sciences, 1985; Payne and Phillips, 1985) by reducing the surface area of the oil. Such interactive processes have been suggested as the major factors limiting the rates of petroleum biodegradation at the IXTOC-I (Atlas et al., 1980a) and Metula (Colwell et al., 1978) oil spills.

2.3. Temperature

Psychrotrophic, mesophilic, and thermophilic oil-degrading microorganisms have been isolated from the environment and microbial biodegradation of hydrocarbons has been observed at temperatures ranging from below 0°C

(Zobell, 1973; Traxler, 1973) to above 70°C (Mateles et al., 1967). In addition to having a direct effect on metabolic processes, changes in temperature alter the physical/chemical state of petroleum within the environment. Low temperatures decrease the rates of volatilization and may increase the solubility of low molecular weight hydrocarbons, some of which are toxic to microflora (Atlas, 1975; Atlas and Bartha, 1972a; Walker and Colwell, 1974). Consequently, in oil spilled under low temperature conditions, the onset of biodegradation may be delayed substantially because toxic fractions of the hydrocarbons are retained within the water column for a considerably longer period than at higher temperatures. Furthermore, the degree of spreading, or slick formation, is reduced by the lower viscosity of the oil at low temperatures and thereby restricts the surface area of the oil available for colonization by hydrocarbon-degrading microorganisms.

An experiment on the combined influence of temperature and chemical composition of crude oils on the selection of hydrocarbon-degrading microorganisms has been conducted by Cook and Westlake (1974). At 4°C, *Achromobacter*, *Alcaligenes*, *Flavobacterium*, and *Cytophaga* were isolated on a substrate of Prudhoe Bay crude oil; *Acinetobacter*, *Pseudomonas*, and unidentified gram-negative cocci on Atkinson Point crude oil; and *Flavobacterium*, *Cytophaga*, *Pseudomonas*, and *Xanthomonas* on Norman Wells crude oil. At 30°C, substantially different major genera were isolated: On the Prudhoe Bay crude oil the major genera were *Achromobacter*, *Arthrobacter*, and *Pseudomonas*; on the Atkinson Point crude oil, they were *Achromobacter*, *Alcaligenes*, and *Xanthomonas*; and on the Norman Wells crude oil, they were *Acinetobacter*, *Arthrobacter*, *Xanthomonas*, and other gram-negative rods.

Environmental adaptation by oil-degrading microbes to low temperature conditions is well documented. For example, Colwell et al. (1978) reported greater degradation of Metula crude oil at 3°C than at 22°C by indigenous populations of microorganisms present in beach sand collected in the Antarctic. In laboratory studies at 0.1% oil concentration, antarctic isolates adapted to the incubation temperatures degraded 48% of the added hydrocarbons at 3°C, and 21% at 22°C. Extensive research on the biodegradation of Prudhoe Bay crude oil in the arctic coastal ecosystem (Atlas, 1985, 1977, 1978; Atlas et al., 1978) has demonstrated that biodegradation potentials for this oil were much lower in ice than in water or in sediment. Natural rates of biodegradation were found to be slow during an arctic summer, with abiotic and biodegradative processes accounting for less than 50% of the oil loss. In the arctic environment, microbial degradation involving indigenous psychrophilic and psychrotrophic microorganisms will occur, albeit much more slowly than during a warm season in a temperate zone. Rates of oil biodegradation in the Arctic appear to be limited by temperature-dependent factors (e.g., physical state of the oil and ice conditions) and the concentration of available nutrients.

Changes in temperature have also been correlated with the induction and/or suppression of metabolic pathways within microorganisms. In studying arctic surface waters during the summer, Horowitz and Atlas (1977) found that the percentage loss of individual components of the oil were similar. It was postulated that cometabolism, in which normally resistant compounds in petroleum were enzymatically attacked by microorganisms sustained on other organic substrates (Horvath, 1972; Perry, 1979; Cerniglia et al., 1980), was enhanced by low temperatures.

2.4. Nutrients

Seawater is far from being an ideal medium for bacterial degradation of oil because the concentration of nutrients necessary for heterotrophic processes is kept low by preferential uptake by other microbial species including phytoplankton. Nitrogen and phosphorus are the nutrients most likely to limit microbial degradation of hydrocarbons in the sea (Gibbs and Davis, 1976; Bergstein and Vestal, 1978; Horowitz and Atlas, 1978). Of the two, in coastal marine environments, nitrogen is generally the nutrient first depleted by phytoplankton (Ryther and Dunstan, 1971). As the result of studies on the effect of nutrient limitation on the biodegradation of hydrocarbons in the oceans, Floodgate (1979) proposed the concept of "nitrogen demand," analogous to biochemical oxygen demand and estimated the nitrogen demand for the biodegradation of Kuwait oil at 14°C to be 0.4 μmol of nitrogen per milligram of oil. A variety of laboratory studies have provided convincing evidence that the slow rates of petroleum degradation often observed in the marine environment are the result of nutrient limitation (Walker and Colwell, 1975; Walker et al., 1976c; Atlas and Bartha, 1972b,c). In particular, nutrient limitation seems to be the predominant factor controlling hydrocarbon biodegradation in the Arctic. Haines and Atlas (1982) reported that biodegradation of Prudhoe Bay crude oil in nearshore sediments of the Beaufort Sea was not apparent until one year after exposure. This initial lack of response by the microbial community was not a direct consequence of low temperatures, since psychrotrophic or psychrophilic hydrocarbonoclastic microorganisms are capable of active growth and metabolism at temperatures below 0°C (Traxler, 1973; Zobell, 1973; Robertson et al., 1973; Morita, 1975). Therefore, it was hypothesized that nutrient and oxygen concentrations were the limiting factors, since Prudhoe Bay crude oil had previously been found not to contain components that were inhibitory to microbial hydrocarbon degradation (Atlas, 1975). Assuming that the optimal rates of hydrocarbon biodegradation occur at C:N and C:P ratios of 10:1 and 30:1 respectively (Atlas and Bartha, 1972b), concentrations of N and P several orders of magnitude higher would have been required to support maximal rates of hydrocarbon biodegradation in arctic sediments. Similarly, Colwell et al. (1978) concluded that hydrocarbons were degraded slowly following the Metula oil spill in the Antarctic because of limitations imposed by relatively low concentrations of nitrogen and phosphorus available in the seawater.

2.5. Oxygen

The oxygen requirement for microbial degradation of oil is substantial. Zobell (1969) estimated that 3 to 4 mg of dissolved oxygen were necessary to oxidize 1 mg of hydrocarbon into CO_2 and H_2O. Oxygen is not a limiting factor for the degradation of hydrocarbons in the open ocean, since the upper parts of the water column, where oil is most likely to be found, are usually well oxygenated. Furthermore the oxygenase enzymes responsible for the biodegradation of oil have a high affinity for oxygen and thus will function even when the concentration of dissolved oxygen is low (Floodgate, 1976). However, low-energy beaches and fine sediments are often depleted in oxygen (Colwell and Walker, 1977; Gibbs and Davis, 1976; Ward et al., 1980) and, therefore, the availability of oxygen may be a limiting factor in these environments. For example, Aminot (1981) reported that some microbial populations in areas polluted by the grounding of the Amoco Cadiz were oxygen as well as nitrogen and phosphorus limited. Johnston (1970) found that the oxygen concentration of interstitial waters within laboratory sand columns decreased rapidly in response to Kuwait crude oil. The mean rate of oxygen consumption over 4 months was 0.45 g m^{-2} d^{-1} at 10°C, corresponding to an oil degradation rate of 90 mg of oil m^{-2} d^{-1}.

Conditions that enhance aeration and the replenishment of nutrients favor biodegradation. After a 3-year study, Rashid (1974) concluded that the degradation of oil spilled on the coast of Nova Scotia from the grounding of the tanker Arrow depended largely on environmental factors, such as wave energy. Degradation was greatest in high-energy environments and lowest in protected embayment areas. In high-wave energy environments, there was a preferential loss of *n*-alkanes, which was attributed to microbial degradation. Presumably, oxygen and nutrient replenishment by wave-driven mixing permitted more extensive degradation. Similarly, biodegradation of oil in sediments has been found to be stimulated by bioturbation (Gordon et al., 1978), presumably through the introduction of oxygen into oil-contaminated sediments by burrowing animals such as polychaete worms.

There have also been reports of microorganisms degrading hydrocarbons, especially the *n*-alkanes, under anaerobic conditions (Chouteau et al., 1962; Parekh et al., 1977; Senez and Azoulay, 1961; Traxler and Bernard, 1969; Iizuka et al., 1969), as well as preliminary reports of anaerobic hydrocarbon degradation in natural ecosystems (Bailey et al., 1973; Pierce et al., 1975; Zobell and Prokop, 1966). Ward et al. (1980) compared the rates of hydrocarbon oxidation in sediments contaminated by the Amoco Cadiz spill under both aerobic and anaerobic conditions. Under strict anaerobic conditions in which methanogenesis was evident, $^{14}CO_2$ production was observed from labeled heptadecane and toluene, but not from hexadecane, although the rates of $^{14}CO_2$ production were orders of magnitude higher under aerobic than under anaerobic conditions. In the absence of oxygen, less than 5% of the added saturated and aromatic hydrocarbons were oxidized to $^{14}CO_2$ in 233 days, compared with

47% in 14 days under aerobic conditions. Thus, anaerobic biodegradation of hydrocarbons in the natural environment appears to proceed at negligible rates.

2.6. Salinity/Pressure

The salinity values observed in the world's oceans are well within the range tolerated by petroleum-degrading bacteria. Although estuaries present salinity, nutrient, and oxygen regimes that are very different from either coastal or oceanic areas, they are not inhibitory to microbial activity. Thus, in neither case is the range of salinities found in the marine environment a controlling factor in the biodegradation of petroleum. Most of the studies on the effects of salinity on microbial biodegradation of hydrocarbons have therefore been concerned with hypersaline environments (e.g., Ward and Brock, 1978)

On the other hand, the rate of biodegradation is affected by depth in the ocean. In addition to hydrocarbons originating from natural seeps, hydrocarbons from the ocean surface may be transported to the sediments of ocean basins via weathering processes that alter their specific gravity. Microbial degradation of organic matter in the deep sea is known to be extremely slow (Jannasch et al., 1971), therefore any oil entering deep-ocean environments will be degraded very slowly and persist for long periods of time. For example, Schwartz et al., (1974a,b, 1975) found that 94% of the ^{14}C-labeled hexadecane in a mixed hydrocarbon substrate was biodegraded by a mixed culture of bacteria isolated from the sediment–water interface at 5000 m depth in 8 weeks at 4°C and 1 atm pressure, whereas 40 weeks were required at 500 atm pressure.

In conclusion, the principal ecological factors that limit the rates of oil biodegradation in the marine environment have been identified. Microorganisms with the capability to degrade petroleum have been found to be ubiquitous in the world's oceans and *in situ* studies have demonstrated that microbial degradation is a major clearing mechanism for removal of petroleum contaminants. Low temperature and/or high pressure conditions have been shown to suppress biodegradation rates; however, results of *in situ* studies suggest that the degradation of petroleum in coastal waters is limited primarily by nutritional factors, e.g., nitrogen and phosphorus, and in some cases by oxygen availability.

3. ENHANCEMENT OF MICROBIAL DEGRADATION

During the last decade, the scientific community has realized that the enhancement of natural biodegradation may provide an effective means of dealing with oil pollution. The first approach to such enhancement is to disperse the oil into the water column to increase the oil/water interfacial area available for attack by indigenous microorganisms. A second approach is to seed the oil with natural or "genetically engineered" microorganisms having degradative capabilities that are superior to those present at the contaminated site. In either case, the rate and extent of biodegradation of the oil may be restricted by the availability of oxygen, nitrogen, or phosphorus, and in such cases the provision of these essential elements at the optimum concentrations and rates may achieve the desired result.

3.1. Chemical Dispersion

Direct microscopic observations show that hydrocarbon-degrading microbes act mainly at the oil–water interface, and therefore biodegradation rates may be accelerated by chemical dispersants that increase the area of the oil–water interface by promoting the rate at which oil droplets are sheared from the slick (Gatellier et al., 1973). Not only is the oil made more available to the microorganisms following dispersion, but, as emulsified droplets move through the water column, oxygen and nutrients become more readily available.

There has long been a controversy regarding the merits of chemical dispersants as a countermeasure to oil spills in the environment. This has been due largely to early experiences, particularly the Torrey Canyon incident (Smith, 1968) in which the solvents in the dispersant formulations caused greater ecological damage than the oil itself. The principal effects of chemical dispersants in the marine environment are the result of enhanced oxidation rates and loss of bacteria due to toxicity, and studies on the biodegradation of hydrocarbons have shown both stimulatory and inhibitory effects on microbial processes following the addition of chemical dispersants. For example, Mulkins-Phillips and Stewart (1974b) found that of four dispersants studied, only one (Sugee 2) increased the degradation of *n*-alkanes in crude oil, and Gatellier et al. (1973) and Robichaux and Myrick (1972) found that some dispersants inhibited populations of hydrocarbon-oxidizing microorganisms, whereas others enhanced them.

The second generation of chemical dispersants have been much less toxic and more effective than the earlier formulations (Gerlach, 1981). However, they are not without adverse biological effects (Lee et al., 1985; Hagstrom and Lonning, 1977; Hsiao et al., 1978; Rogerson and Berger, 1981). Theoretically, all dispersants can inhibit microbial activity by reducing the surface tension to levels that inhibit the growth of bacteria (Colwell and Walker, 1977; Blackman et al., 1978). Griffiths et al. (1981a,b) have reported that the rates of glucose uptake and mineralization in arctic and subarctic marine waters and sediments are suppressed by the presence of the dispersant Corexit 9527 at concentrations as low as 1 ppm. After studying the effects of weathered Lago Medio crude oil on bacteria from nearshore locations in the Northwest Territories of Canada, Bunch et al. (1983b) also concluded that the addition of Corexit to spilled petroleum crude may limit microbial activity. While the activity and numbers of oil-degrading bacteria did not appear to be affected by the oil additions alone, samples treated with Corexit 9527 at a concentration of

0.01% consistently showed lower levels of glutamic acid and hexadecane mineralization than untreated samples.

The behavior of dispersed oil has been studied extensively in the laboratory and under oceanic conditions to assess the possibility that toxic conditions can be established within the water column following the addition of chemical dispersants. McAuliffe et al. (1980) have shown that concentrations of oil in the milligram/liter range may be achieved for a period of some hours under dispersed slicks. The results of biological studies, however, have been contradictory since the higher dilution rates of dispersed oil may reduce toxic effects but highly toxic oil components that would otherwise evaporate from untreated oil may be accommodated in the water by the dispersant. In general, oil emulsions produced by dispersants seem to be more toxic than either the crude oil or the dispersant alone (Swedmark et al., 1973; Parsons et al., 1985). Furthermore, since biodegradation rates of oil under the sea surface are generally slow, dispersed oil droplets may eventually be deposited on the seafloor, particularly in shallow waters (Schwarz et al., 1974a).

Bunch et al. (1983a) reported that addition of Corexit 9527 severely reduced the mineralization of ^{14}C-labeled hexadecane during a 60-day laboratory experiment. Although the dispersant was found to prolong the period prior to mineralization, the addition of nitrate and phosphate rapidly enhanced the rate of hexadecane mineralization. Similarly, Foght and Westlake (1982) observed that enhanced growth of microbial flora in response to preferential use of components within Corexit may reduce nitrate and phosphate levels, causing a delay in the oxidation of hydrocarbons. Thus, in marine environments containing low levels of nitrate and phosphate, biodegradation of petroleum may be severely retarded by the addition of dispersants. Recent studies by Lee et al. (1985) indicate that Corexit 9527, at concentrations similar to those expected in actual field applications, stimulated bacterial production by inducing the release of organic compounds from the indigeneous phytoplankton population, as well as directly serving as a growth substrate. Furthermore, Lee et al. (1985) also attributed the increase in bacterial numbers observed in waters treated with dispersant to an acute toxic response by bacterivores (ciliates and appendicularians).

In practice, evaluating the use of chemical dispersants in the marine environment is complicated by the large number of factors that influence dispersant performance, including dispersant-to-oil volume ratio, temperature, salinity, sea state, physicochemical properties of the oil and the nature of the dispersant–oil mixing process (MacKay and Wells, 1983; MacKay 1981). As an oil slick spreads, weathers, and takes up water to form a water-in-oil emulsion, its viscosity increases. MacKay (1985) has stated that when the viscosity exceeds 3000–8000 mPa.s, the required dispersant dosage becomes so high that the use of chemical dispersants is no longer practical. Hence, there is only a limited period of a few hours to several days, depending on the prevailing spill conditions, within which application of dispersants is effective. Recent statistical analysis by Fingas (1985) indicate that effectiveness of chemical dispersants in field trials was very low (<33%), and consistently effective application of the chemical dispersants was observed only when the dispersants were premixed with the oil.

It is evident, therefore, that while chemical dispersants remove oil from the sea surface and are effective from the cosmetic point of view, they do not eliminate the toxic components of oil. Chemical dispersants merely aid the transport of oil from the surface into the water column. Furthermore, in spite of the well-documented capacity of various chemical dispersants to reduce the size of oil particles and disperse them in the water column, there has been no conclusive study demonstrating that dispersants enhance the extent of biodegradation. All experiments conducted to date suggest that while the rate of mineralization may increase in response to the addition of dispersants, refractory components of petroleum in the environment will persist regardless of the addition of dispersants.

3.2. Seeding

Seeding with oil-degrading bacteria immediately following an oil spill has attracted considerable attention as a method of enhancing petroleum biodegradation and thereby serving as an alternative to mechanical removal for the "cleanup" of oil contaminated sites.

The observation that a large number of *Pseudomonas* had the capability to degrade hydrocarbons encouraged extensive study in genetics and enzymology with regard to hydrocarbon degradation (Chakrabarty, 1972, 1976; Chakrabarty et al., 1973; Williams, 1978). Following the discovery that the hydrocarbon-degradation trait in these organisms was located in the plasmids, genetically engineered strains that contained several plasmids, each carrying genes for the degradation of a major class of petroleum hydrocarbons, were developed (Friello et al., 1976). Seeding of oil-contaminated environments with enriched cultures and/or genetically engineered strains of bacteria was considered a means by which the rate and extent of specific biodegradative processes in the environment could be enhanced (Liu and Dutka, 1972; LaRock and Severence, 1973; Reisfield et al., 1972; Friello et al., 1976; Weinberger and Kollenbrander, 1979; Atlas and Bartha, 1973a; Kado and Lui, 1981). However, in the terrestrial environment, Jobson et al. (1974) found that seeding oil-contaminated soil with oil-degrading bacteria had a much smaller effect in terms of enhanced degradation than the addition of fertilizers (i.e., nitrogen and phosphate). Similarly, studies conducted within the boreal region of the Northwest Territories (Westlake et al., 1978) indicated a rapid increase in bacterial numbers in oiled plots treated with nitrogen and phosphorus followed by a rapid disappearance of *n*-alkanes and isoprenoids and a continuous loss of weight of saturated compounds, but no significant effect on the composition of the recovered oil when the oil-contaminated plots were seeded with oleoclastic bacteria.

There have not been any large-scale *in situ* experiments to assess the effectiveness of seeding as a means of enhancing the biodegradation of petroleum hydrocarbons in the marine environment. Tagger et al. (1983) reported that the hydrocarbon-degradation potential of petroleum-contaminated seawater in 10-m^3 marine enclosures was not enhanced by the addition of a mixed culture of marine hydrocarbon-degrading bacteria. All strains of the mixed culture disappeared from the dominant microflora, while the autochthonous bacteria showed a capacity for adaptation to petroleum degradation approximately four days following the spill. In conclusion, seeding of the experimental spill with a large quantity of a mixed culture of hydrocarbon-degrading bacteria was found to be ineffective, since five months following the spill, the differences between the seeded and unseeded enclosures were not observed.

In addition to the capacity for direct biodegradation, Reisfield et al. (1972) observed that bacteria of the genus *Arthrobacter* appeared to be extremely efficient at emulsifying petroleum hydrocarbons and were able to isolate a strain capable of converting 65% of their test oil into substances that were not extractable by benzene. Following inoculation with *Arthrobacter*, oil in test enclosures was reported to be dispersed in less than one day. However, in this case, experimental evidence indicates that the microorganisms did not degrade components within the oil, but served only to emulsify the oil, which then became resuspended in the water column.

No single microbial species appears to possess the enzymatic capacity to metabolize more than two or three of the compound classes present in petroleum hydrocarbon mixtures. Many different bacterial species in mixed culture may be required for significant degradation of a complex hydrocarbon mixture such as crude oil (Westlake, 1982). Further, there is little evidence to indicate that improvements deriving from the addition of allochthonous microorganisms are sustained under natural conditions where they must compete for survival against the mixed population of natural bacteria (Lee and Levy, in 1987; Tagger et al., 1983; Westlake et al., 1978). Natural bacterial populations have the inductive enzymatic attributes required for hydrocarbon degradation, and the growth potential of bacteria is such that seeding of oil spills offers few advantages since oil degradation in the natural environment is not limited by the need for inocula. Microbial growth in the marine environment appears to be limited primarily by abiotic factors, such as nutrient concentration. Studies on the natural dynamics of bacterial populations by Miget (1973) and Atlas and Bartha (1973a) have clearly demonstrated that when nitrogen and phosphorus concentrations limit bacterial growth following an oil spill, it is futile to treat the contaminated site with additional bacterial species adapted to hydrocarbon degradation.

In conclusion, microbial seeding in the aquatic environment offers promise only in the treatment of contained spills or lagooned effluent where conditions can be more or less controlled. The open sea or the marine environment in general offers a most unlikely situation for successful seeding of oil spills. Furthermore, the hazards of seeding must be appreciated and seeding must be undertaken with caution because of its potential environmental impact, including potential pathogenicity to humans (Cobet et al., 1973).

3.3. Oxygenation

In some environments, such as lagoons or low-energy beaches and sediments, rates of petroleum biodegradation may be limited by oxygen availability, and biodegradation in such cases may be enhanced by the provision of oxygen. For example, Jamison et al. (1975) used forced aeration to enhance hydrocarbon biodegradation in a groundwater supply that was contaminated by a gasoline spill. Since oxygen was the limiting factor in this situation, the addition of nutrients without aeration failed to stimulate biodegradation, but when both nutrients and oxygen were supplied, approximately 1.6×10^5 L of gasoline were removed by microbial degradation.

Enhanced biodegradation was recognized by Halmo (1985) as a method of treating oil-contaminated waste along shorelines. While oxygen was not considered to be the limiting factor for the biodegradation of oil stranded on the shoreline, in situations where oiled seaweed was mixed with oil sorbents and fertilizers, aeration was the primary limiting parameter. A simple composting system of forced aeration from the bottom of the windrows was more efficient and less labor intensive than mechanical mixing at weekly intervals. The porosity of the oily mixtures controlled the degree of aeration and was dependent on the type of oil sorbent used. The most effective combination was obtained by mixing the oily seaweed with pine bark and a nutrient mixture of urea in a microemulsion of oleic acid. Windrow composting of this mixture with forced aeration resulted in approximately 65% of the paraffins in a weathered Statfjord crude oil being degraded within one summer, and 70% over an entire year.

3.4. Nutrient Enrichment

Growth kinetics of microorganisms in the marine environment often approach those of a chemostat wherein growth is limited by a deficiency in one of the nutrients. In oil-contaminated environments, carbon is greatly in abundance and the limiting factors for *in situ* microbial degradation are most likely to be the concentrations of nitrate and phosphate (since petroleum contains virtually no nitrogen or phosphorus compounds that may be used directly by the biota). For example, preliminary laboratory tests (Van der Linden, 1978; Floodgate, 1979), and simulated *in situ* experiments involving flow-through seawater tanks (Atlas and Bartha, 1973c), demonstrated that enhanced oil biodegradation in seawater requires additions of phosphate and nitrate since seawater itself does not usually contain sufficient amounts of these nutrients to sustain continued degradation. Fusey and Oudot (1973) reported that biodegradation of Kuwait crude oil was increased from 10 to 30% after 17 days by the addition of an organic nutrient mixture containing nitrogen and phosphorus.

The optimum C:N and C:P ratios for petroleum biodegradation in marine waters have been determined in laboratory studies. Bridie and Bos (1971), reported that final concentrations of 2.3×10^{-4} M ammonia and 2×10^{-5} M phosphate maintained maximum degradation rates for Kuwait crude in seawater at a concentration of 70 mg L^{-1}. In New Jersey coastal waters, Atlas and Bartha (1972b) observed maximum biodegradation of Sweden crude at nitrogen (as KNO_3 or NH_4NO_2) and phosphorus (Na_2HPO_4) concentrations of 10^{-2} M and 3.5×10^{-4} M respectively. Reisfield et al. (1972) noted that optimum emulsification of Iranian crude oil in Mediterranean seawater (1 g L^{-1}) inoculated with an oil-degrading strain of *Arthrobacter* (RAG-1) was observed when the seawater was supplemented with 3.8×10^{-3} M of nitrogen (as $(NH_4)_2SO_4$) and 3×10^{-5} M phosphate (as K_2HPO_4). However, the addition of nutrients in the form of inorganic salts at concentrations optimized for "closed-system" laboratory tests have little or no effect as single applications in the open sea, since they are diluted rapidly and removed from the oil–water interface where their presence is needed. To maintain optimum nutrient concentrations at the oil–water interface, the use of oleosoluble nitrogen-containing organic nutrients, such as the condensation products of urea and melamine with aldehydes, has been suggested. However, since these organic compounds are still significantly soluble in water, they soon leave the oil–water interface and dissipate into the aqueous phase. To overcome this problem, attempts have been made to encapsulate the water-soluble nutrients in paraffin (Atlas and Bartha, 1973c; Olivieri et al., 1976), or by generating microemulsions of the water-soluble nutrients in a lipophilic phase (Tramier and Sirvins, 1983).

Atlas and Bartha (1973c) conducted tests in the laboratory under simulated marine conditions to demonstrate the effectiveness of an oleophilic fertilizer consisting of paraffinized urea and octylphosphate in enhancing the biodegradation of oil. In contrast to nitrate and phosphate salts, the oleophilic fertilizers were found to supply nutrients to the hydrocarbon-degrading microorganisms selectively and without triggering algal blooms. In a subsequent field trial, in response to a report by Barsdate and Prentki (1972) that indicated that phosphorus limited microbial activity in tundra ponds, Atlas and Busdosh (1976) showed that while oleophilic nitrogen- and phosphorus-containing fertilizers increased the degradation of Prudhoe Bay crude oil in freshwater ponds on the tundra within a 21-day period, addition of inorganic nitrogen and phosphorus (as NO_3 and PO_4) did not enhance the degradation of the oil. Following the preliminary success of these studies, Bartha and Atlas (1976), in conjunction with the U.S. Office of Naval Research, received a patent for the use of fertilizers for stimulating oil-degradation in seawater.

Olivieri et al. (1976) developed a slow release paraffin-supported fertilizer containing $MgNH_4PO_4$ as the active ingredient, and the biodegradation of Sarir crude oil in seawater was found to be sufficiently enhanced by this fertilizer that after 21 days, 63% of the crude oil had disappeared when it was used as compared to 40% in the control area. Results of both field and laboratory experiments suggest that paraffin-supported fertilizers could be used to enhance the degradation of residual oils left behind by mechanical recovery processes. However, the main limitation of paraffin-supported fertilizers is that paraffinized particles become agglomeration centers, especially when applied to viscous oils, and hence decrease the surface area of the oil available to microbial attack. In addition, paraffin coatings of the fertilizer become an additional contaminant which must be degraded. Finally, the release of nutrients in the paraffin-supported fertilizer responds to physical factors (thickness and continuity of the paraffin coat, temperature, etc.), rather than to biological demand. Because of these shortcomings, Olivieri et al. (1978) screened a number of inexpensive nitrogen- and phosphorus-containing compounds with intrinsic lipophilic properties in an attempt to find one that was not soluble in water, and was stable under environmental conditions, biodegradable, and nontoxic. Laboratory tests with Basra crude oil indicated that soya bean lecithin and ethyl allophanate met the above criteria and were suitable sources of phosphorus and nitrogen for the oil-degrading microorganisms. Furthermore, in theory, enzyme-dependent mechanisms would regulate the release of nutrients from these compounds according to the environmental needs. Although preliminary laboratory results were extremely positive, field trials to evaluate the effectiveness of these substances to enhance biodegradation of petroleum hydrocarbons in the marine environment have not been conducted.

Bergstein and Vestal (1978) monitored the degradation of Prudhoe Bay crude oil in arctic tundra ponds fertilized with a commercial oleophilic phosphate fertilizer, Victawet 12 (2-ethylhexyl-dipolyethylene oxide phosphate, Stauffer Chemical Co.) and inorganic phosphate. Bacterial enumeration of water and sediment samples indicated that oil alone did not appear to exhibit either a stimulatory or a toxic effect on the numbers of indigenous heterotrophic and oil-degrading bacteria. In contrast, addition of the oleophilic phosphate mixture at concentrations of 0.1 mM, at weekly intervals, significantly stimulated the indigenous microflora in both the presence and absence of oil. Since equal concentrations of inorganic phosphate did not increase the rate of biodegradation, the stimulation induced by the oleophilic nutrients was attributed to hydrocarbon components within the organic phosphate nutrient mixture. Studies with *n*-[^{14}C]hexadecane and *n*-[^{14}C]naphthalene demonstrated that the microflora mineralized the saturate fraction of the oil before the polyaromatic fraction. The conclusions obtained in this study support those of Atlas and Busdosh (1976), namely that oleophilic fertilizers may be used to enhance the biodegradation of oil in the aquatic environment.

Tramier and Sirvins (1983) and Sirvins and Angles (1986) successfully enhanced biodegradation of Arabian light crude oil in temperate and Antarctic waters with EAP 22, (Elf Aquitaine, France), an oleophilic microemulsion containing a solution of urea in brine encapsulated in oleic acid as the external phase, with laurylphosphate as a cosurfactant. The results of their experiments, which were conducted in the laboratory with cultures and in large outdoor enclosures with indigenous microflora, were quite encouraging. When acclimated species were involved, temperature did not appear to be a limiting factor for the biodegradation of oil even in Antarctic waters. Calculations from 7-day experiments indicated that the rate of oil loss increased from 10–20% to 60–85% in waters supplemented with the oleophilic nutrients, even at temperatures as low as 3°C.

Bronchart et al. (1985) recently suggested that oleophilic nitrogen-containing organic nutrients may be detrimental to pelagic organisms in the open sea since they may become a barrier to the transfer of air and light into the water column, in a manner similar to the oil slick to which they are applied. In response to this limitation, they developed a new generation of oleophilic nutrients with oil-dispersant characteristics. In principle, these nutrient formulations would dissipate the oil and enhance the development of microbial growth around the dispersed oil droplets, since the nutrients remain at the oil–water interface. The proposed nutrient formulations are composed of an oleophilic (normal paraffin or olefin) and a hydrophilic moiety (containing nitrogen and or phosphorus) have the properties of a dispersant as a result of the surfactant-like structure of the nutrient mixtures. Pilot scale trials with Kuwait crude oil in seawater indicated that the "type-B" nutrients (patents pending), containing a nonyl chain and having dodecyl and oleyl radicals as lipophilic moieties, exhibited high dispersion efficiency and initiated biodegradation of the dispersed crude oil shortly after addition.

Recent studies suggest that biodegradation of oil in the littoral zone may be enhanced by water-soluble nutrients. Halmo (1985) monitored the degradation of a weathered Statfjord crude oil emulsion in the supralittoral zone on the coast of Norway. The direct addition of either a commercial grade water-soluble nutrient fertilizer (NH_4^+ and NO_3^-) or an oil-soluble fertilizer originally developed for use offshore (urea in a microemulsion of oleic acid) was effective in enhancing natural biodegradation rates. Biodegradation of the oil stranded on vegetation in the supralittoral zone was significantly increased by the application of nutrients. However, no statistically significant difference in oil biodegradation was observed between use of the oil-soluble and the water-soluble fertilizer. During the initial 4-month period following the spill, 45 to 85% of the paraffins in the crude oil in the fertilized areas were decomposed, while corresponding degradation in the unfertilized reference site was only 10 to 25%. After one year, the *n*-alkanes were completely degraded in both of the fertilized plots, while 20 to 30% of the paraffins were still present at the reference site. Enhancement in the biodegradation of the oil in the fertilized plots was confirmed by time-series observation of the ratios between normal and branched C-17 and C-18 components of the oil. Similar observations were recorded during an extensive study on the biodegradation of Venezuela Lago Medio crude oil stranded on the Arctic coastline (Eimhjellen et al., 1982, 1983; Eimhjellen and Josefsen, 1984). Although a substantial percentage of the viable bacteria in the oiled "control" sediments were potential oil degraders, even after a 2-year period significant changes in the biodegradation index of the oil (ratio of alkanes/isoprenoids) were not observed (Boehm et al., 1984). With the addition of a commercial agricultural fertilizer, an increase in the number of bacteria was observed and degradation of 30–60% of the biologically vulnerable alkane fraction of the oil occurred within one year. Analysis of the alkane/isoprenoid ratio indicated that mechanical mixing of the oil and fertilizer into the sediment supported optimal biodegradation rates, possibly because of improved penetration of oxygen. Furthermore, the effectiveness of these simple fertilizer treatments appears to be dependent on geomorphological conditions, since its effect in coarse-grained sediments, associated with high-energy beaches, was found to be marginal.

In addition to nitrogen and phosphorus, other nutrients may limit the biodegradation of petroleum in the marine environment. The essential mineral nutrient, iron, is one of the most abundant elements in the lithosphere (Deevey, 1970); however, its low concentrations in surface waters may limit biological productivity (Tranter and Newell, 1963; Menzel and Ryther, 1961). Dibble and Bartha (1976) studied the effects of nitrogen, phosphorus, and iron supplements on the biodegradation of South Louisiana crude oil in polluted (10,900 oil degraders per liter) and relatively clean (750 oil degraders per liter) seawater samples collected along the New Jersey coast. Without supplements, biodegradation of the crude oil was negligible in both seawater samples. Addition of nitrogen and phosphorus supported rapid biodegradation (72% in 3 days) in the contaminated seawater. Since the total iron in this contaminated seawater sample was high (5.2 mM), the addition of chelated iron did not increase biodegradation rates. However, in the less contaminated seawater sample with lower concentrations of iron (1.2 uM), biodegradation of the crude oil was considerably slower (21% in 3 days), and the addition of chelated iron had a significant stimulatory effect. Although the effect was far less dramatic than that of nitrogen and phosphorus, this experiment demonstrated that there are many instances in which iron limits petroleum biodegradation in the marine environment. In terms of nutrient enrichment, ferric octoate, an oily substance that is relatively insoluble in water but mixed homogeneously in South Louisiana crude oil, was found to be as effective as chelated iron in stimulating the biodegradation of oil. Dibble and Bartha (1976) suggest that surface oil slicks in temperate waters may be treated with a combination of ferric octoate, paraffin-coated urea, and octyl phosphate.

In general, field studies indicate that following most oil spill incidents the available concentrations of nitrogen and phosphorus severely limit the rate and extent of hydrocarbon biodegradation. Natural rates of nutrient replenishment in the marine environment are generally inadequate to support the rapid biodegradation of large quantities of oil. There is no doubt that fertilization (of isolated spill areas) may have a significant effect and should be seriously considered as a means of enhancing microbial degradation. Based on the data available, optimum N:P:C ratios for the degradation of petroleum can be approximated in the field and the addition of appropriate fertilizers can be used to enhance petroleum biodegradation by the indigenous microflora.

4. SUMMARY

Despite continuous input, the levels of petroleum detected in the oceans have remained low, presumably because of biodegradation processes. In the short term, biodegradation plays a small role in the dispersal and degradation of petroleum following pollution incidents, but over large temporal and spatial scales, it is probably a major mechanism for decontamination.

The hydrocarbons in petroleum can be metabolically degraded by numerous species of bacteria, yeasts, fungi and algae. Such microorganisms have been found to be ubiquitous in the world's oceans, and studies have shown that their numbers and degradation potential are generally higher near sources of chronic petroleum input. However, despite the fact that high number of oil-degrading microorganisms may be present in the marine environment, natural environmental conditions rarely favor optimal rates of biodegradation.

Our knowledge of the interactive and/or sequential events that occur during microbial degradation of petroleum hydrocarbons is incomplete. A succession of species is involved in the biodegradation of the complex mixtures of hydrocarbons present in crude oils and a number of rate-limiting factors have been identified. To enhance predictive capabilities regarding the fate of petroleum in the environment, further research is necessary to improve our understanding of biodegradative mechanisms such as co-metabolism and the microbial formation of persistent contaminants.

During the last decade, considerable environmental research has been focused on the development of techniques that could be used to enhance biodegradation of oil in the environment. *In situ* field studies have shown that low temperatures alone cannot account for the limited rates of hydrocarbon biodegradation. Furthermore, field trials have demonstrated that seeding of oil contaminated areas with oil-degrading bacteria cultures alone offers few advantages since they cannot compete with indigenous microflora. Extensive studies conclude that the degradation of petroleum hydrocarbons in the marine ecosystem is limited primarily by nutritional factors, i.e., nitrogen and phosphorus, and in some cases by oxygen availability. Results of preliminary experiments using oxygenation, chemical dispersion, and nutrient enrichment to enhance the biodegradation of petroleum, indicate that fertilization with nitrogen and phosphorus offers the greatest promise as a countermeasure to oil spills in the marine environment.

With regard to fertilization as an oil-pollution countermeasure, we suggest that biodegradation of oil in the marine environment may be successfully enhanced by periodic addition of nutrients, but only after indigenous microbial populations have adapted to the contaminated site. To maintain maximal petroleum-degradation rates under diverse marine environments, the optimal concentration of various nutrient formulations and their required replenishment rates must be determined. Furthermore, to improve our predictive capabilities regarding the fate of oil in the marine environment, time-series chemical studies must be initiated to monitor the effect of fertilization on the rate and extent of biodegradation of petroleum.

ACKNOWLEDGMENTS

Preparation of this review was supported by the Federal Panel on Energy Research and Development, Canada.

REFERENCES

Aminot, A. (1981). Anomalies in the coastal hydrobiological system following the grounding of the Amoco Cadiz. Qualitative and quantitative considerations on the *in situ* biodegradation of hydrocarbons. In *Consequences of an Accidental Oil Spill*. Le Centre Nationale pour l'Exploration des Oceans, Paris, pp. 223–242.

Arhelger, S.D., B.R. Robertson, and D.K. Button (1977). Arctic hydrocarbon biodegradation. In D. Wolfe (Ed.), *Fate and Effects of Petroleum Hydrocarbons in Marine Ecosystems and Organisms*. Pergamon, Oxford, pp. 270–275.

Atlas, R.M. (1975). Effects of temperature and crude oil composition on petroleum biodegradation. *Appl. Microbiol.* **30**: 396–403.

Atlas, R.M. (1977). Studies on petroleum biodegradation in the Arctic. In D. Wolfe (Ed.), *Fate and Effects of Petroleum Hydrocarbons in Marine Ecosystems and Organisms*. Pergamon, New York, pp. 261–269.

Atlas, R.M. (1978). An assessment of the biodegradation of petroleum in the Arctic. In M.W. Loutit and J.A.R. Miles (Eds.), *Microbial Ecology*. Berlin: Springer-Verlag, pp. 86–90.

Atlas, R.M. (1981). Microbial degradation of petroleum hydrocarbons: An environmental perspective. *Microbiol. Rev.* **45**: 180–209.

Atlas, R.M. (1985). Effects of hydrocarbons on microorganisms and petroleum biodegradation in arctic ecosystems. In F.R. Engelhardt (Ed.), *Petroleum Effects in the Arctic Environment*. Elsevier Applied Science, New York, pp 63–99.

Atlas, R.M. (1986). Fate of petroleum pollutants in arctic ecosystems. *Water Sci. Technol.* **18**: 59–68.

Atlas, R.M., and R. Bartha (1972a). Biodegradation of petroleum in sea water at low temperatures. *Can. J. Microbiol.* **18**: 1851–1855.

Atlas, R.M., and R. Bartha (1972b). Degradation and mineralization of petroleum in seawater: Limitation by nitrogen and phosphorus. *Biotechnol. Bioeng.* **14**: 309–317.

Atlas, R.M., and R. Bartha (1973a). Effects of some commercial oil herders, dispersants and bacteria inocula on biodegradation in seawater. In D.G. Ahearn and S.P. Meyers (Eds.), *The Microbial Degradation of Oil Pollutants*. Louisiana State University Press, Baton Rouge, pp. 283–289.

Atlas, R.M., and R. Bartha (1973b). Abundance, distribution and oil biodegradation potential of microorganisms in Raritan Bay. *Environ. Pollut.* **4**: 291–300.

Atlas, R.M., and R. Bartha (1973c). Stimulated biodegradation of oil slicks using oleophilic fertilizers. *Environ. Sci. Technol.* **7**: 538–541.

Atlas, R.M., and M. Busdosh (1976). Microbial degradation of petroleum in the Arctic. In J.M. Sharpley and A.M. Kaplan (Eds.), *Proceedings, 3rd Internat. Biodegradation Symp.* Applied Science, London, pp. 79–85.

Atlas, R.M., and A. Bronner (1981). Microbial hydrocarbon degradation within intertidal zones impacted by the Amoco Cadiz oil spillage. In *Amoco Cadiz: Fates and Effects of the Oil Spill*. Centre National pour l'Exploitation des Oceans, Paris, pp. 251–256.

Atlas, R.M., A. Horowitz, and M. Busdosh (1978). Prudhoe crude oil in Arctic marine ice, water and sediment ecosystems: Degradation and interactions with microbial and benthic communities. *J. Fish. Res. Board Can.* **35**: 585–590.

Atlas, R.M., G. Roubal, A. Bronner, and J. Haines (1980a). Microbial degradation of hydrocarbons in mousse from Ixtoc-I. In Proceedings, National Oceanic and Atmospheric Administration Symp. on Preliminary Results from the September 1979 *Researcher/Pierce* Ixtoc-I Cruise, June 9–10, 1980, Key Biscayne, Florida, pp. 1–24.

Atlas, R.M., G. Roubal, and J. Haines (1980b). Biodegradation of hydrocarbons in mousse from the Ixtoc-I well blowout. In *Proceedings of Symposium on Assessment of the Environmental Impact of Accidental Oil Spills in the Oceans, August 24–29, San Francisco*. Ann Arbor Science, Ann Arbor, MI.

Azoulay, E., J. Dubrevil, H. Dou, G. Mille, and G. Giusti (1983). Relationship between hydrocarbons and bacterial activity in Mediterranean sediments, 1. Nature and concentration of hydrocarbons in the sediments. *Mar. Environ. Res.* **9**: 1–17.

Bailey, N.J.L., A.M. Jobson, and M.A. Rogers (1973). Bacterial degradation of crude oil: Comparison of field and experimental data. *Chem. Geol.* **11**: 203–221.

Barsdate, R.J., and R.T. Prentki (1972). Nutrient dynamics in tundra ponds. In Proceedings, 1972 Tundra Biome Symp., U.S. Arctic Research Program, pp. 192–199.

Bartha, R., and R.M. Atlas (1976). Biodegradation of oil on water surfaces. U.S. Patent 3939127, May 25, 1976.

Bartha, R., and R.M. Atlas (1977). The microbiology of aquatic oil spills. *Adv. Appl. Microbiol.* **22**: 225–266.

Bergstein, P.E., and J.R. Vestal (1978). Crude oil biodegradation in Arctic tundra ponds. *Arctic* **31**: 159–169.

Blackman, R.A.A., F.L. Franklin, M.G. Norton, and K.W. Wilson (1978). New procedures for the toxicity testing of oil slick dispersants in the United Kingdom. *Mar. Pollut. Bull.* **9**: 234–238.

Boehm, P.C., W. Steinhauer, D. Cobb, S. Duffy, and J. Brown (1984). Chemistry, 2. Analytical Biogeochemistry—1983 Study Results. Baffin Island Oil Spill Working Report 83-2, Environmental Protection Service, Environment Canada, Ottawa.

Bossert, I., and R. Bartha (1984). The fate of petroleum in soil ecosystems. In R. Atlas (Ed.), *Petroleum Microbiology*. MacMillan, New York, pp. 435–474.

Bridie, A.L., and J. Bos (1971). Biological degradation of mineral oil in seawater. *J. Inst. Petroleum (London)* **57**: 270–277.

Bronchart, R.D.E., J. Cadron, A. Charlier, A.A.R. Gillot, and W. Verstraete (1985). A new approach in enhanced biodegradation of spilled oil: Development of an oil dispersant containing oleophilic nutrients. In *Proceedings, 1985 Oil Spill Conf.*, American Petroleum Institute, Washington, DC, pp. 453–462.

Buckley, E.N., R.B. Jonas, and F.K. Pfaender (1976). Characterization of microbial isolates from an estuarine ecosystem: Relationship of hydrocarbon utilization to ambient hydrocarbon concentrations. *Appl. Environ. Microbiol.* **32**: 232–237.

Bunch, J.N., and R.C. Harland (1976). Biodegradation of crude petroleum by the indigenous microbial flora of the Beaufort Sea. Beaufort Sea Tech. Report No.10, Environment Canada, Victoria, B.C.

Bunch, J.N., C. Bedard, and T. Cartier (1983a). Microbiology, 1. Effect of oil on bacterial activity—1981 study results. Baffin Island Oil Spill (BIOS) Project Working Report 81-5.

Bunch, J.N., C. Bedard, and T. Cartier (1983b). Abundance and activity of heterotrophic marine bacteria in selected bays at Cape Hatt, N.W.T.: Effects of oil spills, 1981. Can. Ms. Rep. Fish. Aquat. Sci., No. 1708.

Cerniglia, C.E., and J.J. Perry (1973). Crude oil degradation by microorganisms isolated from the marine environment. *Z. Allg. Mikrobiol.* **13**: 299–306.

Cerniglia, C.E., D.T. Gibson, and C. van Baalen (1980). Oxidation of naphthalene by cyanobacteria and microalgae. *J. Gen. Microbiol.* **116**: 495–500.

Chakrabarty, A.M. (1972). Genetic basis of the biodegradation of salicylate in *Pseudomonas. J. Bacteriol.* **112**: 815–823.

Chakrabarty, A.M. (1976). Plasmids in *Pseudomonas. Ann. Rev. Genet.* **10**: 7–30.

Chakrabarty, A.M., G. Chou and I.C. Gunsalus (1973). Genetic regulation of octane dissimilation plasmid in *Pseudomonas. Proc. Nat. Acad. Sci. U.S.A.* **70**: 1137–1140.

Chouteau, J., E. Azoulay, and J.C. Senez (1962). Anaerobic formation of *n*-hept-1-ene from *n*-heptane by resting cells of *Pseudomonas aeruginosa. Nature* (*London*) **194**: 576–578.

Cobet, A.B., H.E. Guard, and M.A. Chatigny (1973). Considerations in application of microorganisms to the environment for degradation of petroleum products. In D.G. Ahearn and S.P. Meyers (Eds.), *The Microbial Degradation of Oil Pollutants*. Louisiana State University, Baton Rouge, p. 291.

Colwell, R.R., and J.D. Walker (1977). Ecological aspects of microbial degradation of petroleum in the marine environment. *Crit. Rev. Microbiol.* **5**: 423–445.

Colwell, R.R., A.L. Mills, J.D. Walker, P. Garcia-Tello, and V. Campos-P. (1978). Microbial ecology studies of the Metula spill in the straits of Magellan. *J. Fish. Res. Board Can.* **35**: 573–580.

Cook, F.D., and D.W.S. Westlake (1974). Microbiological degradation of northern crude oils. In Environmental-Social Committee; Northern Pipelines, Task Force on Northern Oil Development, Report No. 74-1. R72-12774, Information Canada, Ottawa.

Deevey, E.S., Jr. (1970). Mineral cycles. *Sci. Am.* **223**: 149–158.

Dibble, J.T., and R. Bartha (1976). Effect of iron on the biodegradation of petroleum in seawater. *Appl. Environ. Microbiol.* **31**: 544–550.

Eimhjellen, K., and K. Josefsen (1984). Microbiology, 2. Biodegradation of stranded oil—1983 results. In Baffin Island Oilspill Project Working Report 83-6, Environmental Protection Service, Environment Canada, Ottawa.

Eimhjellin, K., O. Nilssen, K. Josefsen, T. Sommer, E. Sendstad, P. Sveum, and T. Hoddo (1982). Microbiology, 2. Biodegradation of oil—1981 study results. In Baffin Island Oil Spill Working Report 81-6, Environmental Protection Service, Environment Canada.

Fingas, M. (1985). The effectiveness of oil spill dispersants. *Spill Technology Newsletter—Environment Canada* **10**: 47–64.

Floodgate, G.D. (1976). Nutrient limitation. In J.M. Sharpley and A.M. Kaplan (Eds.), *Proceedings, 3rd Internat. Biodegradation Symp*. Applied Science, London, pp. 87–92.

Floodgate, G.D. (1979). Nutrient limitation. In A.W. Bourquin and P.H. Pritchard (Eds.), Proceedings, Workshop on Microbial Degradation of Pollutants in the Marine Environment, Environmental Research Laboratory, Gulf Breeze, FL, U.S. EPA-66019-79-012, pp. 107–118.

Floodgate, G.D. (1984). The fate of petroleum in marine ecosystems. In R.M. Atlas (Ed.), *Petroleum Microbiology*. MacMillan, New York, pp. 355–397.

Foght, J.M., and D.W.S. Westlake (1982). Effect of dispersant 9527 on microbial degradation of Prudhoe Bay oil. *Can. J. Microbiol.* **28**: 117–122.

Friello, D.A., J.R. Mylroie and A.M. Chakrabarty (1976). Use of genetically engineered multi-plasmid microorganisms for the rapid degradation of fuel hydrocarbons. In J.M. Sharpley and A.M. Kaplan (Eds.), *Biodeterioration of Materials*, Vol. 3. Applied Science, London, pp. 205–214.

Fusey, P., and J. Oudot (1973). Note on the acceleration of biodegradation of a crude oil by bacteria. *Material und Organismen* **8**: 157–164.

Gatellier, C.R., J.L. Oudin, P. Fusey, J.C. Lacase, and M.L. Priou (1973). Experimental ecosystems to measure the fate of oil spills dispersed by surface active products. In Proceedings, Joint Conf. Prevention and Control of Oil Spills. American Petroleum Inst., Washington, DC, pp. 497–507.

Gerlach, S.A. (1981). *Marine Pollution*. Springer-Verlag, New York.

Gibbs, C.F., and S.J. Davis (1976). The rate of microbial degradation of oil in a beach gravel column. *Microbial Ecol*. **3**: 55–64.

Gordon, D.C., J. Dale, and P.D. Keizer (1978). Importance of sediment working by the deposit-feeding polychaete *Arenicola marina* on the weathering rate of sediment-bound oil. *J. Fish. Res. Board Can*. **35**: 591–603.

Griffiths, R.P., T.M. McNamara, B.A. Caldwell, and R.Y. Morita (1981a). Field observations on the acute effects of crude oil on glucose and glutamate uptake in samples collected from arctic and subarctic waters. *Appl. Environ. Microbiol*. **41**: 1400–1406.

Griffiths, R.P., T.M. McNamara, B.A. Caldwell, and R.Y. Morita (1981b). A field study on the acute effects of the dispersant Corexit 9527 on glucose uptake by marine microorganisms. *Mar. Environ. Res*. **5**: 83–91.

Gunkel, W., and G. Gassmann (1980). Oil, oil dispersants and related substances in the marine environment. *Helgol. Wiss. Meeresunters*. **33**: 164–181.

Gunkel, W., G. Gassmann, C.H. Oppenheimer, and I. Dundas (1980). Preliminary results of baseline studies of hydrocarbons and bacteria in the North Sea: 1975, 1976 and 1977. In Ponencias del Simposio International en: Resistencia a los Antibiosis y Microbiologia marina, Santiago de Compostela, Spain, pp. 223–247.

Hagstrom, B.E., and S. Lonning (1977). The effects of ESSO Corexit 9527 on the fertilization capacity of spermatozoa. *Mar. Pollut. Bull*. **8**: 136–138.

Haines, J.R., and R.M. Atlas (1982). *In situ* microbial degradation of Prudhoe Bay crude oil in Beauford Sea sediments. *Mar. Environ. Res.* **7**: 91–102.

Halmo, G. (1985). Enhanced biodegradation of oil. In *Proc. 1985 Oil Spill Conf.* American Petroleum Institute, Washington, DC, pp. 531–537.

Horowitz, A., and R.M. Atlas (1977). Continuous open flow-through system as a model for oil degradation in the Arctic Ocean. *Appl. Environ. Microbiol.* **33**: 647–653.

Horowitz, A., and R.M. Atlas (1978). Crude oil degradation in the Arctic: Changes in bacterial populations and oil composition during one year exposure in a model system. *Dev. Ind. Microbiol.* **11**: 517–522.

Horvath, R.S. (1972). Microbial co-metabolism and the degradation of organic compounds in nature. *Bacteriol. Rev.* **36**: 146–155.

Hsiao S.I.C., D.W. Kittle, and M.G. Foy (1978). Effects of crude oils and the oil dispersant Corexit on primary production of arctic marine phytoplankton and seaweed. *Environ. Pollut.* **15**: 209–221.

Hughes, D.E., and S. Stafford (1983). Bacterial degradation of crude oil. In J. Wardley-Smith (Ed.), *The Control of Oil Pollution*. Graham & Trotman, London, pp. 37–46.

Iizuka, H., M. Ilida, and S. Fujita (1969). Formation of *n*-decane-1 from *n*-decane by resting cells of *C. Rosa. Z. Allg. Mikrobiol.* **9**: 223–226.

Jamison, V.W., R.L. Raymond, and J.O. Hudson (1975). Biodegradation of high-octane gasoline in groundwater. *Dev. Ind. Microbiol.* **16**: 305–312.

Jannasch, H.W., K. Eimhjellen, C.O. Wirsen, and A. Farmaian (1971). Microbial degradation of organic matter in the deep sea. *Science* **171**: 672–675.

Jobson, A., M. McLaughlin, F.D. Cook, and D.W.S. Westlake (1974). Effect of amendments on the microbial utilization of oil applied to soil. *Appl. Microbiol.* **27**: 166–171.

Johnston, R. (1970). The decomposition of crude oil residues in sand columns. *J. Mar. Biol. Assoc. U.K.* **50**: 925–937.

Jordan, R.E. and J.R. Payne (1980). *Fate and Weathering of Petroleum Spills in the Marine Environment: A Literature Review and Synopsis.* Ann Arbor Science, Ann Arbor, MI.

Kado, C., and S. Lui (1981). Rapid procedure for the detection and isolation of large and small plasmids. *J. Bacteriol.* **145:** 1365–1373.

Karrick, N.L. (1977). Alterations in petroleum resulting from physico-chemical and microbiological factors. In D.C. Malins (Ed.), *Effects of Petroleum in Arctic and Subarctic Marine Environments and Organisms*, Vol. 1, *Nature and Fate of Petroleum.* Academic, New York, pp. 225–299.

Kator, H., and R. Herwig (1977). Microbial responses after two experimental oil spills in an eastern coastal plain estuarine ecosystem. In *Proceedings, 1977 Oil Spill Conf.* American Petroleum Institute, Washington, DC, pp. 517–522.

LaRock, P.A., and M. Severence (1973). The bacterial treatment of oil spills: Some facts considered. In L.H. Stevenson and R.R. Colwell (Eds.), *Estuarine Microbial Ecology*. University of South Carolina Press, Columbia, pp. 309–328.

Lee, K., and E.M. Levy (1987). Enhanced biodegradation of a light crude oil in sandy beaches. In *Proceedings, 1987 Oil Spill Conf.* American Petroleum Institute, Washington, DC, pp. 411–416.

Lee, K., C.S. Wong, W.J. Cretney, F.A. Whitney, T.R. Parsons, C.M. Lalli, and J. Wu (1985). Microbial response to crude oil and Corexit 9527: SEAFLUXES Enclosure Study. *Microbial Ecol.* **11**: 337–351.

LePetit, J., H.-M. N'Guyen, and S. Tagger (1977). The ecology of a coastal marine zone receiving the effluent from an oil refinery. *Environ. Pollut.* **13**: 41–56.

Liu, D.L.S., and B.J. Dutka (1972). Bacterial seeding techniques: Novel approach to oil spill problems. *Can. Res. Dev.* **5**: 1.

MacKay, D. (1981). Fate and behaviour of oil spills. In J.B. Sprague, J.H. Vandermeulen, and P.G. Wells (Eds.), *Oil Dispersants in Canadian Seas—Research Appraisal and Recommendations.* Environment Canada, Toronto, pp. 7—27.

MacKay, D. (1985). The physical and chemical fate of spilled oil. In F.R. Engelhardt (Ed.), *Petroleum Effects in the Arctic Environment*. Elsvier Applied Science, London, pp. 37–61.

MacKay, D., and P.G. Wells (1983). Effectiveness behavior and toxicity of dispersants. In *Proc. 1983 Oil Spill Conf.* American Petroleum Institute, Washington, DC, pp. 65–71.

Mateles, R.I., J.N. Baruah, and S.R. Tannenbaum (1967). Growth of a thermophilic bacteria on hydrocarbons: A new source of single-cell protein. *Science* **157**: 1322–1323.

McAuliffe, C.D., J.C. Johnson, S.H. Green, G.P. Canevari, and T.D. Searl (1980). Dispersion and weathering of chemically treated crude oils on the ocean. *Environ. Sci. Technol.* **14**: 1509–1518.

Menzel, D.W., and J.H. Ryther (1961). Nutrients limiting the production of phytoplankton in the Sargasso Sea, with special reference to iron. *Deep-Sea Res.* **7**: 276–281.

Miget, R. (1973). Bacterial seeding to enhance biodegradation of oil slicks. In D.G. Ahearn and S.P. Meyers (Eds.), *The Microbial Degradation of Oil Pollutants.* Louisiana State University, Baton Rouge, p. 291.

Mironov, O.G. (1970). Role of microorganisms growing on oil in the self-purification and indication of oil pollution in the sea. *Oceanology* **10**: 650–656.

Morita, R.Y. (1975). Psychrophilic bacteria. *Bact. Rev.* **39**: 144–167.

Mulkins-Phillips, G.J., and J.E. Stewart (1974a). Effect of environmental parameters on bacterial degradation of Bunker C oil, crude oils and hydrocarbons. *Appl. Microbiol.* **28**: 915–922.

Mulkins-Phillips, G.J., and J.E. Stewart (1974b). Effect of four dispersants on biodegradation and growth of bacteria on crude oil. *Appl. Microbiol.* **28**: 547–552.

Mulkins-Phillips, G.J., and J.E. Stewart (1974c). Distribution of hydrocarbon-utilizing bacteria in northwestern Atlantic waters and coastal sediments. *Can. J. Microbiol.* **20**: 955–962.

National Academy of Sciences (1975). *Petroleum in the Marine Environment.* National Academy of Sciences, Washington, DC.

National Academy of Sciences (1985). *Oil in the Sea: Inputs, Fates, and Effects.* National Academy of Sciences, Washington, DC.

Olivieri, R., P. Bacchin, A. Robertiello, N. Oddo, L. Degen, and A. Tonolo (1976). Microbial degradation of oil spills enhanced by a slow-release fertilizer. *Appl. Environ. Microbiol.* **31**: 629–634.

Olivieri, R., A. Robertiello, and L. Degen (1978). Enhancement of microbial degradation of oil pollutants using lipophilic fertilizers. *Marine Pollut. Bull.* **9**: 217–220.

Oppenheimer, C.H., W. Gunkel, and G. Gassman (1977). Microorganisms and hydrocarbons in the North Sea during July–August 1975. In *Proc. 1977 Oil Spill Conference.* American Petroleum Institute, Washington, DC, pp. 593–610.

Parekh, V.R., R.W. Traxler, and J.M. Sobek (1977). *n*-Alkane oxidation enzymes of a *Pseudomonad. Appl. Environ. Microbiol.* **33**: 881–884.

Parsons, T.R., P.J. Harrison, J.C. Acreman, H.M. Dovey, P.A. Thompson, C.M. Lalli, K. Lee, Li Guanguo, and Chen Xiaolin (1985). An experimental marine ecosystem response to crude oil and Corexit 9527, Part 2. Biological effects. *Marine. Environ. Res.* **13**: 265–276.

Payne, J.R., and C.R. Phillips (1985). *Petroleum Spills in the Marine Environment.* Lewis, Chelsea, MI, pp. 1–148.

Perry, J.J. (1979) Microbial cooxidations involving hydrocarbons. *Microbial Rev.* **43**: 59–72.

Pierce, R.H., A.M. Cundell, and R.W. Traxler (1975). Persistence and biodegradation of spilled residual fuel oil on an estuarine beach. *Appl. Microbiol.* **29**: 646–652.

Rashid, M.A. (1974). Degradation of Bunker C oil under different coastal environments of Chedabuto Bay, Nova Scotia. *Estuarine Coastal Mar. Sci.* **2**: 137–144.

Reisfeld, A., E. Rosenberg, and D. Gutnick (1972). Microbial degradation of oil: Factors affecting oil dispersion in sea water by mixed and pure cultures. *Appl. Microbiol* **24**: 363–368.

Robertson, B., S. Arhelger, P.J. Kinney, and D.K. Button (1973). Hydrocarbon biodegradation in Alaskan waters. In D.G. Ahearn and S.P. Meyers (Eds.), *The Microbial Degradation of Oil Pollutants.* Louisiania State University Press, Baton Rouge, pp. 171–184.

Robichaux, T.J., and H.N. Myrick (1972). Chemical enhancement of the biodegradation of crude oil pollutants. *J. Petrol. Tech.* **24**: 16–20.

Rogerson, A., and J. Berger (1981). The toxicity of the dispersant Corexit 9527 and oil dispersant mixtures to ciliate protozoa. *Chemosphere* **10**: 33–39.

Roubal, G., and R.M. Atlas (1978). Distribution of hydrocarbon-utilizing microorganisms and hydrocarbon biodegradation potentials in Alaska continential shelf areas. *Appl. Environ. Microbiol.* **35**: 897–905.

Ryther, J.H., and W.M. Dunstan (1971). Nitrogen, phosphorus, and eutrophication in the coastal marine environment. *Science* **171**: 1008–1013.

Schwarz, J.R., J.D. Walker, and R.R. Colwell (1974a). Hydrocarbon degradation at ambient and *in situ* pressure. *Appl. Microbiol.* **28**: 982–986.

Schwarz, J.R., J.D. Walker, and R.R. Colwell (1974b). Growth of deep sea bacteria on hydrocarbons at ambient and *in situ* pressure. *Dev. Ind. Microbiol.* **15**: 239–249.

Schwarz, J.R., J.D. Walker, and R.R. Colwell (1975). Deep-sea bacteria: Growth and utilization of *n*-hexadecane at *in situ* temperature and pressure. *Can. J. Microbiol.* **21**: 682–687.

Senez, J.C., and E. Azoulay (1961). Dehydrogenation of paraffinic hydrocarbons by resting cells and cell free extracts of *Pseudomonas aeruginosa. Biochim. Biophys. Acta* **47**: 307–316.

Sirvins, A., and M. Angles (1986). Development and effects on marine environment of a nutrient formula to control pollution by petroleum hydrocarbons. In C.S. Giam and H.J.M. Dou (Eds.), *Strategies and Advanced Techniques for Marine Pollution Studies: Mediterranean Sea*, NATO ASI Series G9. Springer-Verlag, Berlin, pp. 357–404.

Smith, J.E. (1968). *Torrey Canyon: Pollution and Marine Life.* Cambridge University Press, London.

Soli, G. (1973). Marine hydrocarbonoclastic bacteria: Types and range of oil degradation. In D.G. Ahearn and S.P. Meyers (Eds.), *The Microbial Degradation of Oil Pollutants.* Center for Wetland Resources, Louisiana State University, Baton Rouge, Publication No. LSU-SG-73-01, pp. 141–146.

Swedmark, M., A. Granmco, and S. Kollberg (1973). Effects of oil dispersants and oil emulsions on marine animals. *Water Res.* 7: 1649–1672.

Tagger, S., L. Deveze, and J. LePetit (1976). The conditions for biodegradation of petroleum hydrocarbons in the sea. *Mar. Pollut. Bull.* 7: 172–174.

Tagger, S., A. Bianchi, M. Juilliard, J. LePetit, and B. Roux (1983). Effect of microbial seeding of crude oil in seawater in a model system. *Marine Biol.* **78:** 13–20.

Tramier, B., and A. Sirvins (1983). Enhanced oil biodegradation: A new operational tool to control oil spills. In *Proc. 1983 Oil Spill Conf.* American Petroleum Institute, Washington, DC, pp. 115–119.

Tranter, D.J., and B.S. Newell (1963). Enrichment experiments in the Indian Ocean. *Deep-Sea Res.* **10**: 1–9.

Traxler, R.W. (1973). Bacterial degradation of petroleum materials in low temperature environments. In D.G. Ahearn and S.P. Meyers (Eds.), *The Microbial Degradation of Oil Pollutants.* Center for Wetland Resources, Louisiana State University, Baton Rouge, Publication No. LSU-SG-73-01, pp. 163–170.

Traxler, R.W., and J.M. Bernard (1969). The utilization of *n*-alkanes by *Pseudomonas aeruginosa* under conditions of anaerobiosis. *Internat. Biodeterior. Bull.* **5**: 21–25.

Van der Linden, A.C. (1978). Degradation of oil in the marine environment. In J.R. Watkinson (Ed.), *Developments in Biodegradation of Hydrocarbons*, Vol. 1. Elsevier Applied Science, London, pp. 165–200.

Walker, J.D., and R.R. Colwell (1974). Microbial degradation of model petroleum at low temperatures. *Microbial Ecol.* **1:** 63–95.

Walker, J.D., and R.R. Colwell (1975). Degradation of hydrocarbons and mixed hydrocarbon substrate by microorganisms from Chesapeake Bay. *Prog. Water. Technol.* **7:** 783–795.

Walker, J.D., and R.R. Colwell (1976a). Measuring the potential activity of hydrocarbon-degrading bacteria. *Appl. Environ. Microbiol.* **31:** 189–197.

Walker, J.D., and R.R. Colwell (1976b). Enumeration of petroleum-degrading microorganisms. *Appl. Environ. Microbiol.* **31:** 198–207.

Walker, J.D., R.R. Colwell, and L. Petrakis (1975a). Degradation of petroleum by an alga, *Prototheca zopfii*. *Appl. Microbial*. **30:** 79–81.

Walker, J.D., R.R. Colwell, Z. Vaituzis, and S.A. Meyer (1975b). Petroleum-degrading achlorophyllous alga, *Prototheca zopfii. Nature* (*London*) **154:** 423.

Walker, J.D., H.F. Austin, and R.R. Colwell (1975c). Utilization of mixed hydrocarbon substrate by petroleum-degrading microorganisms. *J. Gen. Appl. Microbiol.* **21**: 27–39.

Walker, J.D., R.R. Colwell, and L. Petrakis (1976a). Comparison of the biodegradability of crude and fuel oils. *Can. J. Microbiol.* **22:** 598–602.

Walker, J.D., R.R. Colwell, and L. Petrakis (1976b). Biodegradation rates of components of petroleum. *Can. J. Microbiol.* **22:** 1209–1213.

Walker, J.D., P.A. Seesman, T.L. Herbert, and R.R. Colwell (1976c). Petroleum hydrocarbons: Degradation and growth potential of deep-sea sediment bacteria. *Environ. Pollut.* **10**: 89–96.

Ward, D.M., and T.D. Brock (1978). Hydrocarbon biodegradation in hypersaline environments. *Appl. Environ. Microbiol.* **35:** 353–359.

Ward, D., R.M. Atlas, P.D. Boehm, and J.A. Calder (1980). Microbial biodegradation and chemical evolution of oil from the Amoco Spill. *Ambio* **9:** 277–283.

Weinberger, M., and P. Kollenbrander (1979). Plasmid-determined 2-hydroxy pyrimidine utilization by *Arthrobacter crystallopoiedes. Can. J. Microbiol.* **25:** 329–334.

Westlake, D.W.S. (1982). Microorganisms and the degradation of oil under northern marine conditions. In J.B. Sprague, J.H. Vandermulen and P.G. Wells (Eds.), *Oil and Dispersants in Canadian Seas—Research Appraisal and Recommendations*, Environmental Protection Service Canada, Pub. No. EPS-3-EC-82-2, pp. 47–50.

Westlake, D.W.S., A.M. Jobson, and F.D. Cook (1978). *In situ* degradation of oil in a soil of the boreal region of the Northwest Territories. *Can. J. Microbiol.* **24:** 254–260.

Williams, P.A. (1978). Microbial genetics relating to hydrocarbon degradation. In R. Watkinson (Ed.), *Developments in the Degradation of Hydrocarbons*, Vol. 1. Applied Science, London, pp. 135–164.

Wong, C.S., F.A. Whitney, W.J. Cretney, K. Lee, F. McLaughlin, J. Wu, T. Fu, and D. Zhuang (1984). An experimental marine ecosystem response to crude oil and Corexit 9527, Part 1. Fate of chemically dispersed crude oil. *Mar. Environ. Res.* **13:** 247–264.

Zobell, C.E. (1964). The occurrence, effects and fate of oil polluting the sea. *Adv. Water Pollut. Res.* **3:** 85–118.

Zobell, C.E. (1969). Microbial modification of crude oil in the sea. In *API/FWPCA Joint Conf. on Prevention and Control of Oil Spills.* American Petroleum Institute, Washington, DC.

Zobell, C.E. (1973). Bacterial degradation of mineral oils at low temperatures. In D.G. Ahearn and S.P. Meyers (Eds.), *The Microbial Degradation of Oil Pollutants.* Publication No. LSU-SG-73-01, Center for Wetland Resources, Louisiana State University, Baton Rouge, pp. 153–161.

Zobell, C.E., and J.F. Prokop (1966). Microbial oxidation of mineral oils in Barataria Bay bottom deposits. *Z. Allg. Mikrobiol.* **6:** 134–162.

16

PESTICIDES IN FORESTRY AND AGRICULTURE: Effects on Aquatic Habitats

D.C. Eidt

Canadian Forestry Service—Maritimes, Fredericton, New Brunswick, Canada

J.E. Hollebone

Planning and Priorities Division, Pesticides Directorate, Agriculture Canada, Ottawa, Ontario, Canada

W.L. Lockhart

Freshwater Institute, Fisheries and Oceans Canada, Winnipeg, Manitoba, Canada

P.D. Kingsbury

Forest Pest Management Institute, Canadian Forestry Service, Sault Ste. Marie, Ontario, Canada

M.C. Gadsby

Hoechst Canada Inc., Regina, Saskatchewan, Canada

W.R. Ernst

Environmental Protection Service, Environment Canada, Dartmouth, Nova Scotia, Canada

1. Introduction

2. Agricultural pesticides in the aquatic environment: Risks and benefits (J.E. Hollebone)

2.1. Use of agricultural chemicals

2.2. Do pesticides get into the aquatic environment, and what is their impact?

2.3. Can there be agricultural use and environmental safety?

2.4. What are we or should we be doing to allow use of agricultural pesticides, yet ensure safety to aquatic systems?

2.5. Future regulation

3. Risks to aquatic organisms from forestry pesticide use in Canada (P.D. Kingsbury)

3.1. Current forest pesticide use

3.1.1. Use of chemical pesticides

3.1.2. Use of *Bacillus thuringiensis*

3.2. Toxicity and exposure to forest insecticides

3.3. Toxicity and exposure to forest herbicides

3.4. Buffer zones and protective regulations

3.5. Conclusion

4. The role of the pesticide companies in safe pesticide use (M.C. Gadsby)

5. Selected case histories: Effects of pesticides on aquatic habitats (W.L. Lockhart)

5.1. Sample cases

5.2. Regulatory role

5.3. Need for better conceptual models

5.4. Ethical responsibility

6. Assessing and regulating the environmental effects of pesticide use (W.R. Ernst)

6.1. Pesticide registration

6.2. Postregistration evaluation

6.3. Provincial regulation

6.4. Concluding remarks

7. General overview (D.C. Eidt)

References

---

1. INTRODUCTION

In anything we do there are conflicts and trade-offs to be made. The greater the scale of intensity and extent of the activity, the greater the conflicts and the more difficult the adjustments. With demand for higher production of higher quality foodstuffs, and with demand for forest products approaching and, in New Brunswick, exceeding production growth rates, we can no longer afford to share our crops with pest species; thus, the need for pesticides has increased. It is hardly necessary to tell aquatic toxicologists where the conflicts are.

The contributors to this chapter were chosen for their particular insights into the nature of these conflicts. Perhaps we can clarify the problems, even if we cannot solve them. We might even help to achieve the characteristically Canadian solution—compromise. The contributions represent five perspectives, agriculture, forestry, industry, fisheries, and environment, which follow in that order. The first two, agriculture and forestry, are the activities that are considered to be the principal users of pesticides, even though on an area basis a great deal more pesticides are used in urban areas. From the aquatic perspective, they qualify as villains. The pesticide industry is even less popular with those charged with protecting the aquatic environment, even though they must use its services occasionally. However, pesticide companies in Canada are developing an increased awareness that their products must be environmentally safe as well as efficacious to be acceptable in the marketplace. The fisheries are the victim. Whatever we terrestrials do seems to affect water quality one way or another. Pesticides are not the only chemical products that are of concern to aquatic biologists, but they are designed to kill, so they stand high on the list. Fisheries people have been concerned about a number of insecticides used in forestry, including DDT, aminocarb, and fenitrothion; black fly larvicides such as methoxychlor; and a wide range of insecticides and herbicides used in agriculture.

Finally, environmental people tread a thin line between environmental purity on the one hand, and exploitation of the gifts of chemistry on the other. Trying to be impartial makes them seem pawns of industry and government to the environmentalist, and zealots to the grower who lost the only product that worked because it was banned. They are people caught in the middle, perhaps the most difficult position of all.

2. AGRICULTURAL PESTICIDES IN THE AQUATIC ENVIRONMENT: RISKS AND BENEFITS*

Among agricultural practices, the use of chemicals is a recent phenomenon. Extensive chemical use has occurred for less than 35 years but has significantly changed the face of Canadian agriculture. In the golden years, pesticides were regarded as wonder chemicals allowing unprecedented increases in crop yields, improving food quality and production, giving added protection of human health, and improving human welfare. Pesticides have ameliorated the standard of living and provided increased enjoyment of life through improved recreational activities in areas not previously possible. By the 1960s, however, a downside of pesticide use was becoming well documented: Undesirable effects were reported in the environment; in some applications, the chemicals

*Section 2 was written by J.E. Hollebone.

intended to provide economic benefits were found to be jeopardizing human health. In crop management, use of chemicals has sometimes intensified other pest problems intended to be solved, (e.g., development of pest resistance). Great strides have been made since these early years by regulatory agencies to allow needed pesticide uses but to control unacceptable environmental and human safety effects. There is, however, considerable room for improvement.

This section addresses the general subject of agricultural pesticides in the aquatic environment. The subject could perhaps be more pertinently addressed as a question: Can agricultural pesticides be used safely with no or acceptable impact on the aquatic environment?

The question has several sides to explore:

1. Are agricultural chemicals necessary? If they are not, we probably should not be using them.
2. Do chemicals get into the environment? What are the risks if they get into the environment?
3. Can there be an acceptable balance between use and safety?
4. What should we be doing now and in the future?

2.1. Use of Agricultural Chemicals

There is no doubt that for the near future the use of chemicals as an aid to agricultural production is here to stay. Users of pesticides point to the high quality, variety, and volume of food produced with the aid of pesticides. Although new plant varieties, fertilizers, and management practices have played a role in this increase, credit must go in large part to agricultural chemicals for the control of serious diseases, insect pests, and weeds in crops. In agriculture, it is estimated that in the United States for every one dollar invested in pesticides there is a ten-dollar return. This is an average figure. Estimates differ slightly for different situations and different analysts. Pimentel (1982) arrived at an overall return of $2.82. For Canadian field crops the return is also significant. Recent work by Tolman et al. (1986) found that insect-incurred losses in the absence of pesticides were 47–50% for potatoes, 39% for onions, and 58% for rutabagas. In fact, in onions a commercial crop could not be harvested without the use of pesticides (in particular, herbicides are needed for weed control). Stemeroff and Madder (1985) found that apples could not be produced competitively in the absence of fungicides, although, through Integrated Pest Management programs, insecticide use could be reduced considerably in many years. It is not intended to belabor the benefits of chemicals as crop production tools, but merely to point out that many of the perfect fruits and vegetables available to the Canadian consumer today are there because pesticides have allowed high yields and high-quality crops.

2.2. Do Pesticides Get into the Aquatic Environment, and What Is Their Impact?

Agricultural chemicals have been found in the aquatic environment. They are there, sometimes as a result of direct contamination, most often by movement into water due to careless handling (e.g., from disposal of pesticide containers in ditches), accidental spills, or aerial overspray. Or they can reach water indirectly through volatilization and drift from application equipment on nearby treated fields, by runoff in surface water from fields, or from vertical leaching in internal soil drainage systems to groundwater aquifers and lateral transfer to river systems. Finally, natural factors, such as heavy rain, can result in transport to aquatic systems even of pesticides bound tightly to soil particles.

A recent study by R. Frank (1985) illustrates this point for Ontario. Of 61 pesticides studied in Ontario from 1975 to 1977, 18 (30%) appeared in surface waters because of one or all three reasons listed above. Fifteen came from spills, seven from runoff immediately following application of the pesticides, and four from spills plus runoff. Of the 18 pesticides, 14 were not persistent and four were persistent chemicals; the latter appeared in water all year round. The persistent pesticides, such as atrazine, could be divided into 60% in runoff waters, 20% in internal drainage, and 20% from spills. Once in surface water, many of the pesticides degraded rapidly and could not be detected downstream from the point of application. In the Grand River, only three of the 18 appeared at the mouth in concentrations that could be measured, and all three belonged to the persistent category.

The Ontario situation was chosen simply to demonstrate the point that pesticides do occur in water. Results of a number of other surveys carried out by Agriculture Canada, Environment Canada, and Health and Welfare Canada give additional data.

The second part of the question is not as easy to answer. What is the impact on aquatic systems? It depends, of course, upon a large number of factors, such as the amount and nature of the chemical that gets into the aquatic environment, its persistence, its stability, the nature of partitioning in the aquatic environment, and a host of other physical and biological factors. Most important of all is the effect on aquatic organisms and the ability of the chemical to biomagnify or accumulate in food chains.

2.3. Can There Be Agricultural Use and Environmental Safety?

The answer is equivocal: yes and no. Clearly there are instances where the benefits of use do not outweigh the adverse effects. In Canada, as elsewhere in the world, the group of pesticides identified with the most harmful environmental effects were the persistent organochlorine insecticides. DDT, for instance, has been associated with fish kills, unacceptable residues and longevity in the environment, and bioaccumulation in the food chain, resulting in the well-known effects on predatory birds. As a result, and due to other factors,

DDT and other persistent organochlorines, such as aldrin and dieldrin, were deregistered for agricultural uses by the early 1970s.

In other cases, the environmental risks from chemical use may not be perceived to outweigh the possible benefits. In most of these cases the approach taken by regulatory authorities is to try to reduce the risks to acceptable levels by modifying the proposed conditions of use, for example, by specifying use restrictions such as number of applications per season, setbacks, etc. In other cases, decisions may move from the realm of science and objectivity to that of policy and subjectivity. For example, methoxychlor has been used to kill larval black flies in the Athabasca and Saskatchewan rivers for many years. Methoxychlor is known, at the rate used, to have detrimental effects on the aquatic environment. It does not directly kill fish at these concentrations, but residues have been detected at low ppb (parts per billion) concentrations in tissue, and the nonselective kill of invertebrates has been well documented. On the other hand, the unacceptability of the hordes of black flies attacking livestock and humans in the area is also well documented. We have greatly simplified the complexities of this issue. However, the decisions taken have been to allow limited use of methoxychlor to control black flies, yet numbers of treatments have been controlled to reduce environmental impact. Clearly, in this instance the decision has been that limited treatments are necessary to allow livestock production and reasonable human comfort, and that some impact will be tolerated. Meanwhile, however, much effort has gone into exploring alternatives to this broad-spectrum pesticide and recently the biological pesticide *Bacillus thurigiensis israelensis* (*B.t.i.*) has been registered to replace methoxychlor for the majority of treatments. For every decision there are some downsides. While *Bti* has a narrow spectrum of activity and is therefore environmentally much more acceptable, the cost of treatment is higher and the metholodogy of application is still largely experimental and may take some years to perfect.

2.4. What Are We or Should We Be Doing to Allow Use of Agricultural Pesticides, Yet Ensure Safety to Aquatic Systems?

First, we must do the best we can to assess the potential of new chemicals for adverse aquatic impact before they are registered for use. In Canada, pesticides are registered after a presale assessment of their safety under the Pest Control Products Act administered by Agriculture Canada. The review process is a consultative one with other federal departments. Health and Welfare Canada advises on human safety with regard to user and bystander exposure, and sets limits for residues allowed in foods. Fisheries and Oceans Canada and Environment Canada provide advice on the fate of pesticides in the environment and an assessment of possible effects in aquatic systems. In recent years the environmental component of the review process has been greatly strengthened.

Draft Environmental Fate Guidelines have been available since 1982. A revised and updated version is scheduled to be complete by summer 1987. It contains a new section on aquatic fate assessment. The guidelines formalize what has been required for several years, and provide guidance on acceptable protocols. The studies resulting from these guidelines should allow an understanding of what impact a new pesticide will have in the environment. The field of aquatic fate has changed rapidly in the last few years and the new aquatic section is designed to allow better prediction of the fate of a new pesticide in water. Knowing this, plus the inherent toxicity of a pesticide to aquatic organisms, regulators can better decide whether a new candidate pesticide can be used safely and, if it cannot, determine if conditions can be established that would reduce impact to acceptable levels.

Modern analytical chemistry is such that today we can measure concentrations in parts per billion or parts per trillion. This increased sensitivity means that there are very few "perfect" pesticides and that, increasingly, regulators will be assessing benefits of use against risks. The final regulatory position will represent a balanced compromise somewhere between sometimes conflicting needs or points of view.

The synthetic pyrethroid insecticides illustrate the regulatory assessment that is taken in such cases:

1. On the one hand, the pyrethroids are highly effective compounds, with a wide margin of safety to plants, mammals, and birds. They are seen by the farmer as very attractive alternatives to currently registered, more toxic pesticides. In most cases they can be applied without the need for special safety equipment, such as respirators.

2. On the other hand, they are highly toxic to aquatic insects and fish, within the parts per billion range (10^{-9}) to fish and parts per trillion range (10^{-12}) to some aquatic invertebrates. Yet, the synthetic pyrethroids are lipophilic and not particularly mobile. In other words, they do not leach, but are highly adsorptive to soil particles. The argument was made that they would be unlikely to reach aquatic systems if not directly applied to water. No, said the environmental advisors, catastrophic rain events could physically move bound pyrethroids to water where they could dissociate and cause toxicity. Furthermore, direct overspray at operational rates, or spray drift, might cause adverse effects that would be unacceptable, especially in areas considered to be fisheries-productive, such as the salmon streams of British Columbia and New Brunswick, or environmentally sensitive, such as the prairie pothole region of Saskatchewan.

Our approach was to examine drift study data from a number of areas to see if adverse effects could be minimized. We reviewed Environment Canada analyses of research permits in British Columbia that suggested drift beyond 67 m from standard agricultural aerial applications was less than 1%. Then we looked at a pyrethroid study by Ontario Environment, which supported the Environment Canada analysis. Finally, we looked at some data from registrants of these compounds on drift during normal agricultural use.

With the agreement of environmental advisors, a 15-m buffer zone was set for ground-applied pyrethroids and a 100-m buffer zone for aerial application was established around environmentally sensitive areas. Acceptable pyrethroids were registered under temporary registration status, while confirmatory field tests were carried out. In 1985, summer monitoring by Agriculture Canada confirmed that the buffer zones appear to be adequate. Trace amounts, ppt (10^{-12}) concentrations, were found in one prairie slough and low concentrations at a stream mouth adjacent to treated potato fields were reported in an Environment Canada study. The 1985 work was considered preliminary. Further studies were carried out in 1986, again by Agriculture Canada, by Environment Canada, and by the registrants. We do not yet have the results of these studies, but when complete, we hope they will tell us whether an adequate margin of safety has been provided, that is, whether the registrations may continue and farmers will be allowed to use the pyrethroids.

In addition to assessing the aquatic impacts of new chemicals before registration, we try to ensure that pesticides currently registered are being used safely. Fifteen percent of the pesticides detected in water resulted from spills through accidents in handling. One of the most important directions is to raise awareness that pesticides must be used safely. In response to ever increasing queries about pesticides, a Canadian national pesticide call line has been established. The professional staff handled 3271 calls from January to October concerning all aspects of pesticides. Farmer and grower calls accounted for 10% of the enquiries, provincial and extension calls about 30%, and miscellaneous inquiries from such as home gardeners, householders, and environmentally concerned persons accounted for the rest. Also an annual letter is sent to all farmers in Canada to provide advice regarding safe use and handling procedures. Provincial governments have active education programs often associated with licensing. Fisheries and Oceans has produced a cautionary brochure about toxicity of synthetic pyrethroids, and there has been an industry–government filmstrip on container handling and disposal. We see farmer education as being one of our most important means to minimize misuse of pesticides.

Third, we advocate increased monitoring and reevaluation of older chemicals. We do not know the effects of many of the older chemicals in aquatic systems, especially their sublethal effects. A priority scheme for aquatic monitoring has been developed, and a recent pilot study of 684 samples of organophosphates and carbamates in October 1986 found no trace of these compounds in drinking water. Current efforts are underway to coordinate this program with Environment Canada efforts and with the national drinking water monitoring program carried out by the federal and provincial ministries of health.

There are other measures that also can be taken to minimize aquatic impacts. It is not prudent to let pests waste a large proportion of our food. To compete effectively in world and domestic markets, Canadian farmers will be using pesticides. On the other hand, it is not prudent to pollute our environment and it is clear that common sense must be used in the handling of economic poisons.

Agriculture Canada is actively seeking viable alternatives. For instance, there are new thrusts in insect pest management (IPM) endeavors to actively achieve control effectively with less use of pesticides. "Towards an IPM Strategy" (Agriculture Canada, 1986) records a national discussion of the problems, concerns, and research needed to improve effort in this area.

There is a renewed effort toward use of biocontrol organisms to replace, or augment existing chemical control measures. There are risks in this approach, such as the possibility of gypsy moth-like escapes. We must move cautiously in this area. There are also new trends toward use of microbial pesticides and genetically engineered organisms and plants. For these we will need to develop new regulatory guidelines to cover such concerns as their release and fate in the environment.

2.5. Future Regulation

The use of chemicals in the environment will continue to be controversial. By virtue of our increased ability to measure minute deposits and concentrations and increased understanding of cellular processes, the chances of a pesticide without some negative effects in today's world are small. We are becoming increasingly aware of areas where science does not always provide clear directions. We must move from emotional pesticide debate to better cooperation and understanding. We must avoid the polarized positions of pro-agricultural users and those against any pesticide use. There is a need to move toward working more cooperatively to identify and resolve problems and explain solutions, rather than to take rigid accusatory positions on each side. The result will be a more cooperative process in which the interests of one group will be weighed and considered against the interests of another. The solution, which may not be perfect, and may not satisfy everybody, must allow respect for opposing points of view and lead to negotiated settlements. The end product will be a science-driven and balanced decision, which should best protect the interests of Canadians.

3. RISKS TO AQUATIC ORGANISMS FROM FORESTRY PESTICIDE USE IN CANADA*

The effects of forest spray programs on aquatic organisms have been a lively topic of debate for a long time in Canada. Since the time Rachel Carson's classic book *Silent Spring* was published (1962), the effects of forest pesticide use on fisheries resources has generated lively and sometimes heated public and scientific controversy. The effects of DDT used for spruce budworm

*Section 3 was written by P.D. Kingsbury.

control on Atlantic salmon, documented by researchers of the Fisheries Research Board of Canada, constitutes one of the classic examples of the discovery that pesticides are a two-edged sword capable of serious harm as well as substantial benefits. The more recent avalanche of other significant environmental pollution issues, such as acid rain, PCBs, and heavy metals, has caused attention to shift away from forest pesticide use to other environmental issues, particularly as it involves scientific scrutiny and public concern. But the legacy of the serious impacts of DDT still seems to haunt forest managers through the public perception of the risks that current forest pesticide use poses to aquatic organisms. We would like to briefly focus on that risk and attempt to show why such a perception is both outdated and scientifically unfounded.

3.1. Current Forest Pesticide Use

3.3.1. Use of Chemical Pesticides

We must first deal with current pesticide use in forestry and the materials that are being used. There have been some significant changes. First, forestry is a small user of pesticides in Canada (Table 16.1). Forestry use of herbicides, insecticides, and fungicides represents about 1.5% of total pesticide use in Canada, and is not only modest in comparison to agricultural and industrial use, but is also considerably smaller than home and garden use.

There have been some significant trends in forestry, particularly recently, toward less overall use of insecticides, and greater use of less hazardous materials, applied in what we might broadly describe as a more controlled fashion. We have moved away from the organochlorine compounds, DDT in particular, toward organophosphorous and carbamate compounds (Table 16.2). Since the 1970s there has been a substantial decline in the area of eastern Canada sprayed each year to control the spruce budworm, by far the most destructive forest insect pest in the region. Although this is partly in response to the size and intensity of the pest outbreak, there are other factors involved that represent significant changes in policy and practice in forestry.

Table 16.1 Summary of Annual Pesticide Use in Canada, 1977–1982

Use	Amount (kg)	%
Canada—all uses	34,648,000	100.00
Agricultural, industrial, and structural pest control	31,840,000	91.90
Home and garden	2,280,000	6.60
Forestry	504,000	1.43

Source: J.F. Henigman and P.J. Humphreys, British Columbia Ministry of Forests, Protection Branch, Victoria, B.C., personal communication, 1986.

Table 16.2 Operational Use of Insecticides to Control Spruce Budworm in Eastern Canada, 1965–1986[a]

		1000s of Hectares Sprayed[b]					
Insecticide	Period Used	65–69	70–74	75–79	80–84	85	86
DDT	1944–1969	1800	—	—	—	—	—
Phosphamidon	1963–1977	750	2,550	3,950	—	—	—
Fenitrothion	1967–1986	1400	12,800	13,900	7900	352	222
Aminocarb	1970–1986	—	650	8,500	4100	455	197
Bt	1974–1986	—	5	40	520	644	336
Hectares sprayed/yr		790	3,201	5,278	2504	1451	755

[a] Approximate figures compiled from Forest Pest Control Forum reports.
[b] Much of the area sprayed received a second application of the same material.

As forest insect control operations moved away from the use of DDT and phosphamidon, they moved toward materials that are applied in smaller quantities. DDT was generally used at 280 to 1120 g/ha, but fenitrothion is used at 210 or 280 g and aminocarb at 70 to 95 g of active ingredient per hectare, so that even if the toxicities of these materials to aquatic organisms were similar there would be less exposure simply because smaller amounts are used.

Another change is less use of large aircraft on large blocks. This was the type of operation that was relatively common in the early 1970s, in an attempt to suppress insect damage over large areas. Since that time, almost all jurisdictions have deliberately moved away from large aircraft toward smaller aircraft, spraying smaller blocks of forest. One of the main reasons is that there has been a great need to increase the economic and social justification for spraying in terms of what values are to be protected and what the benefit/cost will be. It is expensive to spray large areas of forest.

Almost all jurisdictions have had to use economic analysis to defend, on a long-term basis, that an area has a wood supply critical to the maintenance of an existing industry, and that there will be an economic return on the control operation. All this is part of the current revolution in Canadian forestry which is to get away from forest exploitation and into intensive forest management. The result is a trend toward more intensive use of a smaller portion of the land base. This is particularly true for herbicide use because any site treated with herbicide usually receives other kinds of silvicultural treatments, such as site preparation and planting. All forest herbicide treatments are expensive because the return on investment comes 50 to 80 years later. Herbicides and their application are expensive, so economic justification is needed before money is spent to treat a site.

Table 16.3 Operational Use of Forest Insecticides in Eastern Canada in Recent Years[a]

	1000s of Hectares Sprayed[b]			
Insecticide	1983	1984	1985	1986
Spruce Budworm Spraying Only				
Bt	64 (2%)	376 (19%)	644 (44%)	336 (44%)
Fenitrothion	1553 (56%)	687 (34%)	352 (24%)	222 (29%)
Aminocarb	1162 (42%)	930 (47%)	455 (31%)	197 (26%)
Total	2779	1993	1451	755
All major forest insect pests (spruce & jack pine budworms, gypsy moth, hemlock looper)				
Bt	64 (2%)	376 (19%)	868 (48%)	906 (64%)
Fenitrothion	1553 (56%)	687 (34%)	475 (26%)	301 (21%)
Aminocarb	1162 (42%)	930 (47%)	455 (25%)	197 (14%)
Total	2779	1993	1798	1404

[a] Approximate figures compiled from Forest Pest Control Forum reports.
[b] Much of the area sprayed received a second application of the same material.

3.1.2. Use of Bacillus thuringiensis

For many years the Canadian Forestry Service has carried out research on controlling lepidopterous forest pests with the bacterial pathogen *Bacillus thuringiensis* (*Bt*). Recently there has been a dramatic increase in the use of *Bt* in forestry, both in spruce budworm control and in other insect control programs (Table 16.3). *Bt* was used on only 2% of the area sprayed in 1983; in 1986, 64% of all forest insecticide spraying was based on *Bt* in various commercial preparations. In many jurisdictions, this has been an intentional move and one that is apparently not reversible. It is now policy in the provinces of Nova Scotia and Quebec that *Bt* will be the only acceptable choice for control of insects such as spruce budworm. It is virtually an established practice in other jurisdictions such as Ontario and British Columbia that *Bt* is the only insecticide that politicians are prepared to allow for forest spraying. *Bt* has become and will probably remain the major forest insecticide for those pests against which it is effective. There are obvious benefits in terms of lessened risks to aquatic systems associated with increased used of this highly specific pest control product as an alternative to broad-spectrum chemical insecticides.

3.2. Toxicity and Exposure to Forest Insecticides

A large number of insecticides have been considered and experimented with for forestry use in Canada. Their toxicities to a selection of aquatic organisms

Table 16.4 Toxicity (mg/L), IU/L for *Bt* of Some Forestry Insecticides to Aquatic Organisms[a]

Insecticide	*Daphnia*	Amphipods	Stonefly nymphs	Rainbow trout
Acephate	—	>50	10–12	1100
Aminocarb	0.02[b]	0.01	1.0[b c]	13.5
Bt	—	—	>430 IU/mL[d]	300–1000[e]
Carbaryl	0.006	0.02	0.002–0.005	2.0
DDT	0.005	0.001	0.001–0.007	0.009
Diflubenzuron	0.02	0.03	>2.4[f]	240
Fenitrothion	0.01	0.003	0.004	2.4
Mexacarbate	0.01	0.4	0.01	12.0
Permethrin	0.001	0.001[g]	0.002[c]	0.002[h]
Phosphamidon	0.01	0.01	1.5	7.8
Trichlorfon	0.0002	0.04	0.01–0.04	1.8

[a] 48-h EC_{50} values for *Daphnia* and 96-h LC_{50} values for other organisms except where noted. Data from Johnson and Finley (1980) except where noted.

[b] Holmes and Kingsbury (1980).

[c] 48-h LC_{50} after 1-h exposure (Poirier, 1986).

[d] Eidt (1985).

[e] NRCC (1976).

[f] 14-d EC_{50} after 1-h exposure (Poirier and Surgeoner, 1987).

[g] 24-h $LC_{90\text{-}95}$ after 1-h exposure.

[h] NRCC (1986).

are given in Table 16.4. Of these materials, only fenitrothion, aminocarb, and *Bt* have been used to a significant extent in operational spray programs in the 1980s.

Most of these materials have 96-h LC_{50}'s to rainbow trout under static conditions in the parts per million range. Two exceptions are DDT and the synthetic pyrethroid, permethrin, which are toxic to trout in the parts per billions range 10^{-9}. Not surprisingly, given that we are dealing with insecticides, toxicities to insects and crustaceans are considerably higher than to trout. If we compare these exposures with those that we usually document in forestry situations we see some interesting differences. An extensive study published recently by Morin et al. (1986) summarized peak residues of aminocarb and fenitrothion found in forests, streams, and lakes after spruce budworm sprays in Quebec over a 4-year period. It was based on over 400 water samples collected from various spray blocks immediately after spraying. The median peak concentrations of fenitrothion measured in forest streams were just less than 3 ppb. In terms of the toxicity to fish, these median peak concentrations are one eight hundredth the 96-h LC_{50} value for rainbow trout (Table 16.4). For aminocarb those peak concentrations were less than 1 ppb. The maximum concentrations of fenitrothion found in streams were 127 ppb, a twentieth of concentrations toxic to rainbow trout; for aminocarb they were about 20 ppb. Permethrin, which has a very high toxicity to rainbow trout, has not been used operationally in forestry but the peak concentration found in over a dozen streams directly oversprayed in experimental situations was 2.5 ppb or very close to the concentration toxic to fish.

Peak concentrations refer to the maximum observed pesticide residues in water. These peaks are of short duration in forest streams because there is rapid dissipation and decline in pesticide concentrations once direct spray inputs cease (Fig. 16.1). This is not simply chemical breakdown, but primarily transport and dilution of the material within the type of flowing water systems that we think of as salmonid habitat. Within the first 96 h, the period that we usually use for expressing toxicities, the peak concentration is present for only a short period, declining rapidly so that for most of the 96 h, exposure is to much lower concentrations. Bioassays for 96 h in the laboratory with typical peak concentrations reported in the field expose the fish to a much higher concentrations reported in the field expose the fish to a much higher concentration than that to which they would be exposed in the natural environment. When brief pulses of insecticide are used in bioassays, toxic effects are not seen at concentrations much higher than those producing toxic effects in static bioassays with 96-h exposure.

Some chemical insecticides do pose a hazard to some aquatic invertebrate groups at exposures encountered after normal spray operations. An extensive literature, reviewed by Bart and Hunter (1978) and Trial (1986), suggests that

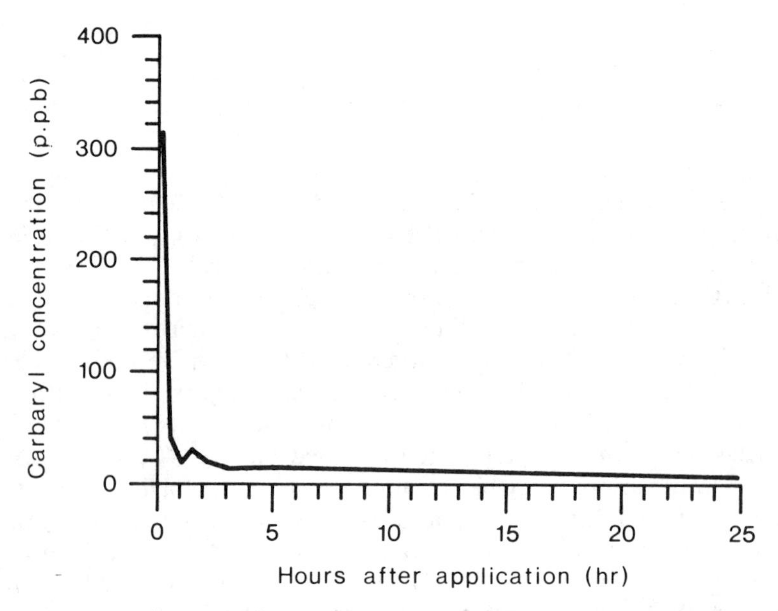

Figure 16.1. Carbaryl residues in stream water following an aerial application of 280 g/ha Sevin-2-oil (after Holmes et al., 1981).

Table 16.5 Toxicity (mg/L) of Some Forestry Herbicides to Aquatic Organisms[a]

Herbicide	*Daphnia*	Amphipods	Shrimp	Stonefly Nymphs	Rainbow Trout
2,4-D					
2,4-D amine	4.0	100.0	0.15	—	100.0
2,4-D ester	1.2	2.9	0.4	2.4	1.0
2,4-D butyric acid	—	—	—	15.0	2.0
Glyphosate					
Glyphosate technical	780.0	—	—	—	130.0
Roundup	5.3	43.0	—	—	8.3
Hexazinone					
Velpar	152.0	—	56.0	—	320.0
Triclopyr					
Triclopyr technical	133.0	—	—	—	117.0
Garlon 3A (Triclopyr amine)	—	—	895.0	—	240.0
Garlon 4 (Triclopyr ester)	10.1	—	—	—	0.74

[a] 48-h EC_{50} values for *Daphnia* and 96-h LC_{50} values for other organisms. Data from Johnson and Finley (1980), Ghassemi et al. (1981), and Weed Science Society of America (1983).

disturbances to stream and pond invertebrate communities often occur with permethrin, carbaryl, and to a lesser extent fenitrothion, although mortality is only partial and recovery occurs within the season of application. There have been cases of more complete and longer lasting impacts of permethrin and carbaryl on the most sensitive invertebrate groups. Aminocarb and *Bt* have little effect on stream invertebrates. Symons (1977) reviewed the literature to 1976, which give data on stream invertebrate drift increases and benthos reductions after fenitrothion sprays, and he concluded that the median response to applications of 210 to 280 g/ha is a 5 to 10% reduction of aquatic invertebrates. In many instances, no effects were noted, and in other cases substantial drift increases and benthos reductions were noted for the most sensitive species, generally mayfly and stonefly nymphs. This reflects variability in spray deposits, protection by overhead forest canopy, and other factors.

3.3. Toxicity and Exposure to Forest Herbicides

The toxicities of the major forest herbicides to fish, as with insecticides, tend to be in the parts per million range (Table 16.5). Herbicides, however, are far less toxic to insects and crustaceans than are insecticides, so that in many cases fish tend to be among the most sensitive organisms on which we have data.

Herbicides enter the aquatic environment by direct application, spray drift, mobilization in ephemeral stream channels, overland flow, and leaching. The relative size of the peak concentration and duration of entry depends on the mechanism, as summarized by Norris (1982):

Entry Mechanism	Relative Size of Peak Concentration	Duration of Entry
Direct	High	Short (during application)
Drift	Moderately High	Short (during application)
Ephemeral channels	Moderate to low	Short (first storms)
Overland flow	Low	Medium
Leaching	Very low	Long

From Norris (1982).

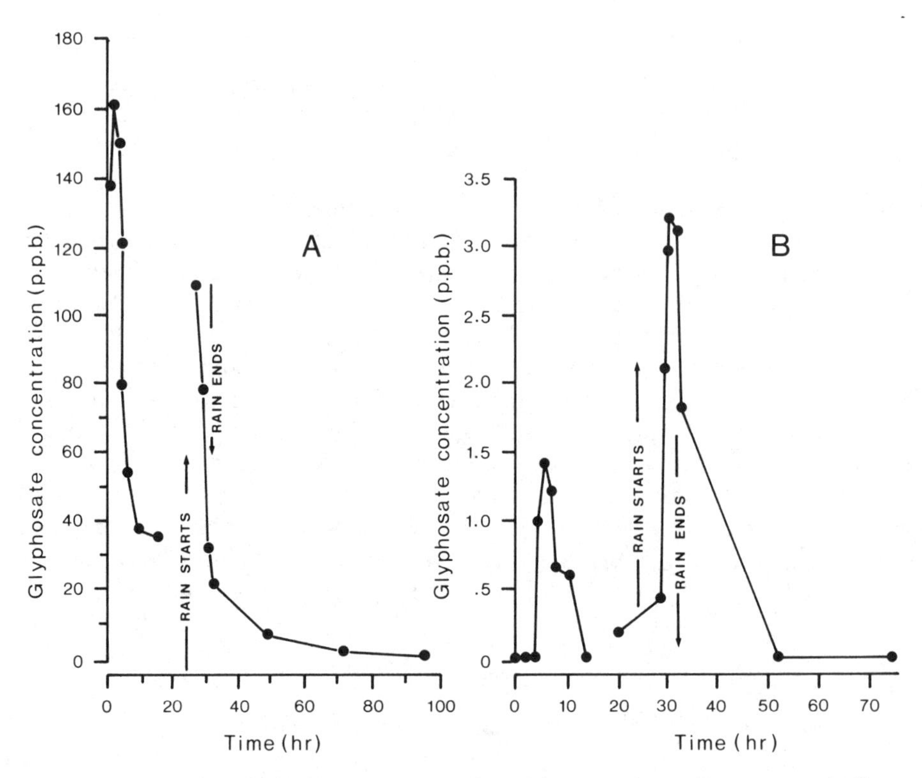

Figure 16.2. Glyphosate resides in stream waters from (A) an experimentally oversprayed tributary and (B) a main channel protected by a 10-m unsprayed buffer zone in a coastal British Columbia watershed (after Feng et al., 1986).

Direct application is the mechanism most likely to introduce significant quantities of herbicide into surface waters and has the potential to produce the highest concentrations.

Herbicides are generally applied at substantially higher rates than are insecticides. In streams, the same type of rapid decline from the initial peaks tends to occur (Fig. 16.2). With herbicides much material is sprayed onto foliage and the forest floor, and when heavy rainfall occurs shortly after application, additional herbicide can enter the aquatic system (Fig. 16.2). Again, this second peak tends to decline rapidly. The peak herbicide residues measured in stream water after forest applications vary with the nature of site, application conditions, and the timing and intensity of subsequent rainfall. Even when streams are directly oversprayed, short-lived peak concentrations in stream water are generally less than 100 ppb, although values approaching 1 ppm are occasionally found (Ghassemi et al., 1981; Norris et al., 1983).

3.4. Buffer Zones and Protective Regulations

Buffer zones are a fact of life in forest spraying. One of the reasons is that aerial application of pesticides to forests falls in the restricted category according to federal legislation. This means that there is a requirement for provincial agencies to grant permits for each application. Provincial regulations are generally accompanied by site inspection by regulatory agencies and their advisors and often by additional stipulation of buffer zone constraints and measures that will limit hazard to aquatic systems. All jurisdictions across Canada have now established buffer zones that are applied to use of pesticides in forestry. They also have regional and local enforcement officers, and because of the high profile of forest pesticide use, enforcement is strict.

Can buffer zones work? There is much research being done on the subject. An example is a recent study by the Forest Pest Management Institute designed to rationalize buffer zones relevant to pyrethroid use (Payne et al., 1986). A number of wading pools containing sensitive test animals were placed at various distances from a permethrin spray line. The data produced (Table 16.6) show that mortality can decrease substantially at distances downwind.

Because of strict control of forest herbicide applications, the herbicides are usually applied to small sites, often from the ground or with helicopters. The data in Fig. 16.2B were generated with sophisticated low-drift application equipment, a microfoil boom, and a 10-m buffer zone. Using such tactics, the peak residue was only a few parts per billion, whereas the peak residue in an intentionally unbuffered tributary (Fig. 16.2A) was about 150 ppb. This occurred despite the fact that the buffered sampling station was in the main stream and received some residue from the small tributary that was intentionally oversprayed.

Table 16.6 Mortality in Populations of *Aedes aegypti* Exposed in Artificial Pools (depth about 7 cm) during Ground and Aerial Applications

	Mortality (%)			
	Ground (35 g of a.i./ha)		Aerial (140 g of a.i./ha)	
Downwind Distance (m)	Predicted	Observed	Predicted	Observed
30	3	2.1	—	—
50	1.2	0.9	39	20
100	<1	1.2	19	2.2
200	<1	0.1	7	0.5
400	—	—	2	0.5

Source: Data from Payne et al. (1986).

Not only are buffer zone restrictions applied to forest pesticide applications, but in some provinces an applicator must have special training before he may apply pesticides to forests. In Ontario, for example, to bid on a contract to apply pesticide to Ontario Crown Land an applicator must have a certain amount of experience in flying in forest situations. The only way he can acquire that experience is to fly under the guidance or direction of an experienced applicator. He must also attend training sessions, including lectures on environmental protection. He is told that if protected waters are sprayed, hazards are greatly increased and furthermore he increases the risk of putting himself and forestry out of business. This is backed up by requirements for performance bonds and liability insurance. The training is taken very seriously and the cavalier attitude that has previously been associated with some pesticide applicators is rapidly becoming a thing of the past. This is particularly true in forestry, where the number of parties involved in and paying for the control operations is small, and those who behave badly soon find no customers for their business.

3.5. Conclusion

There have been extensive studies of the effects on aquatic systems from pesticides that are being used in forestry. Fenitrothion is perhaps the most studied. It was intensively studied by Fisheries Research Board of Canada (now Fisheries and Oceans Canada) researchers at the time that it was introduced as a replacement for DDT. The conclusion of these and other researchers was that the impacts of fenitrothion, as it was used in spruce budworm control at the time, were few and erratically distributed. There was no evidence of direct effects on fish. Sometimes there were increases in stream invertebrate drift and occasionally decreases in benthic invertebrates, but these were sporadic and depended on the nature of the system and the type of exposure. It was felt that not only could these impacts be tolerated but secondary effects on fish were improbable, and given the amount of research that had been done, there was no point to further study the aquatic impacts of fenitrothion under conventional use patterns unless the material or its formulations were made more hazardous. The other two insecticides, aminocarb and *Bacillus thuringiensis*, subsequently introduced and widely used for forest insect control, are even less likely to have impacts on aquatic organisms. Insecticides such as permethrin, which cause dramatic increases in stream invertebrate drift and reduce populations of fish food organisms whenever they are directly introduced into aquatic systems, are not suitable for widespread use in forestry, even though they are still unlikely to cause direct fish mortality. They should only be used in situations where other adequate alternatives for pest control are not available and scientifically established buffer zone requirements have been defined and implemented to protect adjacent water bodies. Forest herbicide use following appropriate buffer zone requirements will have little or no adverse effect on aquatic organisms. The small-scale and controlled nature of herbicide operations and low toxicity and mobility of the herbicides used all contribute to limiting possible effects.

Finally, it should be stressed that forests and aquatic systems are integrally linked. The nature of the forest influences the water quality and yield of the water systems that flow through it (Hynes, 1975). We must manage our forests to maintain the quality of fish habitat. Allowing forest pests like the spruce budworm to ravage extensive forests is not a good way to do it. Drastic changes in cover resulting from tree mortality and salvage logging can drastically alter the stream habitat. For the aquatic environment, the forest's illness left unchecked can be worse than the control measures.

4. THE ROLE OF THE PESTICIDE INDUSTRY IN SAFE PESTICIDE USE*

In tight economic times such as these, it is difficult to formulate a strategy for crop production. Obviously, production efficiency is a paramount concern. Production must be maximized per unit area while minimizing financial drain. This is not merely a matter of controlling input costs but is also a matter of reducing avoidable losses of crop quantity and quality. The only real strategy is intensive management of agricultural and forest resources.

Pesticides are only one small part of the total weaponry available to the producer. How critical to production success are they? Are the needs important or overstated?

The period between detection of a major problem and the time for action is short. Usually no advance warning is available. With literally hundreds of thousands of dollars tied up in equipment, fertilizers, time and seed one must protect that investment. Biological control agents tend to be slow acting. They

*This section was written by M.C. Gadsby.

often require a willingness to maintain low levels of both the pest and the control agent throughout the production period. This is usually not possible. Although very useful, biological control is not the approach for many pest-control situations. Even *Bacillus thuringiensis* and other "bioinsecticides" are not always solutions to critical pest problems. Swifter methods of biological control and biotechnology are still in the future. At present there is little alternative to the use of chemical pesticides.

Pest monitoring reduces the guesswork involved in the timing of pest-control operations. It often makes the difference between effectively treating a real problem and wasting money. Valuable as pest-monitoring is, it cannot by itself control a pest problem, however.

Agriculture and forestry need not be destined to eternal conflict with environmental preservationists. Pesticide use in agriculture and forestry does not guarantee the destruction of sensitive environments. The "Pesticides versus the Environment" issue is not one of irreconcilable conflict. Instead, it is an issue of "essential tension," tension that can fuel "creative compromise." There can be a balance between the benefit from pesticide use and the risk of potential hazard to the environment.

The toxic effects of pesticides must be controlled; there are three basic approaches. The first is applied during pesticide development. Chemicals selected for development are those with toxic effects controlled by the mode of action of the substance, by the method of application, by the timing of application, and by a myriad of other factors. The value of research during the development stage is that it is assessed unemotionally based on the product's performance relative to the projected production costs and market demand. Uncontrollable chemicals are discarded; only reliable, controllable pesticides proceed to market.

The second approach is control of pesticides by legislation; legislation which is determined objectively, based on the real need for pesticides balanced against the potential hazard they represent. Regulations and guidelines based on legislation are established to control uses, dosages, setbacks, waste disposal, and so on, in the same objective way.

Finally, control is exercised over toxic effects of pesticides by careful practices based on responsible guidelines for product use. The decision to use a particular product is objectively made, based on the benefits to the crop, weighed against cost and hazards.

Objective decisions must be based on facts elucidated through competent research. The research must not be a single-minded search for pesticide effect, but a comprehensive look at pesticide behavior. One should ask what normal use patterns pose a real hazard to sensitive environments, how can the chances of pesticide disruption of sensitive aquatic environments be minimized, and how can pesticide effects be mitigated.

These are challenging research questions for which answers will not come easily. Enormous resources are required to respond, and these are not questions that industry or government alone can answer. The real environmental issues can only be elucidated if we work together, using our separate areas of expertise. Any other approach is not realistic or productive. Industry needs access to the experience and knowledge of competent environmental assessors. Environmental scientists need access to agriculture's and forestry's experience in application science, and industry's experience with pure chemistry and goal-oriented research. Together, our expertise is impressive. At present the system is often adversarial, but it need not be. The Canadian system of pesticide registration is one of the best in the world. When adversarial, it works, but cooperatively, it could work a lot better.

5. SELECTED CASE HISTORIES: EFFECTS OF PESTICIDES ON AQUATIC HABITATS*

In terms of the protection of fish and fish habitat, it does not matter whether we discuss pesticides, heavy metals, or other toxic chemicals, since they all represent chemical inputs with which the fish must cope. There is nothing new in this: History and archeology reveal that the release of chemicals into the environment has been with us throughout our cultural existence. What is new is the unprecedented scale with which we can now produce and disseminate chemicals. With ever-growing numbers of people and ever-growing ingenuity in finding new chemicals and new uses of chemicals, there is accelerating growth in the quantities and kinds of chemicals in circulation. Pesticides are just the tip of a chemical iceberg, but pesticides are the only chemicals released into the environment with the deliberate intent of producing biological injury. Public agencies, such as departments of agriculture, forestry, and public health, are all under pressure to explain and justify their uses of pesticides to a public increasingly concerned about pollution. While our primary role in pesticide registration is in the protection of fish and fish habitat, we have some sympathy for other pesticide user groups since fisheries management agencies have frequently found themselves using pesticides, for example, to control undesirable fish species, to control parasites like the sea lamprey, or to control fungal diseases in hatcheries and aquaculture facilities.

The problems of chemical pollution have been recognized at many different levels, as judged by the creation by government departments of environment in many countries. A perceptive statement of the problem was given by Pope John Paul II, reprinted in *Science* (1980): "Man today seems always menaced by what he produces. This seems to constitute the principal act of the drama of human existence today." But for the most part people do not produce chemicals without reason, and certainly not in the case of pesticides. Canadian

*This section was written by W.L. Lockhart. The views expressed in this report represent those of the author and should not be considered the policies of the Canadian Department of Fisheries and Oceans.

farmers are businessmen in a competitive world market, and they do not buy pesticides because they want to spend their money that way; they buy them because they need effective pest control to stay in business. We know that Canadian farmers spent about $740 million (1984/85 figures) on pesticides (74% herbicides) and that they apply about 34 million kg of pesticides annually. In the past it seems that we have used the chemicals for the benefits they promised, and later we have found the menace in biological effects we had not anticipated. Now our registration procedures are aimed trying to evaluate both the risks and the benefits prior to reaching a registration decision.

If we look at world population growth, now some 5 billion (Fig. 16.3), then it is hard to see how the chemical pollution problem will get any better. In terms of our own numbers we continue to create a different kind of world from that of our history. As increasing numbers of people seek improved standards of living, there seems likely to be growth in the demand for more and better pest control, with the accompanying demand on the environment to assimilate greater chemical loadings. Some people will believe almost any scenario about the meaning of those chemicals for the environment and for themselves. In Lewis Carroll's *Through the Looking Glass* the Red Queen said, "Why sometimes I've believed as many as six impossible things before breakfast," and that sentiment sometimes applies to perceptions about environmental chemicals. I have attended public hearings on pesticide uses in Winnipeg, and I have listened to people express their beliefs about the dangers of chemicals in the

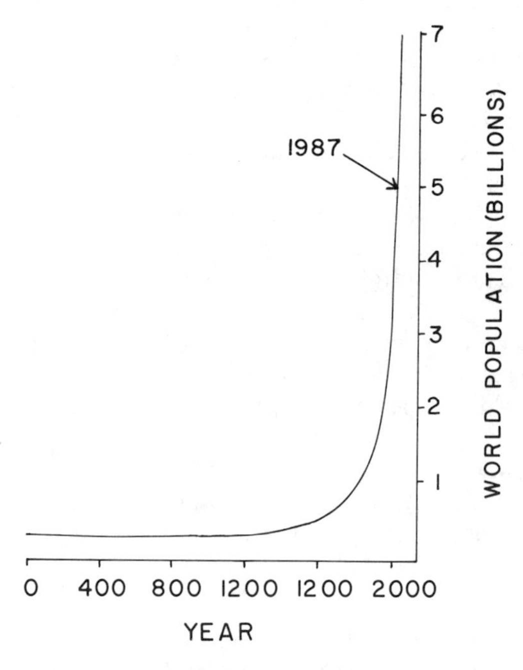

Figure 16.3.

environment. Sometimes the beliefs really are impossible, and sometimes they are sound qualitatively, but amplified out of realistic proportion. I recall one citizen's announcement that 2,4-D was causing cancer in farmers who use that herbicide, and I can remember being unable to find any convincing evidence in support of that. Now I understand that 2,4-D is indeed being investigated for that very effect.

People often complain that they cannot get what they regard as unbiased information about pesticides. It seems that the glossy brochures and instruction booklets are always those from the pesticide manufacturers. One published study of where pesticide users get their information was that of Turpin and Maxwell (1976) who analyzed the way Indiana corn farmers made decisions on their use of soil insecticides. They found that farmers used information from pesticide dealers to define their soil insect problems (45.8% of farmers), to decide what the solution to their problem was (63.2% of farmers), and to select the best insecticide (71.4% of farmers). This seems a clear conflict of interest.

The Fisheries Act contains provisions designed to limit the introduction of harmful chemicals into water, and we have a Fish Habitat Policy (Department of Fisheries and Oceans, 1986) intended to conserve the quality of habitat important for fish. There have been enough convincing cases where pesticide applications have had undesirable side effects on fish and wildlife to cause agencies responsible for the conservation of those species to become cautious about pesticides. The institutional arrangements regulating the use of pesticides have been developed following the scientific understanding of the effects of earlier uses of pesticides before regulations were in place. We recognize how little we can do about pesticide pollution after it has occurred, and we now try to focus our attention on prevention of undesirable side effects. The Department of Fisheries and Oceans has the opportunity to review data submitted by the industry in support of applications to register and market pesticide products. Based on those data a judgment is made on the acceptability of the registration in terms of risk to fish and fish habitat, and that judgment is forwarded as advice to Agriculture Canada where the final decision is made.

Generally, our reviews are based on the data supplied by chemical companies to Agriculture Canada and on data published in the open literature, but sometimes the department also conducts independent field or laboratory research. In these instances private companies and other government agencies have been helpful in providing information and samples for testing, especially samples labeled with radioactive tracers. Agriculture Canada's program of supplying analytical standards of pesticides has been particularly welcome.

5.1. Sample Cases

Most clear examples in which pesticides have become pollution problems occurred before the establishment of the current system of pre-use reviews. Early cases also preceeded the development of gas–liquid chromatography and particularly the electron-capture detector, and this technology has been a key factor in tracing the environmental dynamics of organochlorine pesticides. The presence of residues has in turn been a key factor in supporting cause–effect arguments.

One of the most widely cited cases has been described for Clear Lake, California (Hunt and Bischoff, 1959), which was treated with DDD, starting in 1949, to kill *Chaoborus* gnats. The gnats virtually disappeared for two years and then began to reappear and increase in abundance until the lake was treated again in 1954. Late in 1954 and early in 1955 biologists began to find dead western grebes in the lake and no infectious diseases could be identified. Gnat populations increased again and a third treatment of the lake was carried out in 1957, and again dead grebes were found. On this occasion the fat of some of the birds was analyzed and found to contain high concentrations of DDD. This prompted the collection of fish and they too were found to have high concentrations of DDD. The nesting population of western grebes had historically been over 1000 pairs, but by 1959 it was only 15 pairs and even these were thought to have raised no young. In this instance the pesticide made its way from the water to the fish and then to the fish-eating birds, where the effect was noticed.

Perhaps the clearest cases of effects of pesticides on Canadian fish populations were described in the 1950s and 1960s during widespread use of DDT to control forest pests. Extensive kills of Atlantic salmon, brook trout, suckers, minnows, and invertebrates were reported in New Brunswick streams and rivers, with further delayed mortality noted at the onset of winter (Kerswill, 1967). Comparable kills of young salmon and trout were found in streams on northern Vancouver Island during forest spraying there in 1957 (Crouter and Vernon, 1959). These experiences provided some of the most compelling arguments in favor of replacing DDT with alternative compounds. Since then we have learned that DDT has permeated the planet. In 1969 it was identified in snow samples from the Antarctic (Peterle, 1969) and we know it is now ubiquitous in the biosphere.

More recently toxaphene has been turning up almost everywhere. It has been found in fish from the Gulf of St. Lawrence (Musial and Uthe, 1983) and fish from the Great Lakes (Rice and Evans, 1984). It is the major organochlorine compound in samples of fish from the Mackenzie River drainage (unpublished data). For example, the liver from a burbot (*Lota lota*) from Arctic Red River contained over 5 μg/g (wet weight) of toxaphene. The liver of this species is rich in fat and has historically been consumed by Dene people as a delicacy. At a level of 5 μg/g a single burbot liver would supply several times the "acceptable daily intake" of toxaphene (1.25 to 2.5 μg per kg body weight per day) (*United States National Academy of Sciences, 1977*). The situation is even worse with lake trout from the upper Great Lakes where toxaphene levels around 7 μg/g have been reported by Rice and Evans (1984). Like other organochlorine compounds (DDT, PCB, etc.) toxaphene components move with air currents (Rapaport and Eisenreich, 1986) and contaminate sites remote from any use of the pesticide. Toxaphene has recently been banned from further uses in North America, but that does little to deal with the residues already in place. Chlordane, another organochlorine pesticide, follows toxaphene, and PCBs in fish from the Mackenzie drainage, and it also has been identified in air samples from the Northwest Territories (Hoff and Chan, 1986). How did chlordane find its way into the air in Arctic Canada (Mould Bay)? Did it come from North America or did it come over the Arctic Ocean from Europe or Asia? If there are multiple sources then each can implicate the other. "My brother did it" is always a good defense.

Many of the more recent pesticides are unstable in the environment, and we do not have residues to help trace exposure. We may have fragments of the parent chemical but we may not be able to establish that they came from the pesticide. In these cases it is more difficult to associate any biological responses with exposures to the pesticide. However, given that pesticides do reach beyond the target pests, what does that mean to those unfortunate creatures, the "nontarget organisms"? The toxicological literature on lethal and sublethal effects of pesticides on experimental animals is enormous. We can find these effects at all the different levels of biological organization. At the subcellular level an entire journal (*Pesticide Biochemistry and Physiology*) is devoted to studies describing the biochemical and physiological effects of pesticides. At the anatomical level, structural deformities in fish as a result of experimental exposure to pesticides have been reported (e.g., vertebral damage) (McCann and Jasper, 1972). Attempts have been made to work backward from vertebral damage in a population of fish to explain the deformities as a consequence of earlier pesticide exposure (Baumann and Hamilton, 1984). Subtle behavioral changes have been described in fish exposed to pesticides (Bull and McInerney, 1974) and dramatic changes in the drifting behavior of invertebrates have been described repeatedly following treatments of Canadian rivers with pesticides (Flannagan et al., 1979). The widespread emergence of pest species resistant to pesticides that have been used to control them is a good example of pesticides altering gene frequencies in populations (World Health Organization, 1970). Even at the psychological level an important human reaction to chemical pollution is the fear people express. An editorial in *Science* (Marshall, 1982) described a U.S. court ruling that required the Nuclear Regulatory Commission to deal with the fears of people near the Three Mile Island nuclear plant, whether or not those fears had a rational basis. Citizens' fears of exposure to pesticides have been expressed at several public hearings associated with pesticide use in Manitoba.

Sometimes we may be able to see a clear effect of a pesticide at one level of organization but not understand what that implies for another level. As an example of this, experience is cited with fish in ponds in Winnipeg during a Public Health emergency in which the whole city was sprayed with malathion (Lockhart et al., 1985). Immediately after spraying, the brain cholinesterase activities in walleye fell to only 26% of prespray values and then recovered to near prespray values over about two weeks. A second spray was applied about three weeks after the first, and the cholinesterase response was very similar. However, apparently no fish were killed, and the only ecological response noted was a short-term cessation in growth. Thus, the biochemical observations were clear but their ecological implications were not. This state of knowing something at one level of biological organization but of not knowing how to translate it to another level of organization is not uncommon.

The case of bioaccumulation of DDD in birds in Clear Lake cited above (Hunt and Bischoff, 1959) is an instance of food-chain biomagnification of pesticide residues. The consumption of organochlorine-contaminated fish by mammals can be harmful to mammals, especially to processes of reproduction. For example, mink fed on coho salmon from Lake Erie averaged over four pups per litter, while those fed on coho from Lake Michigan averaged less than one (Aulerich et al., 1973). The salmon from Lake Erie were found to contain much lower levels of DDT and dieldrin than those from Lake Michigan, and these materials may have been the cause of the reproductive impairment in the mink. Marine mammals feeding on fish are particularly vulnerable to accumulation of contaminants from the fish, and organochlorine compounds including DDT have been suggested as the cause for reproductive failure in Baltic seals (Helle et al., 1976) and of premature births in California sea lions (DeLong et al., 1973), although the experiments would not isolate the effect of DDT from that of other chemicals like PCBs. More recently, experimental feeding has shown that a diet of fish taken from the Dutch Wadden Sea, an area receiving the discharge from the Rhine river, can cause reproductive failure in seals (Reijnders, 1986). Low birth weights have been reported in babies born to people residing the Love Canal area of New York (Vianna and Polan, 1984).

In view of the ability of pesticides and other chemicals in fish to cause harmful effects in mammals consuming the fish, it is not surprising that agencies charged with protecting public health set limits on the contaminant levels allowed. However, Canadians (like other people) have become contaminated with a range of stable pesticides (chlorobenzenes, lindane, oxychlordane, *t*-nonachlor, heptachlor epoxide, dieldrin, and DDT) in spite of the efforts to limit exposure (Mes et al., 1982). Fish can accumulate contaminants to some level below a toxic threshold for the fish themselves but still represent a threat to consumers, just as the fish in Clear Lake, California, in the 1950s were hazardous to the western grebes feeding on them. This type of contamination can destroy fisheries by closing markets. Contamination can also require expensive inspection programs to analyze the fish and ensure that only those below the safety limits ever reach markets. Even the perception of a threat of contamination can influence markets, whether the threat is real or not.

5.2. Regulatory Role

The responsibility of the Department of Fisheries and Oceans can be stated very simply. We want to maintain Canada's ability to produce fish, to enhance that ability where we can, and to make sure that the fishery products are wholesome. In making a judgment about the use of a pesticide, a range of socioeconomic arguments come into play. These arguments are difficult to reduce to the required "yes or no" decision on the use of a pesticide, particularly when the costs and the benefits apply to different sectors of society or accrue over different time scales. With a pesticide the cost may occur as some contamination or biological stress on fish or wildlife or other "environmental" components generally and it may occur over decades, while the benefit may be in crop yield or forest growth or public health right now. Our input to the registration decision is generally on the cost or risk side of the balance that must be weighed by Agriculture Canada. We can produce dollar figures, like the value of sales of commercial fish, the value of the sport fishery that might be placed at risk, or the cost of inspection programs, but the party on the benefit side can do the same and those figures may be even larger. It is often easier to associate dollars with the proven and measurable benefits than with uncertainties of possible costs, particularly over short time periods. Consequently it is easy for the "environmental" values to lose short-term socioeconomic arguments.

When our examination of the data supporting a registration is complete, it is given as advice to Agriculture Canada where the final registration decision is made.

5.3. Need for Better Conceptual Models

It seems reasonable to assume that the primary toxic interaction between a pesticide and an organism is at the molecular level. A wide range of sublethal cellular and metabolic effects have been shown to follow from experimental exposures of fish to pesticides, but then it becomes a problem of understanding what the sublethal events mean at higher levels of biological organization. For example, a cell biologist or biochemist might have a good understanding of events within a cell, but have difficulty establishing what the subcellular events mean to the whole organisms. As life scientists we have failed to develop good conceptual models to lead us from the primary interaction between the organism and the chemical at a molecular level through to the responses meaningful to us at population and community levels. When dealing with fish and wildlife, we need to make the jump to ecological levels because individual organisms may have little impact on the population. This is not to say that the injury or death of individuals is meaningless. Experiences with human or veterinary medicine can help us, but these disciplines can often afford to stop at diagnosis and treatment of the individual patient. When dealing with fish and wildlife we have to find ways to proceed to still higher ecological levels of organization.

Concepts of ecosystem stress are beginning to emerge (Rapport et al., 1985), based on Selye's earlier studies, and these may provide the framework needed. However, a valid conclusion that an ecosystem is in a state of stress may offer little help in identifying the components that have given rise to that state. The paradox is that measures with the greatest diagnostic potential have uncertain ecological meaning, and measures with the greatest ecological meaning have uncertain diagnostic capability.

5.4. Ethical Responsibility

Fundamentally the argument is not an economic one based on which set of dollar figures is more persuasive, nor is it a purely scientific one based on an understanding of ecological effects. Rather, the argument is an ethical or religious one. We must protect the environment *because it is right to do so*. No more fundamental argument can be made, and the question is whether Canadian citizens as a voting body accept it. Public opinion polls have suggested that Canadians do care deeply about the environment and will make economic sacrifices to protect it. The widespread distribution of pesticides throughout the biosphere is a good example of "tinkering" on a planetary scale. A comment by Aldo Leopold (Leopold, 1949) bears repeating: "To keep every cog and wheel is the first precaution of intelligent tinkering."

6. ASSESSING AND REGULATING THE ENVIRONMENTAL EFFECTS OF PESTICIDE USE*

There is little doubt that today's commercial agriculture is dependent on the use of pesticides. Modern forestry practice is also wholeheartedly embracing the principles of pest control. The depth of the dependence on pesticides that is necessary for production of food and fiber has been and will continue to be an arguable point. If trends hold, we will continue to increase our standards of living by continually increasing yields per unit area of both food and fiber. This will create a sustained demand for some form of pest control and, for the most part, this control will continue to be provided by chemicals.

This section examines critically how the environmental effects of pesticides are assessed and managed. This is done not with the intent of accusing or laying blame, but rather with the intent of demonstrating where there are opportunities for strengthening our abilities to assess and manage pesticide impacts. This will be my perspective not necessarily that of Environment Canada.

At present there is an elaborate system in place for measuring and controlling pesticide impacts. There is evidence that it is largely successful, in that we are not faced with pesticide-induced, ecological problems of catastrophic proportions (see Section 4.1, however). There are several disturbing trends, however, that indicate the need to increase our efforts in the area of pesticide control.

The first of these trends is that with the introduction of each new type of pesticide compound there has been a decrease in the rate at which that compound is applied. We began in the early 1940s by applying DDT at a rate of

*Section 6 was written by W.R. Ernst.

about 1.5 kg/ha. This was followed by the introduction of organophosphates (such as diazinon and malathion) in the late 1940s and early 1950s, which were applied at rates of 0.5 to 1.0 kg/ha. The carbamates (such as carbaryl and aminocarb), first introduced in the late 1950s and 1960s, further reduced the dosage to 0.25 to 0.5 kg/ha. We are now using synthetic pyrethroid pesticides which are applied at only 0.02 kg or less per hectare (Knüsli, 1985). This means that we are using ever increasingly toxic compounds to do the same job. While the specificity of some of these chemicals has been increased so that they may not be as toxic to birds and mammals, they are still as broadly toxic to 95% of the species in the animal kingdom, namely the invertebrates (Hickman, 1973). This use of ever increasingly toxic compounds means that the margin for error in being able to protect nontarget invertebrate species is being reduced.

Another disturbing trend is that the number of target insect species that are resistant to one or more insecticides is increasing at an exponential rate. In 1955, there were only approximately 25 insect species known to be resistant to control chemicals. As of 1980, there were about 450 insect species displaying some degree of resistance to one or more insecticides (Bottrell and Smith, 1982). From an operational point of view this means that to obtain the same degree of control, either increased application rates of the same type of insecticide must be used, or new insecticides must be developed. This fact, combined with the previously mentioned need to increase yields on a per unit area basis means that we are locked into what has been aptly called the "pesticide treadmill"(Van den Bosch, 1979). In other words, once you are on the treadmill, you no longer retain the option of getting off and you must continually move faster just to stay in the same spot.

The fact that total pesticide sales in Canada skyrocketted from about 55 million dollars in 1970 to about 190 million dollars in 1977 (Agricultural Institute of Canada, 1981) is a forceful indication of just how fast this treadmill is moving. Unfortunately, the collection of data on comprehensive national pesticide use in Canada ceased in 1977, so we have no idea of how much is sold now.

This information is not presented with the intent of suggesting that we try to get off the pesticide treadmill, because we do not have the technology to do so. The principles of integrated pest management (IPM) have been reasonably well established for over 25 years (Stern *et al.,* 1959) and yet they still offer only marginal hope of reducing our dependence on pesticides. Likewise, the development of more pest-resistant varieties of crops through biotechnological advances offers some hope for reducing pesticide needs (Worthy, 1984), but, this will probably come slowly.

The purpose in discussing the depth of our dependence on pesticide chemicals is to demonstrate the magnitude of the problem in assessing and regulating their environmental impacts and to indicate that these problems will probably get worse before they get better. The public is also beginning to develop an appreciation of the problems that are involved in assessing and regulating pesticides. I would agree with those who believe that we now live in a chemophobic society (Bolker, 1986) and that there has been considerable overreaction to the pesticides issue. Nonetheless, the vocal public is exerting considerable pressure on politicians and regulators to ensure that all toxic chemicals, but particularly pesticides, are properly regulated. Unfortunately, in the eyes of many members of the public and, to be fair, in the eyes of many of the government regulators themselves, we are doing a less than creditable job of managing these toxic chemicals (Hall, 1986). This view has been supported during the recent review of the federal Department of Environment by the Nielson Task Force (Anonymous, 1986), which indicated that the Environment Canada record in toxic chemicals management is poor and is primarily a consequence of weak and fragmented environmental protection legislation in Canada.

6.1. Pesticide Registration

The principal mechanism for controlling pesticide use in Canada is registration of all pesticides in commerce under the authority of the Pest Control Products Act. This act is administered by Agriculture Canada, but, during the preregistration review of a particular product, Agriculture Canada solicits the opinion of technical experts from Health and Welfare Canada, the Department of Fisheries and Oceans, and Environment Canada. Therein lies one of the first problems we encounter with controlling the environmental effects of pesticides. While Health and Welfare Canada has a formal "Memorandum of Understanding" defining the nature of the relationship between these two departments, Environment Canada's relationship with Agriculture Canada has no formalized agreement. There has been an effort to establish a "Memorandum of Understanding" between the two departments for over four years; however, it has not come to pass. This means that Agriculture Canada calls upon Environment Canada to review information only as Agriculture Canada sees fit and is under no formal obligation to accept the advice that is offered. To be fair, the working relationship between the two departments is good, but there have been instances when Environment Canada's advice has been ignored.

The preregistration evaluation of pesticide products is probably more thorough than for any other chemicals in commerce. A measure of this thoroughness is the often quoted fact that to obtain the information necessary for a complete modern pesticide registration package may take 9 to 10 years and cost upward of 30 million dollars. There are several qualifiers that must be mentioned to put this fact in perspective. The first is that most of the data for agricultural pesticides are generated primarily for registration in the U.S. market. (The situation is different for forestry uses, where most of the data are generated under Canadian situations.) Much of the environmental fate and effects information is therefore generated under U.S. conditions with species that may not be endemic to Canadian environments. Where this is judged to be a problem, a request is usually made to prospective registrants to generate more locally relevant information that may be used to review purposes (Millson, 1986). A specific set of guidelines for the environmental fate and effects information needs to be established for registration. These guidelines should demand data relevant to the Canadian situation. Guidelines have recently been drafted for environmental chemistry and microbial and biological pesticides but they are not yet satisfactory to Environment Canada advisors. Guidelines have yet to be established for environmental toxicology requirements.

It is also important to recognize that the information reviewed prior to registration is generated by the prospective registrant (in many instances with their own personnel, but generally by contract investigators). The Industrial BioTest Laboratories experience taught us that information generated by these so-called "independent" contract investigators should not always be accepted as face value. Furthermore, because these testing facilities exist in other countries, Canada has no access to them and, therefore, is in no position to evaluate their capabilities for producing high-quality data. This would seem to argue for an audit or verification capability by those agencies responsible for evaluating submitted data. Health and Welfare Canada has decided, as has the Environmental Protection Agency in the United States, that departmental audit of certain pivotal information is necessary before registering a pesticide product. Unfortunately, Environment Canada has not begun to audit *any* of the information it reviews prior to registration of the product. It is now almost 10 years since the Industrial BioTest Laboratories affair and we still have not come to grips with the problem of verifying environmental information. This is a serious fault in the integrity of the registration process.

It is also worth noting that few of the presently registered pesticides have had anything that approaches an intensive review, since most were registered for use in the years when environmental impacts and human health effects were not adequately considered. It has been estimated that only 15% of the pesticide active ingredients presently registered have ever been reviewed by Environment Canada, let alone been subjected to the testing detail we now know is necessary to predict environmental fate and behavior (Millison, 1984). At present, there is an attempt on the part of Environment Canada, in cooperation with Agriculture Canada, to systematically evaluate this large number of currently registered pesticides which have had no environmental review. A high priority needs to be put on the systematic reevaluation of all currently registered pesticides.

6.2. Postregistration Evaluation

Once a pesticide has been evaluated it receives a registration, which dictates the way in which it may be used. Certain individuals, mostly the lay public, seem to have difficulty in accepting the scientifically logical fact that it will never be possible to prove the negative, namely that a pesticide will not produce adverse effects during operational use. Until the time of registration, however, the burden of proof has been to demonstrate that once the chemical is put into commerce, its use will not result in unacceptable environmental or health impacts. The chemical, until it is registered, is judged to be "guilty" until the weight of evidence demonstrates that it is innocent of being a potential environmental problem. After the pesticide is registered, however, that "burden of proof" shifts. The chemical is now judged to be *reasonably* safe and it must be demonstrated *positively* that the product causes harm before changes in its registration status will be considered.

This change in burden of proof is really the only logical approach that can be taken in registering pesticides. This creates a problem, however. Simply stated, it is after registration that the truly independent investigation into a chemical's environmental fate and behavior begins. This is the time that government agencies and academic researchers begin fate and effects studies. Because of this shifted burden of proof, the genuinely independent researchers must now demonstrate serious negative impacts before regulatory action will be taken.

Postregistration evaluation is also somewhat hampered by the fact that academic and many government researchers do not have access to the fate and effects information that is submitted to regulatory authorities. This is because such information is classed as confidential business information under Canadian patent laws. This means that the independent researchers must start from square one and repeat most of the fate and effects studies just to begin to focus in on potential problem areas—a process that takes years in some instances. Much valuable research time and effort could be saved if registration information was not classed as confidential. There are some efforts underway by the National Pest Management Advisory Board (Versteeg, 1986) to initiate changes in the rules regulating access to pesticide information. These efforts should be supported and nothing less than total access to all information should be tolerated.

The postregistration monitoring of pesticide impacts is one of the most important parts of the overall pesticide management process. Most of the preregistration evaluation is laboratory toxicity work (primarily acute lethality studies) and in some instances small-scale controlled field studies, which, with the exception of forestry pesticide, are usually conducted in another country (Rawn, 1986). It is only after the chemicals are in commerce that their effects under actual operational conditions can be measured. Field studies conducted under operational conditions do have the disadvantage of being much less controlled than laboratory or small-plot research trials. There are regimes of weather and application techniques to contend with, and no two receiving environments are the same. Well-conducted field studies, under operational conditions, have the advantage of measuring the end result of pesticide use after all the modifying factors have had a chance to operate.

Unfortunately, there are few investigators in Canada who are now monitoring field impacts of pesticides. Almost all of these are government scientists in environmental or resource agencies. It has been estimated (Environment Canada, 1984) that in 1984 within Environment Canada, which has a total staff of over 10,000 people, there were only 16 person-years dedicated to assessing pesticide use after registration. It is likely that only half of these people are involved in environmental fate and effects monitoring. This indicates an inadequate dedication of resources to the issue.

One of the principal problems in assessing pesticide impacts is determining what constitutes an adverse effect. It must come as no surprise to anyone that virtually all pesticide use causes nontarget mortality. The real trick is being able to determine when these impacts become ecological problems. Some ecologists (Hall, 1986) have argued that a severe impact is indicated by measurable changes in ecosystem structure or function, while others are of the opinion that this measurable change must continue for a period of time, such as one annual cycle. This is not the time to explore this concept. The point is that determining what constitutes a significant ecological impact is within the realm of scientific judgment. Unfortunately when it comes to pesticide impacts, researchers have been reluctant to make and put forward these judgments. Furthermore, if such judgments are to become part of the decision-making process they must be properly directed and defended vigorously.

6.3. Provincial Regulation

Provincial authorities are responsible for authorizing the use of pesticides within their jurisdictions. They may add stipulations to the use permit and, therefore, in principal, are able to act as a second-tier mechanism for pesticide review. Unfortunately, in most cases, they are hampered in this activity by two factors. The first is that they are not allowed access to the registration review information and must base their decisions on the limited data in the open literature. Second, they do not have the staff resources to review the environmental implications of each pesticide on a case-by-case basis. Accordingly, provincial pesticide use monitoring generally consists of an observance of compliance with permit restrictions and there is not much effort directed toward measuring environmental impacts.

6.4. Concluding Remarks

It is incorrect to leave the impression that pesticide assessment and management is in a state of disarray—it is not. Pesticide chemicals as a group are probably more intensively and extensively evaluated for potential problems both before and after they are introduced into commerce than other commercial chemicals. The rigor of these evaluations has restricted the registration of new active ingredients to an average of about five per year over the past eight years. Making it increasingly difficult to register pest control products can be double-edged sword that must be wielded carefully. There could come a point when, in an honest but zealous attempt to prevent the introduction of a chemical into commerce, one could prevent the displacement of a currently registered chemical with environmental effects that may be substantially greater. In other words, we should not become so rigorous in our evaluation demands that we economically impede the development of newer, possibly more environmentally benign, products.

Some steps that could be taken to greatly improve pesticide management abilities in Canada are as follows:

1. To ensure that environmental advice is given due respect, a "Memorandum of Understanding" should be required between Environment Canada and Agriculture Canada.
2. Preparation of guidelines for detailed environmental fate and toxicity data requirements should be expedited. These guidelines must be specific to the Canadian situation and be satisfactory to Environment Canada advisors.
3. The authenticity of submitted environmental registration data should be verified by selectively repeating certain pivotal tests in independent laboratories.
4. A high level of effort needs to be directed toward the systematic reevaluation of all presently registered pesticides according to existing evaluation standards.
5. There needs to be freer access to the data submitted in support of registration.
6. There needs to be an increased level of effort dedicated to evaluating environmental impacts of pesticides after they are registered.
7. Finally, scientists, particularly those from independent research facilities, must be more willing to make clear judgments on the ecological importance of measured impacts. Also, they must not hesitate to direct these judgments, by whatever means or channels they feel necessary, to the proper people.

Many people are indicating that, with new developments in biotechnology and biorational control agents, we are on the brink of a new era in pest control (Science Council of Canada, 1985). If that is the case, we will be hard pressed to keep up with these developments with the control strategies we now have. Environment Canada has just come to the realization that our legislative mandate for toxic chemicals is inadequate, and it is in the final stages of implementing new toxic chemicals legislation. Efforts to find alternative control strategies are to be encouraged, but this is proving to be a catch-up game and we need to move much faster if we are ever going to provide an assessment and regulatory process that is satisfactory to all the stakeholders involved.

7. GENERAL OVERVIEW*

The registration process for pesticides in Canada is strict, even stricter than that for pharmaceuticals. That is as it should be because pharmaceuticals present a risk only to the treated individuals, whereas pesticides present a risk to entire

*Section 7 was written by D.C. Eidt.

communities. Strict as the regulations may be, they are subject to the simple principle already cited, that, stated another way, says that if something is convincingly demonstrated it is proven, but something not demonstrated is not disproven. Thus, the constant vigilance and cooperation of all organizations concerned about the benefits and hazards of pesticides is needed. There is everything to be gained or maintained, because environmental protection is in everybody's best interest. Agriculture and forestry cannot risk the loss of pesticides as critical tools of production; industry cannot risk the loss of products in which it has invested millions of dollars.

A sixth element in the pesticide controversy is public opinion and in particular the environmentalists who purport to represent it. That is not to say that Environment Canada is insensitive to public opinion, but the response is indirect through government and departmental policymakers. Environmentalists have been keen watchdogs whose concerns have led to many changes in the way pesticides are regulated. Once ruled more by emotion than rationality, they have become increasingly sophisticated and, like scientists, businessmen, and resource managers in fisheries, agriculture, forestry, the chemical industry, and government, are becoming increasingly aware of the benefits and costs and conflicts of interests involved in pesticide usage.

In summary there is general agreement on the following grounds pertaining to pesticide use:

1. Pesticides will be with us into the indefinite future, and we will have to cope with them in aquatic habitats because of the ubiquitous nature of water.
2. We have a good pesticide regulatory process in Canada that is being constantly improved.
3. National regulation may not be good enough for us and our neighbors. We have to think and regulate hemispherically or even globally.
4. The regulatory process is not perfect, but it will continue to improve. A system based on antagonism is not the best way to improve it.
5. The environment is not negotiable. One cannot put a price on some things, and the trade-offs will be judgmental rather than cost-benefit exercises.

Much of the body of data in a registration package is not available for postregistration scrutiny. Therefore, persons concerned about possible environmental (or health) effects must start many of their investigations from the beginning, and are compelled to repeat work already done. Thus there are philosophical differences in approach to the study of environmental side effects: before registration, emphasis is on environmental safety; after registration, emphasis is on environmental hazard. Furthermore, evidence of the invalidity of a registration is taken seriously before registration and may delay or prevent it (guilty until proven innocent), but after registration, similar evidence will be vigorously opposed, has little chance of resulting in suspension of the registration, and may require a protracted conflict before deregulation or suspension occurs (innocent until proven guilty).

REFERENCES

Agriculture Canada (1986). Towards an IPM strategy. Agriculture Canada Research Branch Publication.

Agricultural Institute of Canada (1981). Pesticides, agriculture, and the environment. *Agronews*, January 1981: 1–12.

Anonymous (1986). Programs of the Minister of the Environment: A study team report to the Task Force on Program Review. Dept. Supply Services, Ottawa.

Aulerich, R.J., Ringer, R.K., and Iwamoto, S. (1973). Reproductive failure and mortality in mink fed on Great Lakes fish. *J. Reprod. Fert., Suppl.* **19**: 365–376.

Bart, J.B., and Hunter, L. (1978). Ecological impacts of forest insecticides: An annotated bibliography. New York Cooperative Wildlife Research Unit, Cornell University, Ithaca, NY.

Bauman, P.C., and Hamilton, S.J. (1984). Vertebral abnormalities in white crappies, *Pomoxis annularis* Rafinesque, from Lake Decatur, Illinois, and investigation of possible cause. *J. Fish Biol.* **25**: 25–33.

Bolker, H.I. (1986). Scaring ourselves to death. *Can. Chem. News.* Feb.: 19–22.

Bottrell, D.G., and Smith, R.F. (1982). Integrated pest management. *Environ. Sci. Technol.* **16**: 283A–288A.

Bull, C.J., and McInerney, J.E. (1974). Behavior of juvenile coho salmon (*Oncorhynchus kisutch*) exposed to Sumithion (fenitrothion), an organophosphate insecticide. *Fish. Res. Board Canada* **31**: 1867–1872.

Carson, R.L. (1962). *Silent Spring*. Houghton Mifflin, Boston.

Crouter, R.A., and Vernon, E.H. (1959). Effects of black-headed budworm control on salmon and trout in British Columbia. *Can. Fish Cult.* **24**: 23–40.

Delong, R.L., Gilmartin, W.G., and Simpson, J.G. (1973). Premature births in California sea lions: Association with high organochlorine pollutant residue levels. *Science* **181**: 1168–1169.

Department of Fisheries and Oceans. (1986). *Policy for the Management of Fish Habitat*. Department of Fisheries and Oceans, Fish Habitat Management Branch, Ottawa.

Eidt, D.C. (1985). Toxicity of *Bacillus thuringiensis* var. *kurstaki* to aquatic insects. *Can. Ent.* **117**: 829–837.

Environment Canada. (1984). Pest management: Environment Canada's three year plan of action. Unpubl. Environment Canada Report, Ottawa.

Feng, J.C., Reynolds, P.E., and Klassen, H.P. (1986). Persistence of glyphosate in streamwater following aerial application with a microfoil boom at Carnation Creek, British Columbia, 1984. Forest Pest Management Institute File Report 73.

Flannagan, J.F., Townsend, B.E., deMarch, B.G.E., Friesen, M.K., and Leonhard, S.L. (1979). The effects of an experimental injection of methoxychlor on aquatic invertebrates: Accumulation, standing crop, and drift. *Can. Entomol.* **111**: 73–89.

Frank, R. (1985). Farm Water Quality and Pesticides—Highlights of Agricultural Research in Ontario 8, 18.

Ghassemi, M., Fargo, L., Painter, P., Quinlivan, S., Scofield, R., and Takata, A. (1981). Environmental fates and impacts of major forest use pesticides. TRW Environmental Division, Redondo Beach, California.

Hall, R.H. (1986). Why we need an alternative to Environment Canada. *Probe Post* Spring 1986: 26–29.

Helle, E., Olsson, M., and Jensen, S. (1976). DDT and PCB levels and reproduction in ringed seal from the Bothnian Bay. *Ambio* **5**: 188–189.

Hickman, C.P. (1973). *Biology of the Invertebrates*, 2d ed. Mosby, St. Louis.

Hoff, R.M., and Chan, K.W. (1986). Atmospheric concentrations of chlordane at Mould Bay, N.W.T., Canada. *Chemosphere* **15**: 449–452.

Holmes, S.B., and Kingsbury, P.D. (1980). The environmental impact of nonyl phenol and the Matacil formulation, Part 1. Aquatic ecosystems. Forest Pest Management Institute, Sault Ste. Marie, Ontario. Report FPM-X-35.

Holmes, S.B., Millikin, R.L., and Kingsbury, P.D. (1981). Environmental effects of a split application of Sevin-2-oil. Forest Pest Management Institute, Report FPM-X-46.

Hunt, E.G., and Bischoff, A.I. (1959). Inimical effects on wildlife of periodic DDD applications to Clear Lake. *Calif. Fish Game* **46**: 91–106.

Hynes, H.B.N. (1975). The stream and its valley. Edgardo Baldi Memorial Lecture. *Verh. Internat. Verein. Limnol.* **19**: 1–15.

Johnson, W.W., and Finley, M.T. (1980). *Handbook of Acute Toxicity of Chemicals to Fish and Aquatic Invertebrates.* U.S. Dept. of the Interior, Fish and Wildlife Service, Resource Publication 137, Washington, DC.

Kerswill, C.J. (1967). Studies on effects of forest sprayings with insecticides, 1952–63, on fish and aquatic invertebrates in New Brunswick streams: Introduction and summary. *J. Fish. Res. Board Can.* **24**: 701–708.

Knüsli, E. (1985). Avenues of the safe handling of plant protection agents. UNEP Industry and Environment, Fall 1985.

Leopold, A. (1949). *A Sand County Almanac.* Oxford University Press, New York.

Lockhart, W.L., Metner, D.A., Ward, F.J., and Swanson, G.M. (1985). Population and cholinesterase responses in fish exposed to malathion sprays. *Pesticide Biochem. Physiol.* **24**: 12–18.

Marshall, E. (1982). Fear as a form of pollution. *Science* **215**: 481.

McCann, J.A., and Jasper, R.L. (1972). Vertebral damage to bluegills exposed to acutely toxic levels of pesticides. *Trans. Am. Fish. Soc.* **101**: 317–322.

Mes, J., Davies, D.J., and Turton, D. (1982). Polychlorinated biphenyl and other chlorinated hydrocarbon residues in adipose tissue of Canadians. *Bull. Environ. Contam. Toxicol* **28**: 97–104.

Millson, M.F. (1984). The EPS involvement in the pesticide registration process: A review. Unpubl. Environment Canada Report, Ottawa.

Millson, M.F. (1986). Conservation and protection, Environment Canada, Ottawa (personal communication).

Morin, R., Gaboury, G., and Mamarbachi, G. (1986). Fenitrothion and aminocarb residues in water and balsam fir foliage following spruce budworm spraying programs in Quebec, 1979 to 1982. *Bull. Environ. Contam. Toxicol.* **36**: 622–628.

Musial, C.J., and Uthe, J.F. (1983). Widespread occurrence of the pesticide toxaphene in Canadian east coast marine fish. *Intern. J. Environ. Anal. Chem.* **14**: 117–126.

Norris, L.A. (1982). Behavior of herbicides in soil and water in the forest environment. *Northeastern Weed Society* **36** (supplement): 44–62.

Norris, L.A., Lorz, H.W., and Gregory, S.V. (1983). Influence of forest and rangeland management on anadromous fish habitat in Western North America, Part 9. Forest chemicals. USDA Forest Service General Technical Report PNW-149, Pacific Northwest Forest and Range Experiment Station, Portland, Oregon.

NRCC (1976). *Bacilllus thuringiensis*: Its effects on environmental quality. Associate Committee on Scientific Criteria for Environmental Quality, National Research Council of Canada, Ottawa, Ontario. NRCC No. 15385.

NRCC (1986). Pyrethroids: Their effects on aquatic and terrestrial ecosystems. Associate Committee on Scientific Criteria for Environmental Quality, National Research Council of Canada, Ottawa, Ontario. NRCC No. 24376.

Odum, E.P. (1971). *Fundamentals of Ecology*, 3d ed. Saunders, Toronto.

Payne, N., Helson, B., Sundaram, K., Kingsbury, P., Fleming, R., and de Groot, P. (1986). Estimating the buffer required around water during permethrin applications. Forest Pest Management Institute Information Report FPM-X-70.

Peterle, T.J. (1969). DDT in Antartic snow. *Nature (London)* **224**: 620.

Pimentel, D. (1982). Perspectives of integrated pest management. *Crop Protection* **1**: 5–26.

Poirier, D.G. (1986). The toxicity of four insecticides used in forest pest management in Canada to selected aquatic invertebrates. M.Sc. thesis, University of Guelph.

Poirier, D.G., and Surgeoner, G.A. (1987). Toxicity of the insect growth regulator diflubenzuron to stream invertebrates. Dept. of Environmental Biology, University of Guelph, PRUF contract report to Forest Pest Management Institute, Sault Ste. Marie, Ontario.

Pope John Paul II. (1980). Einstein session of the Pontifical Academy. Address of Pope John Paul II. *Science* **207**: 1165–1167.

Rapaport, R.A., and Eisenreich, S.J. (1986). Atmospheric deposition of toxaphene to eastern North America derived from peat accumulation. *Atoms. Environ.*, **20**: 2367–2379.

Rapport, D.J., Regier, H.A., and Hutchinson, T.C. (1985). Ecosystem behavior under stress. *Am. Nat.* **125**: 617–640.

Rawn, G. (1986). Conservation and protection, Environment Canada, Ottawa (personal communication).

Reijnders, P.J.H. (1986). Reproductive failure in common seals feeding on fish from polluted coastal waters. *Nature (London)* **324**: 456–457.

Rice, C.P., and Evans, M.S. (1984). Toxaphene in the Great Lakes. In J.O. Nriagu and M.S. Simmons (Eds.), *Advances in Environmental Science and Technology*, Vol. 14, *Toxic Contaminants in the Great Lakes*, Wiley, New York, pp. 163–194.

Science Council of Canada (1985). Seeds of renewal: Biotechnology and Canada's resource industries. Report 38, Science Council of Canada, Ottawa.

Stemeroff, M., and Madder, D.J. (1985). The productivity of agricultural pesticides in B.C. and eastern Canada. Working Paper, Development Policy Branch, Agriculture Canada.

Stern, V.M., Smith, R.F., van den Bosch, R., and Hagen, K.S. (1959). The integrated control concept. *Hilgardia* **29**: 81–101.

Symons, P.E.K. (1977). Dispersal and toxicology of the insecticide fenitrothion: Predicting hazards of forest spraying. *Residue Rev.* **38**: 1–36.

Tolman, J.H., McLeod, D.G.R., and Harris, C.R. (1986). Yield losses in potatoes, onions and rutabagas in southwestern Ontario, Canada—The case for pest control. *Crop Protection* **5**: 227–237.

Trial, J.G. (1986). Environmental monitoring of spruce budworm suppression programs in the Eastern United States and Canada: An annotated bibliography. Maine Agricultural Expt. Station. Miscellaneous Report 312, University of Maine, Orono

Turpin, F.T., and Maxwell, J.D. (1976). Decision-making related to use of soil insecticides by Indiana corn farmers. *J. Econ. Entomol.* **69**: 359–362.

United States National Academy of Sciences (1977). Cited from J.R. Sullivan and D.E. Armstrong, Toxaphene status in the Great Lakes. University of Wisconsin Sea Grant Institute Priority Pollutant Status Report No. 2, Nov. 1985, p. 33.

Van den Bosch, R. (1979). The pesticide problem. *Environment* **21**: 13–42.

Versteeg, H. (1986). Access to proprietary data. Memorandum to all interested parties Aug. 25, 1986, Pest Management Advisory Board, Ottawa.

Vianna, N.J., and Polan, A.K. (1984). Incidence of low birth weight among Love Canal residents. *Science* **226**: 1217–1219.

Weed Science Society of America (1983). *Herbicide Handbook*, 5th ed. Weed Science Society of America, Champaign, IL.

World Health Organization (1970). Insecticide resistance and vector control. WHO Tech. Rep. Ser. No. 443.

Worthy, W. (1984). Pesticide chemists are shifting emphasis from kill to control. *Chem. Eng. News* July: 22–26.

INDEX

Acidification:
  episodic, 24–25
  reproductive failure, 24
  runoff water, spawning site effects, 30–31
Acute bioassays, 8–10, 19
Adsorption, Nafion membranes, 78–79,
Adsorption coefficient, 195
Aedes aegypti, mortality, 261–262
Aeromonas salmonicida, transfers of triple sulfa resistance, 135, 138
Agricultural pesticides, 247–253
  benefit/risk, 250–251
  black fly control, 250
  buffer zone, 252
  Draft Environmental Fate Guidelines, 250–251
  education programs, 252
  entry into water, 249
  environmental safety, 249–250
  future regulation, 253
  impact on aquatic systems, 249
  monitoring and reevaluation of older chemicals, 252
  return on investment, 248
  steps to ensure aquatic safety, 250–253
  use, 248
Algae:
  bioaccumulation, 206
  hydrocarbon-degrading, 219
  place in food web, 206
  versatility in ecotoxicological applications, 206
Algal bioassays:
  bioaccumulation protocol, 208–209
  bioanalytical determination of metal uptake, 209
  chemical determination of metal uptake, 209
  culture and growth conditions, 207–208
  metal uptake, *C. variabilis,* 209–210
  photooxidation, 209
Aluminum:
  runoff water, 25
  lethality tests, lake charr, 31–32, 36–38
Aminocarb, 255, 257, 263
Anemia, 50–51
Anodic stripping voltammetry, 74
Antibiotic resistance:
  transfer, 134–135, 137
  in vitro transfer, 130–131
  patterns
    bacterial strains, 130, 132
    gram-negative bacteria, 132–135
Antibiotics:
  preparation, 129
  sensitivity tests, speckled trout, 132–135
  suppliers, 139–140
  therapy, speckled trout, 126–127
Application factor, 9
Arthrobacter, hydrocarbon degrading, 229
Assimilative capacity, 17
Aurora WP Lake, 28, 38
Avoidance tests, 116, 118
  apparatus, 112–116, 123
    channel, 115
    data acquisition facilities, 115–116
    flow diagram and plan view, 113
    holding facilities, 115
    side view, 114
    temperature control, 115
    water treatment, 115
  avoidance curve, 119
  avoidance reaction representation, 119, 124
  data analysis and interpretation, 118–120
  fluviarium system, 112
  lateral distribution, 122–123
  longitudinal distribution, 120–121
  main avoidance, 119
  rainbow trout, 116, 118
  secondary avoidance, 119
  shallow concentration gradients, 112

Avoidance tests (*Continued*)
  steep gradient method, 112
  sublethal threshold concentration, 123
  vertical distribution, 121

Bacillus thuringiensis, 259, 263–264
  forestry pesticide, 256
  toxicity, 257
Bacillus thuringiensis israelensis, 250
Bacteria; *see also* specific bacteria
  hydrocarbon-degrading, 218–219
  oil-degrading, seeding, 228–229
Baltic seals, 270
Bauer et al. agar dilution method, 129
Beaverlodge Lake, 49, 51–52
  blood parameters, 59
  hematology, 55–57
  U-series radionuclides, 49
Behavioral tests, 111–112
  sublethal effects testing, 13
Benthos, fjord, 14
Bioaccumulation:
  algae, 206, 208–209
  ecotoxicants, 177–178
Bioassay; *see also* Algal bioassays
  acute, 8–10, 19
  copper, 75
  medium-term sublethal ecotoxicity, 174
  nematode, 148–149
  sediments, 146–148
  short-term sublethal ecotoxicity, 172–173
  sublethal effects testing, 14
Biochemical effects, sublethal effects testing, 12
Bioconcentration factor, 195
Biocontrol organisms, 253
Biodegradation, ecotoxicants, 176–177. *See also* Petroleum, biodegradation
Biological control, 263–264
Biological material, variability, 15
Biological systems, toxic material entrance, 143–144
Biomonitoring, 144–145
  field organisms, 145
Bioturbation, 224
Birds, DDD bioaccumulation, 270
Black Rock Harbor, dredging project, 16
Blood parameters:
  fish, size and sex correlations, 59–60
  food effects, 61
  radionuclides, 53
Bowland Lake, 27–28
Brain cholinesterase, 269–270
Brook charr, 28
Buffer zone:
  agricultural pesticides, 252
  forestry pesticide, 261–262
Burbot, 28

Cadmium, speciation, as function of pH, 83
Carbamates, 273
Carbaryl, 259
  residues, stream water, 258
Carcinogens, 144
Carp, 50
Catostomus commersoni, see White suckers
Cause–effect relations, 18
cDNA:
  9-S mRNA, 101, 103
  synthesis, 97
Cell free translation, metallothionein messenger RNA, 96
Cell proliferation, radionuclides, 57
Cellular systems, functioning, 20
CEPEX bags, 17
Chalmydomonas variabilis, 207–208
  metal uptake, 211
Chaoburus gnats, 268
Chelex 100, 74
Chemical analysis, compared to bioassay, sediments, 148
Chemical dispersants, biodegradation effects, 226–228
Chemophobia, 274
Chlopyrifos, 197
Chlordane, 269
Communities, effects of disturbance, 20
Contaminants, impact, 19
Convention for the Prevention of Marine Pollution by Dumping from Ships and Aircraft of 1972, 16
Convention on the Prevention of Marine Pollution by Dumping of Wastes and Other Matter, 8, 16
Copepods, 14
Copper:
  bioassay studies, 75
  speciation, 75
    in presence of humic colloids, 84–85
Coregonus clupeaformis, see Lake whitefish
Corexit 9527, 226–227
Cumulative effect, 17
Cyanide:
  analytical procedures, 66
  determination
    cyanometallates, 69–70
    distillation from complex matrix, 67

in presence of transition metal cations, 68
procedure, 67
reagents and equipment, 66–67
spiked surface waters, 69–70
recovery from pure and saline aqueous samples, 69
sources, 66
uses, 65
Cyanobacteria, hydrocarbon-degrading, 219
Cyanometallates, 69–70
Cypermethrin, 197
Cyprinus carpio, 50
[^{35}S]Cysteine, incorporation, poly A RNA effects, 99–100
Cytosol, gel filtration profiles, 97–98

2,4-D, 267
Darwinian evolution, paradigm, 2
DDD, 268, 270
DDT, 249–250, 270
effects on fish, 268
forest use, 253–255
Desorption, sediments, 143
Dialysis, 75
Dieldrin, 270
Digestive glands, biochemical composition, 11, 19
Dinoseb, 199, 201
Dissolution, organic materials, 157
Dissolved oxygen, role, sediments, 154
Donnan exclusion, 85

EC_{50}, sediments, 147, 150
Ecological effects, sublethal effects testing, 13–14
Ecological stage, 171
Ecosystemic pathology, 2
Ecosystem stress, pesticides, 271
Ecotoxicants:
bioaccumulation, 177–178
persistence and biodegradability, 176–177
Ecotoxicity:
chronic, 179–180
global assessment, 182–185
holistic view, 206
levels, 186
Ecotoxicity spiral, 162–165
bioaccumulation, 177–178
chronic ecotoxicity, 179–180
ecological stage, 171
exposure, 171
genotoxicity, 178–179
integration, 180–181
medium-term sublethal ecotoxicity, 174–175
microbiological characteristics, 170–171
operational procedure, 182–188
persistence and biodegradability, 176–177
physicochemical characteristics, 166–169
sampling, 168
second-level analyses, 171–175
short-term sublethal ecotoxicity, 172–173
third-level analyses, 175–180
Effluents:
bacterial quality, 170
chronic ecotoxicity, 179–180
ecotoxicological characterization, 180
physicochemical quality, 166–169
possible toxic effects with and without treatment, 165
potential lethal ecotoxicity, 171
potential sublethal ecotoxicity, 173, 175
simultaneous integrated ecotoxicological assessment, 187–188
Eh levels, 154–155
Eldorado Nuclear Ltd., 49
Elutriation, 146
Embryo, encapsulated, acidification effects, 29, 33–34
Environmental conditions, variability, 10
Environmental protection:
levels, 162, 164
second-level analyses, 171–175
third-level analyses, 175–180
Equilibrium distribution model, size of compartments, 195
EROD, 12
Erythrocytes, fish, 56
ET_{50}, sediments, 151
Eutrophication, oxygen depletion, 158
Evolution, characterization, 2
Exposure, 171

Fate guidelines, agricultural pesticides, 250–251
Fenitrothion, 255, 257, 259, 262–263
Fenvalerate, 197
Ferene reagent, 66, 68
Field toxicity tests, lake charr, *see* Lake charr
Fish: *see also specific fish*
acidification effects, 24
blood parameters, size and sex correlations, 59–60
body zinc levels, 92
bone uranium levels, 57–58
chronic radiation studies, 50

Fish (*Continued*)
  metal-binding proteins, 92–93
  metallothioneins, 91–92
  parasite survey, 54–55, 59
  protection, 265
  tumor survey, 54, 57
Fisheries Act, 267
Fish habitat, protection, 265
Fish Habitat Policy, 267
Forestry pesticide, 253–254
  Bacillus thuringiensis, 256
  benefit/cost, 255
  buffer zones and protective regulations, 261–262
  carbaryl residues, stream water, 258
  glyphosate residues, stream waters, 260–261
  herbicide, 255, 260
  toxicity and exposure:
    herbicides, 259–261
    insecticides, 256–259
  training requirements, 262
  use, 254–256
Fredette Lake, 49, 51–52
  hematology, 55–57
Fugacity, definition, 194
Fugacity model:
  freshwater, 194–195
  LC_{50}, 195
  levels, 194
  MacKay model, 201
  materials and methods, 194–196
  seawater, 194–195
  use, 201
Fungi, hydrocarbon-degrading, 219

Genetic effects, sublethal effects testing, 12–13
Genotoxicity, effluents, 178–179
Gram-negative bacteria:
  antibiotic resistance patterns, 132–135
  SxT and RO5 effects, 138
1978 Great Lakes Water Quality Agreement, 4–5
Guidelines, pesticides, 275

Hazard assessment schemes, 206
Hazard index, 195
Hematology, U-series radionuclides, 55–57
Hematopoietic system, 50
Herbicides:
  forestry pesticide, 255
  toxicity:
    exposure and, 259–261
    mobility and, 262–263
Hexadecane, mineralization, 227
H^+ lethality tests, lake charr, 31–32, 36–38
Hydrocarbon-degrading microorganisms, 218–221

Insecticides:
  peak concentrations, 258
  toxicity and exposure, 256–259
Integrated pest management, 253, 273
Intergovernmental Oceanographic Commission, 15–16
  Group of Experts on Methods, Standards and Intercalibration, 15–16
  Group of Experts on the Effects of Pollutants, 15, 19
  Scientific Committee on GIPME, 15
  Workshop on Biological Effects Measurements, 12–13, 15
    scope for growth, 19
Interjurisdictional negotiation, science role, 4
International Council for the Exploration of the Sea, 14–15
  Advisory Committee on Marine Pollution, 14–15, 17
    causal ralations criteria, 18
  Study Group on Biological Effects Techniques, 15
Ion exchange, free metal ions, 74. *See also* Nafion dialysis
Iron:
  hydrocarbon-degrading effects, 234
  release, 157

Joint Groups of Experts on the Scientific Aspects of Marine Pollution, biological variable ranking, 11

Katepwa Lake, sediments, 156–157

Laboratory experiments, using natural water, lake charr, 32–33, 38–39
Laboratory toxicity tests, lake charr, 31–33, 36–39
Lake Athabasca, 51–52
Lake charr, 25, 27
  chemical characteristics of study lakes and laboratory test water, 27
  chronic effects of acid and/or metals, 40
  combination of chronic and episodic toxic conditions, 40
  delayed hatching, 36–37
  embryos, 25, 27
    encapsulated, 29, 33–34
    sensitivity periods, 37

survival following exposure to low pH and Al, 36
episodic acidification effects, 25
fertilization, 29, 33–34
field toxicity tests, 28–31, 33–36
H^+ and Al lethality tests, 31–32, 36–38
laboratory toxicity tests, 31–33, 36–39
mortality of different life stages, 37–38
natural populations, acidic water effects, 39–41
radionuclide content, 48–49
site-specific mortality, 29, 31, 34–36
study sites, 26–28
tests using natural water, 32–33, 38–39
Lake whitefish, 59
radionuclide content, 48–49
white blood cells, 61
LC_{50}, 9, 195
effluents, 171
pesticides, 198–201
sediments, 147, 151
Lead:
β radiation, 50
hematopoietic system effects, 50
Nafion dialysis, 79–80
Lead-210, 49–50, 61
Lethal ecotoxicity, effluents, 171
Lethality index, 195
London Dumping Convention, 8, 16
Lota lota, 28

Malathion, 197, 269
Mammals, consumption of contaminated fish, 270
Manganese, release, 157
Mass-balance evaluation, 17
Medium-term sublethal ecotoxicity, 174–175
Mercury, desorption, 154
dissolved oxygen levels, 155–157
rate constants, 155–156
temperature effects, 155–156, 158
Mesocosms, 17
Metal-binding proteins, fish, 92–93
Metal ions, free
Nafion dialysis, 83
speciation, 74
Metallothionein; *see also* mRNA
amino acid composition, 92–93
fish, 91–92
heavy metal metabolism role, 105
hepatic, winter flounder, 97–98
induction, 94
polarographic techniques, 105
purification, 94–95
role, 104–105
structures and functions, 91
synthesis, 104–105
winter flounder, 92–93
Metal uptake:
bioanalytical expression, 211–213
C. variabilis, 209–210
Methidathion, 197
Methoxychlor, 250
Microcosms, 17
Microorganisms, hydrocarbon-degrading, 235
chemical dispersant toxicity, 226–227
environmental adaptation, 222
Microtox bacterial bioluminescence procedure, 207
Microtox test, 150–151, 212–213
Milliken Lake, 49, 51–52
hematology, 55–57
Mill tailings, 48
Mineralization, hexadecane, 227
Monitoring, 17–18
assessment phase, 17–18
biological effects techniques, 17
phases, 17, 20
Morphological effects, sublethal effects testing, 12
mRNA:
analysis of translation products, 96
cadmium chloride effects, 104
cDNA, 97, 101, 103
cell free translation, 96
hepatic, isolation, 95–96
isolation, 99
polyacrylamide gel electrophoresis, 100–101, 103–104
polyadenylated, isolation, 95–96
sucrose:
density gradients, 101–103
gradient centrifugation, 96
in vitro translation, 99–100
Mueller-Hinton culture media, 128, 139
Mussel watch, 19

Nafion dialysis, 75–76
advantages, 76
binding capacity, 80–82
dialyzate analysis, 78
free metal ion, 83
multication studies, 81–84
binding capacity titrations, 81–82
pH speciation, 83–84
simple studies, 80–81
speciation procedure, 76–77
studies involving natural ligands, 84–86

Nafion membranes:
permselectivity, 79–80
sorption characteristics, 78–79
Natural water, laboratory experiments, lake charr, 32–33, 38–39
Nematode bioassay, 148–149
Nitrogen, microbial hydrocarbon degradation limit, 223
Nutrients
biodegradation effects, 223
enrichment, hydrocarbon degradation, 230–234
littoral zone, 233
oleophilic fertilizers, 231–232
optimum C:N and C:P ratios, 230
paraffin-supported fertilizer, 231–232

Ocean dumping, predictive methods for assessing effects, 16
Octanol/water partition coefficient, 195
Oil, physical/chemical properties, 221
Oleophilic fertilizers, 231–232
Organisms:
biocontrol, 253
hydrocarbon-degrading, 218–221
Organochlorine insecticides, 249–250
Organochlorine pesticides, 254, 269
contaminated fish in food-chain, 270
environmental dynamics, 268
Organophosphates, 273
Ormetoprim, 127
solvent, 140
Oslo Convention, 16
Oxygen; *see also* Dissolved oxygen
biodegradation effects, 224–225
depletion, eutrophication, 158
Oxygenation, hydrogen-degrading effects, 230

Panagrellus redivivus, 148–149, 151
Paraffin-supported fertilizer, 231–232
Parasites:
bacterial and viral infection effects, 61
fish survey, 54–55, 59
Pathological effects, sublethal effects testing, 12
PCBs, 269
Permethrin, 258, 259, 261, 263
Permselectivity, Nafion membranes, 79–80
Persistence, ecotoxicants, 177
Pest Control Products Act, 274
Pesticide industry, role in safe use, 263–265
Pesticides; *see also* Agricultural pesticides; Forestry pesticide
aquatic habitat effects, 265–272
assessment, 272–274
behavioral effects, 269
benefit/cost, 271
benefit/risk, 264, 266, 279
burden of proof, 275–276
control by legislation, 264
ecosystem stress, 271
environmentalists, 279
ethical responsibility, 272
food-chain biomagnification, 270
guidelines, 275
human contamination, 270
impacts, 277–278
information, 267
LC_{50}, 198–201
legislation, 278
lethal and sublethal effects, 269
need for conceptual models, 271–272
postregistration evaluation, 275–277
preregistration evaluation, 274–276
properties, 198–201
provincial regulation, 277
psychological fear, 269–270, 274
public opinion, 279
registration, 267, 271, 274–275, 278
regulation, 272–274, 279
regulatory role, 270–271
relative hazard indices, 196
research, 264–265
resistance of pests, 269, 273
sales, 273
steps to improve management, 278
sublethal effects, 271
trends, 272–273
verification, 275
Pesticide treadmill, 273
Pest monitoring, 264
Pest-resistant crops, 273
Petroleum, biodegradation:
anaerobic, 224–225
chemical dispersion, 226–228
enhancement, 225–226
hydrocarbon-degrading microorganisms, 218–221
nutrients, 223
enrichment, 230–234
oxygen, 224–225
oxygenation, 230
physical/chemical properties of oil, 221
rates, 220–221
salinity/pressure, 225
seeding, 228–229
temperature, 221–223

Phorate, 197
Phosphamidon, 255
Phosphorus:
  microbial hydrocarbon degradation limit, 223
  oil-degrading bacteria effects, 231–232
Photobacterium fisherii, 150
Photooxidation:
  algal bioassays, 209
  toxicity, 210–212
Physicochemical excess, degrees, 168–169
Polonium-210, 61
Polyacrylamide gel electrophoresis, metallothionein messenger RNA, 100–101, 103–104
Polyadenylated mRNA, isolation, 95–96
Poly A RNA, isolation, 95–96
Potentiometry, free metal ions, 74
Preference-avoidance reactions, 112
Pressure, biodegradation effects, 225
Prince Edward Island, 193
Priority chemical lists, 144, 148–149
Protein, fish, 56
Pseudomonas, hydrocarbon-degrading, 228
Pseudopleuronectes americanus, see Winter flounder
Pyrethroid insecticides, 251–252, 261, 273

Quantifying effects, 18

Radiation:
  α, β, and γ, 50
  chronic, 50
    low-level, 61–62
Radionuclides, 48
  blood parameters, 53, 55–57
  cell proliferation, 57
  field methods, 51, 53
  gross necropsy, 53
  literature review, 50–51
  water quality, 61–62
  white blood cell effects, 61
Radium-226, 49–50, 61
Rainbow trout, 116, 118
  chronically acidic conditions, 41
  forestry pesticide, toxicity, 257
  multiple bioassay approach, 150–151
Recruitment:
  acidification effects, 24
  chronically acidic conditions, 40–41
Redox potential, 154–155
Regan Lake, 28–29, 31
Relative hazard index, 195–196
Reproduction:
  acidification effects, 24
  contaminant effects, 19
  state, variability, 12
RNA; *see also* mRNA
  isolation, winter flounder, 95
  separation by centrifugation, 102
RO5, intestinal flora effects in trout, 131, 133, 138
Royal Society of London discussion, 10
Runoff water:
  acidification, spawning site effects, 30–31
  aluminum, 25
Ruth-Roy Lake, 28, 39

Salinity, biodegradation effects, 225
Salmo gairdneri, see Rainbow trout
Salmon, aquaculture, 126
Salmonella typhimurium, 149
Salvelinus fontinalis, see Speckled trout
Salvelinus namaycush, see Lake charr
Saskatchewan, uranium mining, 49
Saskatchewan River, sediments, 149–150
Science:
  in interjurisdictional negotiation, 4
  reliance on authority, 3
  roles in aquatic toxicity and ecotoxicology, 3–4
Scientific knowledge:
  growth, 2
  syntheses, 4
Scope for growth, 11, 16, 19
Seawater, quality trends, 17
Sediments, 143–145
  analytical methods, 155
  bioassay, 146–148
    compared to chemical analysis, 148–151
    end point, 147–148
    multiple bioassay approach, 150–151
    primary indicator, 151–152
  desorption properties, 143
  dissolved oxygen role, 154
  EC_{50}, 147, 150
  elutriates, 151–152
  elutriation, 146
  ET_{50}, 151
  extraction, 146
  freshwater, 154
  Katepwa Lake, 156–157
  LC_{50}, 147, 151
  sample preparation, 145–146
  sampling, 154
  Saskatchewan River, 148–150
  seed germination test, 151

Sediments (*Continued*)
toxic threshold criteria, 147
treatment, 154–155
Seed germination test, 151
Seeding, oil-degrading bacteria, 228–229
Selenastrum capricornutum, 207, 209
Selenastrum capricornutum micro test, 211–212
Shield lakes, 51
Sodium phosphate buffer, 140
Sodium sulfate, 49
Speciation, 74. *See also* Nafion dialysis
copper, 75, 84–85
free metal ions, 74
ionic strength effect, 84–85
methods involving kinetic parameter, 74
methods involving size parameter, 75
Speckled trout, 28
antibiotic sensitivity tests, 128–129, 132–135
antibiotic therapy, 126–127
bacterial growth conditions, 130, 132
bacterial strains isolated from, 137
postmortem examination, 127–128
SxT and RO5 effects on intestinal flora, 131, 133, 138
transfer of antibiotic resistance in vitro, 130–131
Spruce budworm, insecticide control, 255
Standardization, 19–20
Statistical Analysis System, 53
Strontium-90, 50
Student–Newman–Keuls test, 53, 55
Sublethal ecotoxicity:
medium-term, 174–175
short-term, 172–173
Sublethal effects testing, 10–11
behavioral effects, 13
bioassay, 14
biochemical effects, 12
ecological effects, 13–14
genetic effects, 12–13
morphological and pathological effects, 12
physiological effects, 11
Suckers, white blood cells, 61
Sucrose:
density gradients, 101–103
gradient centrifugation, 96
Sulfadimethoxine, 127
Sulfamethoxazole, 127
SxT, intestinal flora effects in trout, 131, 133, 138

Temperature, biodegradation effects, 221–223
Tetracycline, resistance transfer, 135
Torque test, 13
Toxaphene, 268–269
Toxicity:
determination, target organism, 144
priorities of biological effects, 145
Toxic threshold criteria, sediments, 147
Translation products, analysis, 96
Trimethoprim, 127, 140
Triple sulfa:
components, 140
resistance transfer, 135, 138
TU_1, effluents, 171
Tumor, fish survey, 54, 57

Ultrafiltration, 75
U.S. Corps of Engineers/Environmental Protection Agency Field, Verification Program, 16
Uranium, *see* U-series radionuclides
Uranium mining, 48–49
U-series radionuclides, 48
Beaverlodge Lake, 49
bone levels, 57–58
hematology, 55–57

Water quality, radionuclides, 61–62
Western grebes, 268
White blood cell:
Beaverlodge Lake fish, 61
fish, 57–58
radiation effects, 61
types, 53
Whitepine Lake, 28
overwinter test of mortality, 29, 31, 34–36
surface water chemistry monitoring, 31
White suckers, 28
radionuclide content, 48–49
white blood cell types, 53
Winter flounder, 90–91, 94
hepatic metallothionein, 97–98
metallothionein, 92–93
Workshop on Biological Effects of Marine Pollution and the Problems of Monitoring, 10–11, 13

Zero discharge, 3
Zinc:
binding capacity titrations, 81–82
fish body levels, 92
Nafion dialysis, 80–81
winter flounder metabolism, 91